PHOTOELECTRIC EFFECTS IN SEMICONDUCTORS

FOTOÉLEKTRICHESKIE YAVLENIYA V POLUPROVODNIKAKH

ФОТОЭЛЕКТРИЧЕСКИЕ ЯВЛЕНИЯ В ПОЛУПРОВОДНИКАХ

PHOTOELECTRIC EFFECTS IN SEMICONDUCTORS

by

Solomon Meerovich Ryvkin

Translated from Russian by
Albin Tybulewicz
Editor, *Soviet Physics—Semiconductors*

CONSULTANTS BUREAU
NEW YORK
1964

First Printing — October 1964
Second Printing — March 1969

The original Russian text, published by the State Press for Physics
and Mathematical Literature, FIZMATGIZ, in Leningrad in 1963,
has been brought up to date by the author for the English edition.

Соломон Меерович Рывкин.
Фотоэлектрические явления в полупроводниках.

Library of Congress Catalog Card Number 64-25832

ISBN-13: 978-1-4684-1559-9   e-ISBN-13: 978-1-4684-1557-5
DOI: 10.1007/978-1-4684-1557-5

PREFACE

Investigations of photoelectric effects occupy an important place in studies of semiconductors.

Recently investigations of photoconductivity and photoelectromotive forces have been intensified in step with the general development of semi-conductor physics.

An important feature of current work is an approach which combines several experimental methods of investigating photoconductivity by meas-uring a number of the parameters which govern it (e.g., quantum yield, lifetime, etc.). In other words the study of steady-state photoconductiv-ity is being replaced by the study of photoconductivity kinetics.

Another important feature is the extension of the studies beyond pure-ly photoelectric phenomena by the use of radiations other than light. Such an extension is justified not only by practical requirements but also by the close similarity between the ionization processes produced by different radiations. It can be shown that the charge carriers liberated by light and other radiations are not unique as far as their behavior in a crystal lattice is concerned. This can be seen from the following considerations.

Processes occurring during ionization in semiconductors may be di-vided into two categories: intrinsic ionization processes, i.e., formation of free carriers, and processes of motion and recombination of the lib-erated carriers. The former are governed by the interaction of ionizing radiations with matter and are characterized by parameters such as the absorption (or attenuation) coefficient or the yield. The latter depend on the interaction of "nonequilibrium" carriers with matter and are character-ized by parameters such as the relaxation time (or the mobility) and the lifetime.

The intrinsic ionization processes vary with the type of ionizing ra-diation, but the subsequent processes of carrier motion and recombination are usually independent of the radiation which liberated these carriers. This is due to the very strong interaction between carriers and the lat-tice. As a result of this interaction the carriers produced by ionization in bands very quickly lose their excess energy over the average energy of thermal motion (this excess varies with the type of radiation) by collisions with the lattice ("cooling" of carriers) and become indistinguishable from the "equilibrium" carriers liberated by thermal ionization.

Subsequently, i.e., during most of the time of their free existence in a band, the carriers liberated by ionizing radiations have normal mobility and recombination coefficient and show no unique properties which might be connected with the nature of the radiation which was responsible for the ionization. This often makes it possible to use identical methods for the study and analysis of the conductivity induced by radiations differing by a factor of several million in the energy of their particles or quanta.

These reasons are responsible for the transformation of the relatively narrow subject of photoelectric effects in semiconductors into a wider study of the internal ionization in semiconductors, produced by very differing radiations. However, even this wider study does not cover completely all the investigations, similar in their experimental methods and analysis of the results, currently proceeding in this branch of semiconductor physics.

Until recently all methods of generating nonequilibrium carriers in semiconductors (with the exception of very strong electric fields) involved internal ionization. Nowadays, however, nonequilibrium carriers are frequently generated by the injection of "minority" nonequilibrium carriers by means of a rectifying contact (this is particularly effective at the contact between two semiconductors with different signs of conduction).

In semiconductors with carriers of high mobility and long lifetime, and consequently long diffusion length, this injection method makes it possible to raise the carrier density in a sufficiently large region near the injecting contact.

The behavior of injected carriers and those liberated by ionization is the same (here again the reason is the strong interaction with the lattice). Thus, injection is simply a new method of generating nonequilibrium carriers in a semiconductor.

Including this new method we now have an even wider field of study of nonequilibrium electron processes in semiconductors, or, in other words, the behavior of nonequilibrium carriers in semiconductors.

The extension of photoelectric studies by the use of various radiations and injection in conjunction with the methods of combined investigation of nonequilibrium conductivity has widened enormously the experimental means and given new results important both from the scientific and practical points of view.

This can easily be understood because nonequilibrium carriers are the working substance of the most important semiconductor devices, such as diodes, transistors, photodiodes, phototransistors, photoresistors, p-n junction convertors of optical into electrical energy, crystal and semiconductor counters, etc.

Nonequilibrium processes may be divided into three main groups:

Generation of nonequilibrium carriers;

Motion of such carriers (diffusion; drift in an electric field;
   motion in a magnetic field; motion in homogeneous semiconductors, including those containing p-n junctions);

Recombination of nonequilibrium carriers.

All these phenomena are to some extent covered by the present book.
However, the greatest attention is given to the recombination and motion
of nonequilibrium carriers.

The processes of carrier generation are briefly analyzed in Chap. IV.
Their motion is dealt with in Chaps. XII-XV. Recombination is covered
in Chaps. V-XI.

The first three chapters form an introduction which gives a phenomenological description of photoconductivity and formulates preliminary
definitions of some quantities and concepts (Chap. I). It describes the
principles of some methods used to study photoconductivity (Chaps. II
and III).

The title of the book indicates that we are concerned principally with
the phenomena connected with o p t i c a l ionization. However, in view of
what was said above, it is obvious that with the exception of the nature of
the excitation process itself, the rest of the treatment (especially in the
case of recombination) applies also to the phenomena occurring when nonequilibrium carriers are generated by other means.

The present book does not claim to be complete or rigorous in the
description of all photoelectric processes. It does not deal with some
important problems which have not yet received sufficient attention.* The
selection of the material reflects to some extent the problems investigated
in the Laboratory for Nonequilibrium Processes in Semiconductors of the
A. F. Ioffe Physics Institute, USSR Academy of Sciences.

The principal aim of the author is to give a quantitative but lucid description of some of the important phenomena. For this reason the treatment frequently begins with the particular and leads to the general, which
is not strictly logical, but illustrates the phenomenon more clearly.

Actual experimental data for various semiconductors are used very
sparingly† and only to illustrate some phenomenon.

The book does not give the history of photoelectric studies. Each
chapter has a chronological list of selected references. Some of the
items in these lists are specifically referred to in the text, but there are
others which are recommended as additional reading on the problems discussed.

*Among them are the problems of the mechanism of radiationless recombination, the role of
excitons in photoconductivity, etc.

† There are monographs which give these data in full. See, for example, T. S. Moss, Photoconductivity in the Elements (London, 1952); T. S. Moss, Optical Properties of Semiconductors
[Russian translation](IL, 1961); R. H. Bube, Photoconductivity of Solids [Russian translation](IL,
1962).

The author is grateful to his colleagues at the Laboratory for Non-equilibrium Processes in Semiconductors, in particular to L. G. Paritskii, I. D. Yaroshetskii, E. N. Arkad'eva, N. B. Strokan, A. A. Grinberg, R. Yu. Khansevarov, B. M. Konovalenko, F. M. Berkovskii, and A. A. Rogachev, for reading the manuscript and valuable advice.

The author is grateful also to D. G. Ryvkina and I. V. Andrianova for carrying out various calculations and plotting some graphs.

S. M. Ryvkin

CONTENTS

Chapter I

PHENOMENOLOGICAL DESCRIPTION OF PHOTOCONDUCTIVITY

Chapter II

METHODS OF MEASURING STEADY-STATE PHOTOCONDUCTIVITY

Chapter III

DETERMINATION OF THE PRINCIPAL PHENOMENOLOGICAL PARAMETERS β AND τ FROM STUDIES OF PHOTOCONDUCTIVITY KINETICS

Chapter IV

PROCESSES OF GENERATION OF NONEQUILIBRIUM CARRIERS

Chapter V

RECOMBINATION THROUGH SIMPLE LOCAL CENTERS

Chapter VI

PROCESSES OF NONEQUILIBRIUM CARRIER TRAPPING

Chapter VII

RECOMBINATION THROUGH MULTIPLY-CHARGED CENTERS

CONTENTS

Chapter VIII

DIRECT (INTERBAND) RECOMBINATION

Chapter XI

MEANING OF THE CONCEPT "LIFETIME"

Chapter XII

DIFFUSION AND DRIFT OF NONEQUILIBRIUM CARRIERS
(UNIPOLAR CASE)

Chapter XIII

DIFFUSION AND DRIFT OF NONEQUILIBRIUM CARRIERS
(AMBIPOLAR CASE)

Chapter XIV

SOME PHOTOMAGNETOELECTRIC
AND PHOTOMAGNETODENSITY EFFECTS

(i) Photomagnetoelectric Effects

(ii) Magnetodensity Effects

Chapter XV

PHOTOELECTROMOTIVE FORCES
IN INHOMOGENEOUS SEMICONDUCTORS

Chapter I

PHENOMENOLOGICAL DESCRIPTION OF PHOTOCONDUCTIVITY

§1. EQUILIBRIUM AND NONEQUILIBRIUM CARRIERS; NONEQUILIBRIUM CONDUCTIVITY

The process of formation of free charge carriers requires energy for overcoming the energy gaps between allowed bands or between local impurity levels and these bands.

Under normal conditions this energy is drawn from the reservoir of thermal energy present in the crystal.

Electrons in a crystal interact very strongly with the atoms (or ions) of the crystal lattice and therefore under normal conditions the lattice and electron temperatures are practically identical.

Consequently, heating a semiconductor amplifies the intensity of the thermal vibrations of atoms (or ions) at lattice sites and simultaneously changes the electron energy distribution. At the same time the numbers of electrons in the conduction band and of holes in the valence band are altered.

Free electrons and holes (liberated by "thermal ionization") having densities corresponding to thermal equilibrium (i.e., equilibrium between electrons and the lattice) will be called "equilibrium" carriers.

Apart from thermal ionization, other means may serve to generate free electrons and holes, for example, optical ionization (internal photoeffect) or ionization caused by other radiations. Moreover, excess carriers may appear in a semiconductor due to "injection" at a rectifying contact, application of strong fields (impact ionization), etc.

The generation of excess carriers requires energy (for example, the energy of light quanta absorbed in the internal photoeffect). In contrast to the thermal energy, the energy used to generate excess carriers is retained mainly by the electrons, and the average thermal energy of the crystal lattice remains practically unaffected. Consequently the thermal equilibrium between the lattice and electrons is disturbed. Therefore carriers formed in some way other than by thermal ionization are called "nonequilibrium."

1

Obviously when the external excitation (light, injection, strong field) ceases to act, the thermal equilibrium between the lattice and electrons will (after a time) be re-established. Usually the number of nonequilibrium carriers is not very large and their excess energy is small compared with the thermal energy of the crystal lattice. Therefore the process of re-establishing equilibrium between the lattice and electrons reduces to the recombination of nonequilibrium electrons and holes, and the temperature of the lattice and the whole crystal (and consequently the density of equilibrium carriers) remains practically unchanged.

Taking this into account we may assume that the application or removal of an excitation (for example, illumination of a semiconductor or placing it in darkness) changes the density of nonequilibrium carriers without affecting the density of equilibrium ones, and therefore the total density (n or p) is simply the sum of the equilibrium (n_0, p_0) and non-equilibrium (Δn, Δp) carrier densities:

$$n = n_0 + \Delta n,$$
$$p = p_0 + \Delta p.$$

The presence of nonequilibrium carriers alters the conductivity of a semiconductor, and in the general case this can be written in the form

$$\sigma = e\,(\mu_n n_0 + \mu_p p_0 + \mu_n \Delta n + \mu_p \Delta p), \tag{1.1}$$

where e is the electronic charge, and μ_n and μ_p are respectively the electron and hole mobility.

Consequently the excess (nonequilibrium) conductivity can be written in the form

$$\Delta\sigma = e\,(\mu_n \Delta n + \mu_p \Delta p). \tag{1.2}$$

In writing down Eqs. (1.1) and (1.2) we ascribed the same mobilities to equilibrium and nonequilibrium carriers. Later (§3) we shall show that such an assumption may be made in the great majority of actual cases.

Let us now see which factors affect the values of the nonequilibrium densities Δn and Δp governing the nonequilibrium conductivity $\Delta\sigma$.

To make the case definite we shall assume that the nonequilibrium conductivity appears as the result of illumination (internal photoeffect or photoconductivity). The relationships obtained may be easily extended to any other form of excitation.

When a semiconductor is illuminated with light of wavelengths lying in the fundamental absorption region, the absorption of light quanta produces electron transitions from the valence band to the conduction band and consequently generates nonequilibrium electrons and holes.

It is reasonable to assume that the numbers of electrons and holes generated per unit time in a unit volume, which we shall denote respectively by $\Delta n'$ and $\Delta p'$, should be proportional to the optical energy absorbed in that time in the unit volume.

Let the light intensity be I. Then the amount of light energy absorbed per unit time in 1 cm^2 of a layer of thickness dx (x is the direction of propagation of light) is proportional to I and the layer thickness dx:

$$- dI = kI\,dx, \qquad (1.3)$$

where k is a coefficient of proportionality, known as the optical absorption coefficient.

The optical energy absorbed per unit time in unit volume is

$$- \frac{dI}{dx} = kI. \qquad (1.4)$$

Thus $\Delta n'$ and $\Delta p'$ should be proportional to the quantity kI:

$$\Delta n' = \Delta p' = \beta kI, \qquad (1.5)$$

where β is a coefficient of proportionality. If I represents the number of quanta per second, then β represents the "quantum yield," i.e., the number of pairs formed by a single quantum. *

We shall assume that illumination began at a certain moment. If no other processes took place except carrier liberation, then obviously the density of nonequilibrium carriers would increase with time without limit according to the law

$$\Delta n = \Delta p = \beta kIt. \qquad (1.6)$$

Figure 1 shows such a dependence for Δn by a dashed straight line. It is known from experiments that in fact after a certain time from the commencement of illumination a constant (steady-state) photoconductivity $\Delta\sigma_{st}$ is established corresponding to steady-state values of the nonequilibrium carrier densities Δn_{st} and Δp_{st}. It follows therefore that as well as the process of generation of free carriers there must be a converse process of carrier annihilation and the rates of both processes must be equal when the steady state is reached.

This converse process is the recombination of nonequilibrium electrons and holes. Obviously the rate of recombination is directly related to the nonequilibrium carrier densities, since at the commencement of illumination when there are still few nonequilibrium carriers this rate is

---

*Usually β cannot exceed unity. However, if the energy of the quantum exceeds double the width of the forbidden band, then the carriers generated may possess sufficient kinetic energy to form additional nonequilibrium pairs by impact ionization. Then $\beta > 1$.

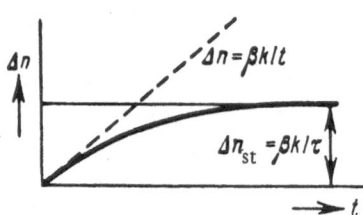

Fig. 1. Variation of carrier density with time during illumination.

low, but later, as the number of such carriers increases, the rate of recombination increases becoming equal to the rate of generation. This corresponds to the steady-state nonequilibrium conductivity. Thus the true curve of the variation of nonequilibrium density with time exhibits "saturation" after a certain time from the commencement of illumination (Fig. 1, continuous curve).

Let us determine how the value of the steady-state density may be expressed in terms of parameters representing the process of the interaction of light with matter. For this purpose we shall introduce the concept of the average lifetime of nonequilibrium carriers. Each nonequilibrium carrier liberated by light exists in the free state for a certain time until it recombines (this is the so-called free-state lifetime). This lifetime may of course be different for different carriers. Therefore we introduce the concept of the "average lifetime," denoted by τ. The value of this lifetime varies within wide limits (usually from $\sim 10^{-2}$ to 10^{-7} sec) depending on the substance.

Obviously the steady-state electron density Δn_{st} may be written in the form of the product of the number of carriers liberated by light per unit time in unit volume ($\Delta n' = \beta kI$) and the average time of their existence in the band before recombination τ_n:

$$\Delta n_{st} = \beta kI \tau_n. \tag{1.7}$$

Similarly, for holes

$$\Delta p_{st} = \beta kI \tau_p. \tag{1.8}$$

Substituting Eqs. (1.7) and (1.8) into Eq. (1.2), we obtain

$$\Delta \sigma_{st} = \Delta \sigma_n + \Delta \sigma_p = e\beta kI \, (\mu_n \tau_n + \mu_p \tau_p). \tag{1.9}$$

If one of the terms in parentheses in the above equation is much greater than the other term (because of a large difference in the mobilities or the lifetimes of electrons and holes), then we have "unipolar" nonequilibrium conductivity due to carriers of one sign only [2, 4, 5]:

$$\Delta \sigma_{st} = e\mu \tau \beta kI. \tag{1.10}$$

In this case the steady-state nonequilibrium conductivity is governed by four parameters: μ, τ, β, and k. Two of these (k and β) represent the interaction of light with matter and govern the process of the generation of nonequilibrium carriers, while the other two (τ and μ) describe the interaction of carriers with matter and represent the processes of motion and recombination of nonequilibrium carriers.

§2. SOME CHARACTERISTICS OF EQUILIBRIUM CONDUCTIVITY

Before we consider in greater detail the problem of nonequilibrium conductivity, let us briefly describe the equilibrium processes, since an understanding of these will be essential in the later treatment of nonequilibrium carriers.

The probability of the occupation of energy levels with energy $\mathscr{E}$ by electrons is given by the Fermi function

$$f = \frac{1}{e^{\frac{\mathscr{E}-F}{kT}} + 1},$$ (2.1)

the nature of which is given for a nonzero temperature on the right of Fig. 2. In the same figure, the Fermi level F is shown by a chain line (the

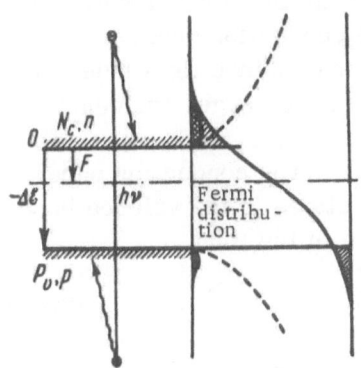

probability of occupation of a state with the energy F is exactly 1/2). We are interested in the probabilities of occupation of states in the allowed bands. We shall consider first the case when the Fermi level in the forbidden band is sufficiently far from the edges of this band ($|F| \gg kT$ $|-\Delta\mathscr{E} - F| \gg kT$), i.e., the case of no degeneracy.

In this case the probability of the occupation of the electron states in the conduction band and of the hole states in the valence band is small (cf. the lightly hatched "tails" of the Fermi distribution within the bands, shown in Fig. 2) and consequently the Pauli exclusion principle cannot greatly affect the nature of the distribution. Therefore the Fermi distribution in the allowed bands reduces to the Maxwell distribution. For electrons in the conduction band

Fig. 2. Distribution of electrons and holes over the energy states in the conduction and valence bands.

$$f = e^{\frac{F-\mathscr{E}}{kT}},$$ (2.2)

and for holes in the valence band

$$f' = e^{\frac{\mathscr{E}-F}{kT}}.$$ (2.3)

To find the density of free electrons (holes), it is necessary to multiply the number of states in the energy interval $d\mathscr{E}$ in the appropriate band by the probability of occupation f (or f') and to sum all possible states in the band.

The dashed curves on the right of Fig. 2 show the dependence of the number of states on the energy and the heavily hatched regions represent the electron distribution over the states in the bands.

Calculations show that the total densities of free electrons and holes are, respectively,

$$n_0 = \frac{2}{h^3} (2\pi m_e kT)^{3/2} e^{\frac{F}{kT}} = N_c e^{\frac{F}{kT}}, \qquad (2.4)$$

$$p_0 = \frac{2}{h^3} (2\pi m_h kT)^{3/2} e^{\frac{-F-\Delta\mathscr{E}}{kT}} = P_v e^{\frac{-F-\Delta\mathscr{E}}{kT}}, \qquad (2.5)$$

(here m_e and m_h are the effective masses of an electron and a hole). Expression (2.4) is the product of a quantity with the dimensions of a density N_c and a multiplier $\exp(F/kT)$ which can be treated as the probability of the occupation of a level at a distance F from the Fermi level, i.e., the level corresponding to the lower edge of the conduction band.

Consequently, in calculating the density we can replace the whole conduction band by energy states coinciding with its lower edge, the density of such states being assumed to be N_c. For this reason N_c is frequently called the effective density of the electron states in the conduction band. Similarly P_v is the effective density of the hole states in the valence band.

The equilibrium conductivity may be written in the form

$$\sigma = e \, (\mu_n n_0 + \mu_p p_0). \qquad (2.6)$$

The electron and hole mobilities μ_n and μ_p are determined by the nature of the interaction of carriers with the lattice, and they depend, in particular, on the energy distribution of the carriers in the bands. At equilibrium the mobility values correspond to the equilibrium energy distribution of the carriers in the bands shown in Fig. 2. For this reason such mobilities may be called "equilibrium" mobilities.

§3. ENERGY DISTRIBUTION OF NONEQUILIBRIUM CARRIERS

The considerations set out in the preceding section indicate that equilibrium electrons and holes occupy some of the states near the band edges, and they have a definite energy distribution at a given temperature.

We shall now determine the energy distribution of nonequilibrium electrons and holes.

Immediately after ionization the energy of a nonequilibrium electron and hole, formed by a photon of sufficient energy $h\nu$, may be considerably greater than the average energy of the equilibrium carriers (Fig. 2), which is of the order of kT.

However, as the result of interaction with phonons and lattice defects, the kinetic energy of the nonequilibrium carriers rapidly decreases to the equilibrium value (cf. wavy arrows), i.e., nonequilibrium carriers attain the temperature of the lattice. A calculation has shown [1] that an electron with energy of about 1 eV loses its excess energy after $\sim 10^3$ collisions. This means that an electron having a mean free path of $\sim 10^{-6}$ cm and a velocity of 10^7 cm/sec (in fact the average velocity is considerably higher) loses its excess energy in 10^{-10} sec.

We have already pointed out that the lifetime of nonequilibrium carriers in the bands amounts to 10^{-2}-10^{-7} sec, which is considerably longer than the time necessary to lose the excess energy. Consequently a nonequilibrium carrier exists for a large part of its lifetime in a state with the same average kinetic energy (temperature) as the equilibrium carriers. In the great majority of cases therefore we may assume that the energy distributions of the nonequilibrium and equilibrium carriers in the band are identical (it is assumed that the nondegenerate state is retained when the nonequilibrium carriers appear in the bands).

Thus the generation of nonequilibrium carriers in semiconductors simply alters the density of free carriers, leaving unaffected the energy distribution of these carriers in the bands and the average kinetic energy per free carrier.

We saw [cf. Eqs. (2.4) and (2.5)] that the equilibrium carrier density is determined uniquely (at a given temperature) by the position of the Fermi level.

In the absence of equilibrium it is frequently convenient to represent the total electron and hole densities (equilibrium and nonequilibrium) by means of certain energy levels which enter the expressions for the total densities in the same way as the Fermi level in Eqs. (2.4) and (2.5), i.e.,

$$n = n_0 + \Delta n = N_c e^{\frac{F_n}{kT}}, \tag{3.1}$$

$$p = p_0 + \Delta p = P_v e^{\frac{-F_p - \Delta \mathscr{E}}{kT}}. \tag{3.2}$$

The levels with the energies F_n and F_p are called Fermi quasi-levels. It should be stressed that at equilibrium the Fermi level is the same for electrons and holes in the bands or at local states but the Fermi quasi-levels are different for the conduction and valence bands and have separate values for each type of local energy state.

It follows that the introduction of Fermi quasi-levels is a purely formal assumption and the method of giving the distance from these quasi-levels to a given level is simply another method of representing the electron density at this level.

We established that the energy distribution of the nonequilibrium carriers in a band does not differ from the equilibrium value for the largest part of their lifetime. Consequently the mobility of the nonequilibrium carriers does not differ from the mobility of equilibrium ones, which has already been used in writing down the expression for the nonequilibrium conductivity [Eq. (1.2)]. Obviously a similar conclusion can be drawn about any other parameter of carriers. In particular, the equilibrium and nonequilibrium carriers have the same average probability of recombination, and in the absence of degeneracy not only the equilibrium carriers but also the nonequilibrium ones satisfy the Einstein relationship between the mobility μ and the diffusion coefficient D:

$$\frac{D}{\mu} = \frac{kT}{e} .$$
(3.3)

§4. LIFETIME OF NONEQUILIBRIUM CARRIERS

Each nonequilibrium carrier, for example an electron, takes part in thermal motion, and consequently while moving in a band it has a definite probability of meeting a hole and of recombining with it or, as it is sometimes expressed, of being captured by a hole.

In general, a given semiconductor may contain several "types" of holes (free holes in the valence band, holes localized at various impurity levels or structure-defect levels, etc.).

Obviously the probability of an electron meeting holes of a given type is proportional to the density of these holes and to the average relative velocity of motion of the electron and the holes v_n. If we introduce a concept of "capture cross section" (or "recombination cross section") q_n for the capture of an electron by holes of a certain type the density of which is p, then obviously the average time between two meetings of the electron with these holes is

$$\tau_n = \frac{\cdot \ 1}{q_n v_n p} .$$
(4.1)

Since in fact each such meeting ends with the capture of the electron by a hole, the time τ_n is essentially the average electron lifetime before capture by holes of the given type. If the semiconductor contains k "types" of hole which capture electrons, then for each of them we can introduce the lifetime τ_{nk}:

$$\tau_{nk} = \frac{1}{p_k q_{nk} v_{nk}} .$$
(4.2)

In this case the effective electron lifetime in the conduction band may be deduced from the following considerations. The quantity in the denominator of Eq. (4.1), $q_n v_n p$, governs the number of possible collisions of an electron with holes per unit time. In fact, the first collision ends with the capture of the electron, but this is unimportant for our calculations.

If there are several types of hole which capture electrons, and their densities, cross sections, and average relative velocities of motion are p_k, q_{nk}, and v_{nk} respectively, then the total number of possible collisions per unit time with all types of hole is obviously given by the sum

$$\sum_k p_k v_{nk} q_{nk}. \tag{4.3}$$

Consequently, the effective electron lifetime is

$$\tau = \frac{1}{\sum_k p_k v_{nk} q_{nk}}. \tag{4.4}$$

Expression (4.4) may be rewritten in the form:

$$\frac{1}{\tau} = \sum_k \frac{1}{\tau_{nk}}. \tag{4.5}$$

Thus, complex processes involving several types of "capture center" may be considered in terms of effective recombination characteristics obtained by summation of the "reciprocal lifetimes."

All the considerations given above apply to the calculation of the electron lifetime τ_n in the conduction band, but obviously they are also fully applicable to the calculation of the hole lifetime τ_p in the valence band. Thus, in place of Eq. (4.2) we can write for the hole lifetime:

$$\tau_{pk} = \frac{1}{n_k q_{pk} v_{pk}}, \tag{4.6}$$

where n_k is the density of electrons of a given type capable of capturing holes, and q_{pk} is the cross section for the capture of a hole by these electrons.

We shall now find how we can express the rate of recombination, i.e., the number of recombination (capture) acts per unit time in unit volume. As stressed above, the quantity in the denominator of Eq. (4.2) [or Eq. (4.6)],

$$p_k q_{nk} v_{nk} = \frac{1}{\tau_{nk}} \tag{4.7}$$

determines the number of collisions (which end with capture or recombination) of one nonequilibrium electron per unit time. Therefore multiplying this number by the electron density and averaging $q_{nk} v_{nk}$ we obviously obtain the recombination (capture) rate

$$\overline{q_{nk}v_{nk}}\, p_k\, \Delta n = \frac{\Delta n}{\tau_{nk}}. \tag{4.8}$$

Similarly, for holes the recombination (capture) rate is

$$\overline{q_{pk}v_{pk}}\, n_k\, \Delta p = \frac{\Delta p}{\tau_{pk}}. \tag{4.9}$$

Thus the recombination (capture) rate is governed by a fraction the denominator of which contains the lifetime for a given transition.

The product of the cross section and the velocity, averaged for all the carriers in the band,

$$\overline{q_{pk}v_{pk}} = \gamma_{pk} \tag{4.10}$$

is called the recombination (capture) coefficient and represents, as can easily be seen from Eq. (4.9), the recombination rate when the densities n_k and Δp are equal to unity. Similarly

$$\overline{q_{nk}v_{nk}} = \gamma_{nk}. \tag{4.11}$$

From Eqs. (4.2) and (4.6) we find that the values of the average lifetime of all the carriers in the bands (for which we retain the previous notation τ_{nk}, τ_{pk}) can be written in the form

$$\left.\begin{aligned} \tau_{nk} &= \frac{1}{\gamma_{nk}p_k}, \\ \tau_{pk} &= \frac{1}{\gamma_{pk}n_k}. \end{aligned}\right\} \tag{4.12}$$

Concluding, we note that since the values of the lifetimes τ_{nk} and τ_{pk} depend on the densities p_k and n_k, they are not only not constant for a given material, but can also vary with time (under non-steady-state conditions), with temperature, with intensity of illumination, etc. Consequently the lifetimes in Eqs. (4.8) and (4.9) cannot in general be regarded as constants.

§5. RELAXATION OF NONEQUILIBRIUM CONDUCTIVITY [2, 3, 5]

If illumination begins at a certain moment (cf. Fig. 1), then the steady-state value of the nonequilibrium conductivity of a semiconductor is reached only after a finite time. Similarly, the nonequilibrium conductivity does not decay instantaneously when the illumination ceases (Fig. 3). The rise or decay curves of the nonequilibrium conductivity for any variation of the illumination intensity are known as the relaxation curves of the nonequilibrium conductivity. We must determine how the nonequilibrium conductivity depends on time during relaxation, what is the ef-

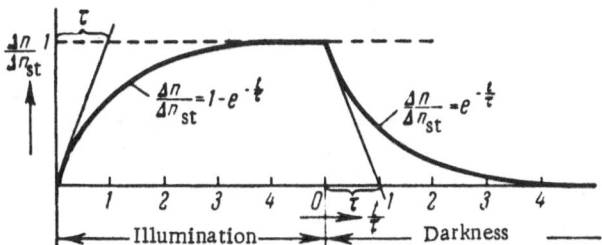

Fig. 3. Relaxation of the nonequilibrium density on excitation with a square light pulse. Linear recombination case.

fective time for the establishment of the steady-state values of the conductivity, and what is the relationship between the nature of the relaxation
processes and the mechanism of nonequilibrium conductivity.

Since at any given time the electron and hole components of the nonequilibrium conductivity $\Delta\sigma$ are proportional to the nonequilibrium densities Δn and Δp, we shall be interested in the relaxation of the density
Δn (or Δp). The change in the number of carriers (for example, electrons) per unit time in unit volume is the difference between the number
of carriers liberated, which can be written in the form βkI [cf. Eq. (1.5)],
and the number of carriers which recombine. In the present section we
shall consider two important special cases without going into the details of
the recombination mechanism.

1. The rate of recombination (capture) is proportional to the first
power of the nonequilibrium carrier density. [This case is realized, for
example, when there are holes of only one type which recombine with the
nonequilibrium electrons, and the density of these holes p is very high
and practically independent of the illumination. Then, in accordance with
Eq. (4.12), $\tau_n = 1/\gamma_n p$ = const and the rate of recombination (capture) of
electrons, equal to $\Delta n/\tau_n$, is directly proportional to the density.] This
case will be called linear recombination.

2. The rate of recombination is proportional to the square of the nonequilibrium carrier density. This happens, for example, when the density of equilibrium carriers is zero and on ionization the electrons are
transferred from a lower to an upper band; then the densities of nonequilibrium electrons and holes are equal and the rate of recombination, ac-
cording to Eqs. (4.8) and (4.10), is $\gamma\Delta n\Delta p = \gamma (\Delta n)^2$. This case will be
called quadratic recombination.

A. Linear Recombination

The change of the nonequilibrium carrier density per unit time may
be written in the form *

---

*All these considerations are given for the nonequilibrium electrons. Obviously they also apply
to the nonequilibrium holes.

$$\frac{d}{dt}(\Delta n) = \beta kI - \frac{\Delta n}{\tau}.$$ (5.1)

Let us assume that the illumination of the sample with light of constant intensity [i.e., I in Eq. (5.1) is constant] begins at a time t = 0. Then, separating variables and integrating, using the initial condition $\Delta n = 0$ when t = 0, we obtain

$$\Delta n = \tau\beta kI\left(1 - e^{-\frac{t}{\tau}}\right).$$ (5.2)

Thus the relaxation of the equilibrium density (and the corresponding conductivity) in the linear recombination case is exponential. As $t \to \infty$

$$\Delta n = \tau\beta kI = \Delta n_{\text{st}}.$$ (5.3)

This conclusion is the same as that obtained earlier [cf. Eq. (1.7)]. If the sample is not illuminated, then in place of Eq. (5.1) we can write

$$\frac{d(\Delta n)}{dt} = -\frac{\Delta n}{\tau}.$$ (5.4)

Assume now that the illumination of the sample ceased at a time t = 0 and that the nonequilibrium density reached the value $\Delta n = \Delta n_{\text{st}} = \tau\beta kI$ as the result of illumination up to this time. Then from Eq. (5.4) we find that the dependence of Δn on time has the form of a decreasing exponential function:

$$\Delta n = \tau\beta kI e^{-\frac{t}{\tau}}.$$ (5.5)

The relaxation (rise and decay) curves of the nonequilibrium conductivity in the case of illumination by a "square" light pulse are shown in Fig. 3.

It is important to note that the "time constant" of the exponential curves is equal to the lifetime. This gives a simple method of determining τ directly from the relaxation curves in the case of linear recombination.*

B. Quadratic Recombination

In this case in the presence of illumination we have, instead of Eq. (5.1),

*We note that the quantum yield β can be determined from the initial portion of the rise curve [for example, from Eq. (5.2)]. Expanding the expression in the parentheses of Eq. (5.2) as a series we find that for small t we have $\Delta n = \beta kIt$, from which, knowing kI, we can determine the quantity β (cf. Chap III).

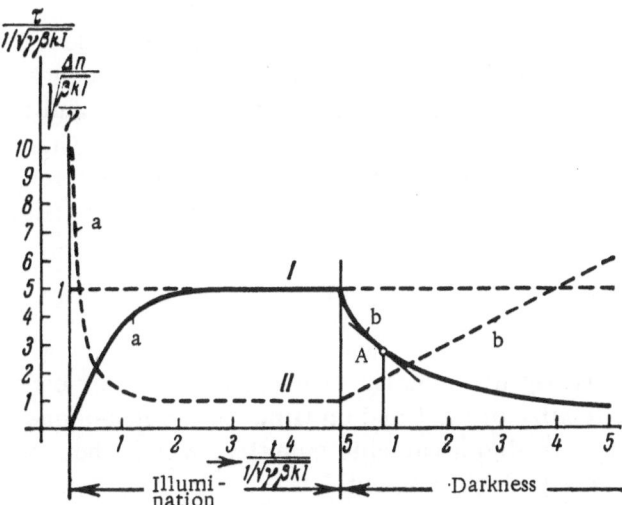

Fig. 4. Relaxation of the nonequilibrium density on excitation with a square light pulse. Quadratic recombination case. Curve I: a) Rise [cf. Eq. (5.8)]; b) decay [cf. Eq. (5.9)]. Curve II, [instantaneous lifetime]: a) Rise [cf. Eq. (5.11)]; b) decay [cf. Eq. (5.12)].

$$\frac{d}{dt}(\Delta n) = \beta kI - \gamma (\Delta n)^2. \tag{5.6}$$

When the illumination ceases we have

$$\frac{d}{dt}(\Delta n) = - \gamma (\Delta n)^2. \tag{5.7}$$

Using the initial conditions, similar to the conditions in the linear recombination case, we find that on illumination with a square light pulse of sufficient duration the relaxation curves of rise and decay of the nonequilibrium density (conductivity) are given by the expressions:
for rise

$$\Delta n = \sqrt{\frac{\beta kI}{\gamma}} \ \tanh \ t \sqrt{\gamma \beta kI} , \tag{5.8}$$

for decay

$$\Delta n = \sqrt{\frac{\beta kI}{\gamma}} \frac{1}{t \sqrt{\gamma \beta kI} + 1} . \tag{5.9}$$

Now the rise and decay are no longer symmetrical, in contrast to the linear recombination case where the rise and decay curves are exponentials with the same time constants.

Here the rise is described by a hyperbolic tangent, and the decay (when t is large) by a much slower hyperbola (Fig. 4).

It is particularly important to note that in the present case (in contrast to the linear recombination case) we cannot introduce the concept of constant lifetime independent of the intensity of light and retaining its value with time during the whole relaxation process. This however by no means implies that the concept of lifetime is altogether inapplicable.

In accordance with Eq. (4.12) the average electron lifetime is in the present case

$$\tau = \frac{1}{\gamma \Delta p}.$$
(5.10)

Since the quantity Δp (the density of holes in the lower band) itself depends on the intensity of light and on time (in the non-steady-state case), it is clear that τ is also a variable quantity, which, however, has a definite meaning at any given moment.

C.   Instantaneous Value of the Lifetime [5, 6]

Bearing in mind the considerations above, we shall introduce the concept of instantaneous lifetime τ_{inst} * (this is completely analogous to the concept of instantaneous velocity in nonuniform motion). In fact it was the instantaneous value of the lifetime which was given in expressions (4.2), (4.6), (4.12), etc. Only in the special case of linear recombination does the instantaneous value of the lifetime remain the same for the whole process of relaxation.

Thus in the general case $\tau = f(I, t)$. We shall determine this function for the quadratic recombination case. For the rise curves we find from Eqs. (5.10) and (5.9) that

$$\tau = \frac{1}{\gamma \Delta p} = \frac{1}{\gamma \Delta n} = \frac{1}{\sqrt{\gamma \beta k I}} \coth t \sqrt{\gamma \beta k I}.$$
(5.11)

For the decay curves we have

$$\tau = \frac{1}{\gamma \Delta n} = \frac{1}{\sqrt{\gamma \beta k I}} (t \sqrt{\gamma \beta k I} + 1).$$
(5.12)

Figure 4 shows by dashed curves the variation of the instantaneous value of τ for a relaxation process. It is not suprising that when t = 0 in the rise case and t → ∞ in the decay case, i.e., when $\Delta n \to 0$, the instantaneous value of τ tends to infinity [cf. Eqs. (5.11) and (5.12)]. This is because when the number of electrons is very small the number of holes is also very small. Therefore the probability of recombination is very

_____

*In subsequent treatment we shall use the subscript "inst" only when it is necessary.

small and the lifetime increases very considerably (although not to infinity, which is obtained only because we have ignored the discrete nature of the electron gas and considered it to be a continuum).

We have now found the dependence of the instantaneous lifetime τ on time for the relaxation processes in the case of quadratic recombination [Eqs. (5.11), (5.12)]. We can, however, easily obtain a general expression for the instantaneous lifetime τ for recombination of any description. In fact the differential equations

$$\frac{d\,(\Delta n)}{dt} = \beta kI - \frac{\Delta n}{\tau} \qquad (5.13)$$

and

$$\frac{d\,(\Delta n)}{dt} = - \frac{\Delta n}{\tau} \qquad (5.14)$$

are valid for relaxation in the most general case, when τ is understood to be the instantaneous value of the lifetime.

Consequently we obtain from Eqs. (5.13) and (5.14) the instantaneous lifetime τ for the rise curve:

$$\tau = \frac{\Delta n}{k\beta I - \frac{d}{dt}(\Delta n)}, \qquad (5.15)$$

and for the decay curve

$$\tau = - \frac{\Delta n}{\frac{d}{dt}(\Delta n)}. \qquad (5.16)$$

It is very convenient to determine the instantaneous value of τ experimentally from the decay curve, by measuring at a given point (for example, that denoted by the point A in Fig. 4) the distance from the abscissa and the slope of the tangent and then take the ratio of these two quantities.

Thus, in the nonlinear case, a relaxation process has an infinite number of values of τ. However, one of these values has special meaning: this is the lifetime under steady-state conditions τ_{st}. From Eq. (5.15) we find that in the steady-state case $\left(\frac{d}{dt}(\Delta n) = 0\right)$

$$\tau_{st} = \frac{\Delta n_{st}}{\beta kI}. \qquad (5.17)*$$

---

* The same expression for τ_{st} follows directly from Eq. (1.7).

The steady-state lifetime can be calculated relatively simply even in the case of complex recombination processes because all that is needed is the solution of the problem under steady-state conditions. This is why τ_{st} is used as an important characteristic of recombination processes.

Obviously in the linear case we have

$$\tau_{inst} = \tau_{st} = \text{const.} \qquad (5.18)$$

In conclusion we shall deal with the problem of the dependence of steady-state nonequilibrium conductivity (density) on the intensity of illumination (the so-called "lux-ampere characteristic"). The steady-state density is, in general,

$$\Delta n_{st} = \tau_{st}\,\beta k I. \qquad (5.19)$$

In the linear case τ_{st} is a constant independent of the intensity of illumination, and consequently the steady-state nonequilibrium density is proportional to the illumination intensity:

$$\Delta n_{st} \sim I. \qquad (5.20)$$

In any nonlinear recombination case τ_{st} depends on the illumination intensity and therefore there is no longer a linear relationship between Δn_{st} and I. In particular, in the quadratic recombination case, we find from Eq. (5.11) that

$$\tau_{st} = \frac{1}{\sqrt{\gamma \beta k I}} \qquad (5.21)$$

and consequently

$$\Delta n_{st} \sim \sqrt{I}. \qquad (5.22)$$

Figure 5 shows the dependences of Δn_{st} on the intensity of illumination for both cases (linear and quadratic recombination).

Thus the linear recombination case is characterized by a linear dependence on the intensity of illumination and the nonlinear recombination case by a nonlinear characteristic. Consequently the concepts of "linear" and "nonlinear" nonequilibrium conductivity are frequently used.

§6. PHOTOCONDUCTIVITY OF FINITE SAMPLES

The measured total change of the conductivity of a sample cannot, in general, be written in the form of a simple combination of the quantities e, μ, k, β, τ, and I, as was done in the case of unipolar photoconduc-

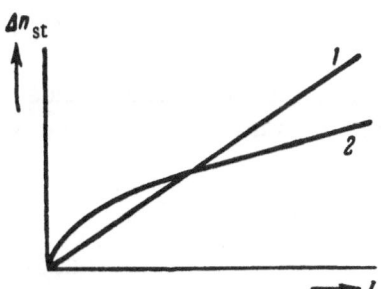

Fig. 5. Dependence of the steady-state nonequilibrium density on the illumination intensity. 1) Linear recombination case; 2) quadratic recombination case.

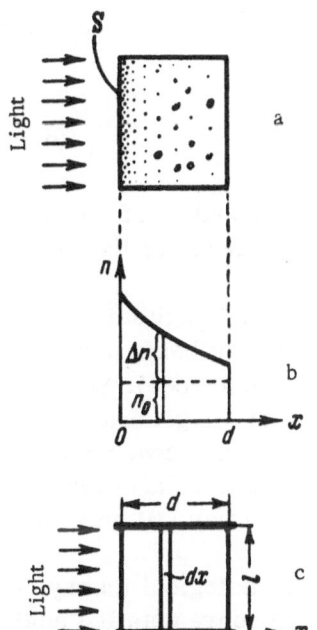

Fig. 6. a) Decrease of the nonequi-librium-carrier generation rate with depth in a sample. b) Dependence of the carrier density on the distance from the illuminated surface. c) Calculation of the transverse photo-conductivity.

tivity, because the value of I is not constant for the whole sample, due to absorption. This difficulty may be overcome in the following two ways:

1. By using sufficiently thin samples and weakly absorbed light we can ensure approximately the same illumination intensity at all points of the sample. Naturally this method is inconvenient because it restricts the range of wavelengths available for study.

2. Knowing the dependence of the photoconductivity on the illumination intensity, and the absorption law, we can express the measured total photoconductivity in terms of the principal phenomenological parameters.

The second method may be used in a relatively simple fashion only in the case of linear photoconductivity and normal orientation of the light beam with respect to the electric field applied to the sample ("transverse" photoconductivity).

Consider a sample in the form of a rectangular parallelepiped of dimensions $l \times h \times d$, which is illuminated with monochromatic light as shown in Fig. 6. The intensity of light decreases with depth in the sample according to the law $I = I_0 e^{-kx}$. The steady-state density is $n_0 + \Delta n_x$, and Δn_x decreases with depth according to the same law:

$$\Delta n_x = \Delta n_0 e^{-kx} = \beta k I_0 \tau e^{-kx}. \qquad (6.1)$$

(In the linear unipolar case considered here the effect of diffusion on the carrier density distribution along the X axis may be neglected for semiconductors with sufficiently high conductivity.) Let us divide the sample into layers of thickness dx. The photoconductivity of one such layer is

$$e\mu \, \Delta n_x \frac{h \, dx}{l} \, . \tag{6.2}$$

Since all the layers dx are "connected" in parallel, the photoconductance of the whole sample $\Delta\sigma^*$ may obviously be found by summation over the whole thickness d†

$$\Delta\sigma^* = \frac{h}{l} \, e\mu \int_0^d \Delta n_x \, dx . \tag{6.3}$$

Substituting Eq. (6.1) into Eq. (6.3), we obtain

$$\Delta\sigma^* = \frac{h}{l} \, e\mu\beta\tau I_0 (1 - e^{-kd}) . \tag{6.4}$$

For sufficiently thick samples, when the effective depth of penetration of light 1/k is considerably less than the sample thickness d (i.e., $kd \gg 1$), we obtain from Eq. (6.4)

$$\Delta\sigma^* = \frac{h}{l} \, e\mu\beta\tau I_0 . \tag{6.5}$$

Thus, the photoconductivity of "thick" samples is independent of the absorption coefficient and is governed only by the total amount of light energy I_0 penetrating into the sample.

As well as the "transverse" photoconductivity, the "longitudinal" photoconductivity is sometimes investigated with the light beam parallel to the field (illumination through a semitransparent electrode). Without analyzing the longitudinal case we should point out that, in general, it has a more complex relationship between the steady-state photoconductivity and the phenomenological parameters (k, β, τ). Simple relationships of the type of Eq. (6.4) or (6.5) are obtained (cf. §12 in Chap. II) only when $\Delta\sigma \ll \sigma_0$.

We have considered some aspects of the phenomenological theory of nonequilibrium conductivity and of recombination. A more detailed analysis of these problems is dealt with mainly in Chaps. IV-XI.

Before considering these problems we shall deal first with the experimental methods of investigating photoconductivity.

---

† The superscript • will be used to distinguish the conductances σ_0^*, $\Delta\sigma^*$ for the whole sample from the conductivities σ_0 and $\Delta\sigma$.

Chapter II

METHOD OF MEASURING STEADY-STATE
PHOTOCONDUCTIVITY

In studies of photoconductivity one needs to measure both the value of the steady-state photoconductivity and the parameters (for example, in the simplest case, τ, β, etc.)* which require a study of the photoconductivity kinetics (cf. §5 in Chap. I). However, the present chapter deals only with the principal methods of measuring the steady-state photoconductivity; methods of investigating the photoconductivity kinetics are discussed in Chap. III.

§7. METHODS USING CONTINUOUS AND MODULATED ILLUMINATION

In the case when $\Delta\sigma_{st} \gg \sigma_0$† the value of $\Delta\sigma_{st}$ may be measured by means of a very simple circuit connecting in series a voltage source, the sample, and a measuring instrument (for example, a galvanometer, as shown in Fig. 7). In this circuit the current rise on illumination is measured on the "background" of the dark current. Therefore, when $\Delta\sigma_{st} \ll \sigma_0$, which is usually true for semiconductors with high conductivity, other circuits are used to eliminate the dark current in some way.

As a rule this is done by one of the following two methods:

Using a standard compensation circuit, as shown in Fig. 8, in which the voltage drop across the load R in darkness is first compensated with a potentiometer P and then the change of the voltage (and therefore of the current) on illumination is measured;

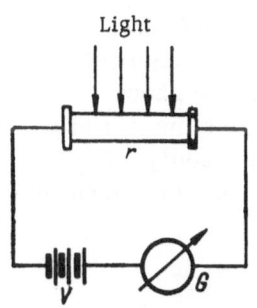

Fig. 7. Simple circuit: r is the sample, V — voltage source, G — galvanometer.

* In complex cases the number of such parameters is much larger (cf., for example, §15 in Chap. III).

† Here and later (§7-10), we shall understand σ_0, $\Delta\sigma_{st}$, r_0, etc., to mean the conductance (resistance) of the sample; the superscript • will be omitted.

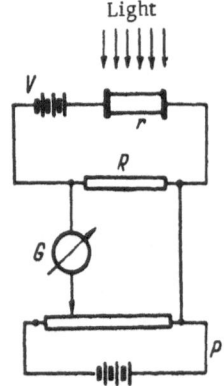

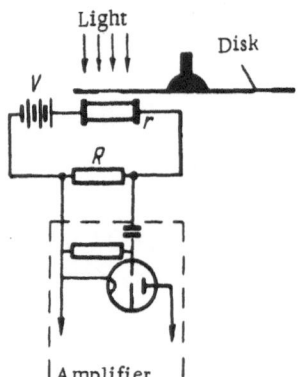

Fig. 8. Compensation cir-
cuit used for measure-
ments.

Fig. 9. Measuring circuit using
chopped illumination.

Using chopped (or, in general, intensity-modulated) illumination, and distinguishing the dark current and the photocurrent by
by their "frequency" [1].

If the circuit in Fig. 9 is used with a constant-current source, then
the dark current has the frequency $f = 0$ and the photocurrent varies with
the frequency of modulation of the light (the modulation is achieved, for
example, by passing the light through apertures in a rotating disk*). If
the measuring instrument is such that it is sensitive to the frequency of
modulation of the light but does not react to $f = 0$ (for example, an ac
amplifier with a tube voltmeter at the output), then obviously the alternating component of the signal represents the value of the photocurrent.

The modulated-illumination method for measuring $\Delta\sigma_{st}$ has certain
practical advantages. Moreover, it may be used, as will be shown later
(cf. Chap. III), not only to measure $\Delta\sigma_{st}$ but also to determine τ and β.
Because of this the method is used very widely. The sensitivity and the
procedure of calculating $\Delta\sigma_{st}$ from the experimental results are discussed
below with this method in mind.

§8. CALCULATION OF PHOTOCONDUCTANCE
FROM EXPERIMENTAL DATA

The circuit using chopped illumination is shown in Fig. 9. Light
interrupted by a rotating disk with apertures falls on a sample whose dark

---

* It should be noted that, as shown in Chap. III, the modulation frequency in measurements of
$\Delta\sigma_{st}$ should not exceed a certain limit.

resistance is r_0. The sample is connected in series with a battery supplying a voltage V and a load resistance R. When the disk rotates, interrupting the light, the sample resistance varies with the chopping frequency because of the photoconductivity appearing in it. Consequently the current in the circuit has a constant component and an alternating one. The voltage drop across the resistance R (or more correctly its alternating component v) is amplified by an ac amplifier, and a recording instrument is connected to its output.

We shall now find the relationship between the alternating voltage v, received at the amplifier input, and the variation of $\Delta\sigma_{st}$ under the action of light. We shall denote the change of the sample resistance on illumination by Δr, the current in darkness by i_d and the current during illumination by i_i. Then

$$v = (i_i - i_d)\,R,$$

where

$$i_i = \frac{V}{R + r_0 - \Delta r}, \quad i_d = \frac{V}{R + r_0},$$

$$v = VR\,\frac{\Delta r}{(R + r_0 - \Delta r)\,(R + r_0)}. \tag{8.1}$$

Hence

$$\Delta r = \frac{v\,(R + r_0)^2}{VR + v\,(R + r_0)}. \tag{8.2}$$

In order to deduce $\Delta\sigma$ from Δr we must find the relationship between these quantities:

$$\Delta\sigma = \sigma_i - \sigma_0 = \frac{1}{r_0 - \Delta r} - \frac{1}{r_0} = \frac{\Delta r}{r_0\,(r_0 - \Delta r)}.$$

Hence

$$\Delta r = \frac{r_0^2\,\Delta\sigma}{1 + r_0\,\Delta\sigma} \tag{8.3}$$

$$\left(\text{when }\ \Delta r \ll r_0 \quad \text{or} \quad r_0\,\Delta\sigma \ll 1 \quad \Delta r = \Delta\sigma r_0^2\right).$$

From Eqs. (8.2) and (8.3) we find $\Delta\sigma$:

$$\Delta\sigma = \frac{v\,(R + r_0)^2}{r_0^2 VR - v r_0 R\,(r_0 + R)}. \tag{8.4}$$

Thus, in general, the relationship between the signal v and the photoconductance $\Delta\sigma_{st}$ is seen to be nonlinear and quite complex (to calculate $\Delta\sigma$ we need to know, in addition to V and R, also the dark resistance of the sample r_0).

In some special cases this relationship is simpler. We can distinguish three regimes which are usually employed in experiments and which differ essentially in the method of selecting the load resistance R.

A. Constant-Field Regime

If a small load resistance is used, $R \ll (r_0 - \Delta r)$, then Eq. (8.4) transforms into

$$\Delta\sigma = \frac{v}{VR}. \tag{8.5}$$

The small load-resistance conditions when the relationship between the photoconductivity and the signal is linear may be called the "constant-field regime," because when R is small the illumination does not greatly alter the electric field distribution in the sample and in the load resistance. Consequently, the field in the sample remains constant.

B. Constant-Current Regime

As well as the "constant-field regime," the "constant-current" one is frequently used. In the latter case the load resistance is sufficiently large that $R \gg r$ and the circuit current $I = \frac{V_0}{R+r} \approx \frac{V_0}{R}$ is practically unaffected by illumination. Then

$$\Delta\sigma = \frac{vR}{r(Vr - vR)}. \tag{8.6}$$

Equation (8.6) shows that the constant-current regime does not imply proportionality between $\Delta\sigma$ and v, and moreover, it has no special advantages. * However, in some cases it is used in practice.

C. Maximum-Sensitivity Regime

The sensitivity of a photoresistor may be defined as the ratio of $\Delta\sigma$ (or $\Delta\sigma/\sigma$) to the intensity of incident (or absorbed) light. The sensitivity of the system containing the photoresistor connected in a certain circuit is defined as the ratio of the recorded electrical signal to the intensity

---

* We can easily see that in this regime there is a linear relationship between v and Δr: v = V · $\Delta r/r$. However, one can hardly use the quantity Δr as the characteristic of nonequilibrium processes. In fact, since nonequilibrium processes appear as changes of the carrier density, we should use the quantity $\Delta\sigma$, which is proportional to the nonequilibrium carrier density. The quantity Δr has a complex relationship with $\Delta\sigma$ [this follows from Eq. (8.3)] and therefore with the nonequilibrium carrier density.

of light. To judge the advanges of a circuit one may define the sensitivity of the circuit as the ratio of the recorded electrical signal to the absolute photoconductance $\Delta\sigma$.

We can easily show that for a given sample (i.e., for given σ) we
can select the load resistance to make the signal v a maximum. Thus
we can distinguish a regime of maximum circuit sensitivity (maximum v),
in addition to the constant-field and constant-current regimes.

Differentiating Eq. (8.1) with respect to R and equating dv/dR to zero,
we find the load resistance corresponding to the maximum signal is

$$R_m = \sqrt{r_0(r_0 - \Delta r)} = \frac{1}{\sigma_0}\frac{1}{\sqrt{1 + \frac{\Delta\sigma}{\sigma_0}}}. \tag{8.7}$$

When $R = R_m$, the signal is

$$v_m = \frac{V}{2}\frac{\Delta\sigma}{\sigma_0}\frac{1}{1 + \sqrt{1 + \frac{\Delta\sigma}{\sigma_0}}}. \tag{8.8}$$

D. Case of Weak Relative Photoconductance

Comparison of Eqs. (8.5), (8.6), and (8.8) shows that only under the
constant-field regime is the signal directly proportional to the photoconductance.

However, if the relative change of the photoconductance is small
($\Delta\sigma/\sigma \ll 1$), then a linear relationship between the signal and the photoconductance is obtained for any load resistance and consequently for any
of the regimes mentioned above.

In fact when $\Delta\sigma/\sigma \ll 1$ the second term in the denominator of the
general expression (8.4) may be neglected * and we obtain

$$\Delta\sigma = \left(1 + \frac{R}{r_0}\right)^2\frac{v}{RV} = \text{const } v. \tag{8.9}$$

Figures 10 and 11 show the dependence of the signal, in relative units,
on the ratio of the sample and load resistances when the photoconductance

* In fact if [cf. Eq. (8.4)]

$$\frac{\Delta\sigma}{\sigma} = \frac{v(R+r)^2}{r_0RV - vR(R+r_0)} \ll 1,$$

then the numerator of this expression is much less than the denominator and therefore much less
than the first term in the denominator: $v(R+r_0)^2 \ll r_0VR$. On multiplying both terms by the
same constant we have $vr_0R(R+r_0)\frac{R+r_0}{R} \ll r_0^2VR$. Since $\frac{R+r_0}{R} \geqslant 1$ then the condition

$vR(R+r_0) \ll r_0VR$ is satisfied to an even greater extent.

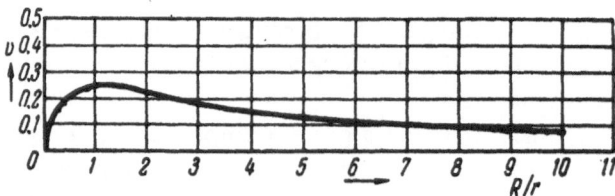

Fig. 10. $v = f(R)$.

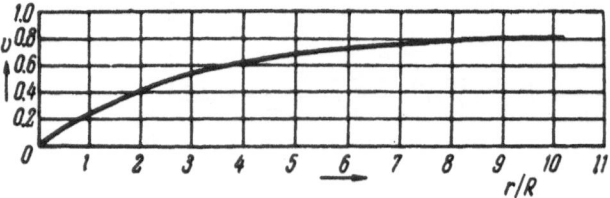

Fig. 11. $v = f(r)$.

$\Delta\sigma$ is small and constant. Figure 10 shows that when $R/r = 1$ the signal is a maximum, as expected. It is apparent from Fig. 11 that, for a given load resistance, the signal increases with increase of the sample resistance.

In the regime of maximum sensitivity when $R = \sqrt{r_0(r_0 - \Delta r)} \approx r_0$, we obtain from Eq. (8.8) [or Eq. (8.9)]

$$v = \frac{V}{4} \frac{\Delta\sigma}{\sigma_0}, \qquad (8.10)$$

i.e., the signal is proportional to the relative change of the conductance on illumination.

§9.   SENSITIVITY THRESHOLD

When small signals v and the corresponding small values of the photoconductance are measured, various types of noise superimposed on the measured signal become important.

The noise may be noise in the sample or in the first stages of the amplifier, or it may be due to stray signals induced in the circuit. If the noise signal is comparable with v or greater than v, we cannot measure the photoconductance with the circuit described above. Therefore, the "sensitivity threshold," which is the minimum light intensity (excitation) to give a measurable effect (conventionally it is equal to the noise signal), is an important characteristic of the performance of a given measuring system (including the sample and the circuit) when small values of $\Delta\sigma$ are measured.

When small signals are measured near the sensitivity threshold the selection of the measuring method and elements of the measuring circuit is governed by the need to achieve a high signal-to-noise ratio. The noise signal (including the fluctuation noise) may be expanded into a frequency spectrum. As a rule this signal has roughly the same intensity over a wide range of frequencies.

Obviously to increase the ratio of the useful signal to the noise, the measuring device should be sensitive in the narrowest possible frequency band around the light-modulation frequency. *

For this reason weak signals are detected and measured by means of narrow-band amplifiers, tuned to the light-modulation frequency.†

If the noise signal v_n at the amplifier input is known, the minimum detectable (threshold) value of the photoconductance can be determined. Thus, in the maximum sensitivity regime we find from Eq. (8.10)

$$\Delta\sigma_n = 4\,\frac{v_n}{V}\,\sigma_0. \tag{9.1}$$

If v_n is, for example, 10^{-7} V, and $V = 10$ V, then

$$\Delta\sigma_{\text{threshold}} = 4 \cdot 10^{-8}\,\sigma_0. \tag{9.2}$$

Thus, the minimum detectable photoconductance is governed by the value of the equilibrium conductance σ_0 and, other conditions being equal, decreases with increase of the sample resistance. ‡

Concluding, we note that it may be impossible to satisfy the maximum sensitivity condition $R = r$ in the case of poorly conducting (insulating) samples because R may be shunted by the input impedance of the amplifier which for many reasons cannot be increased without limit. One may then find that the method using constant illumination, a bridge circuit and an electrostatic instrument (for example, an electrometer) as a null indicator (Fig. 12) may be more sensitive.

§ 10. OPTIMUM SAMPLE DIMENSIONS FOR PHOTOCONDUCTANCE MEASUREMENT

We shall consider briefly the problem of what sample dimensions are most convenient for photoconductance measurements.

---

* It is desirable also that the noise be minimum near this frequency.
† Other methods: synchronous detection, frequency conversion, etc., will not be considered here.
‡ It follows that although the equilibrium conductance is eliminated from the results of measurements when modulated illumination is used, its value still affects the sensitivity and the threshold of sensitivity.

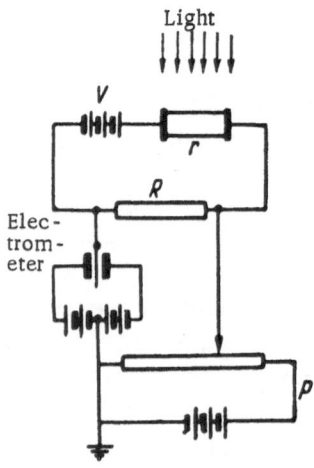

Fig. 12. Circuit for measuring photoconductivity of high-resistivity samples.

We have shown above that under optimum conditions R = r, the sensitivity v is proportional to $\Delta\sigma/\sigma$

$$v = \frac{V}{4}\frac{\Delta\sigma}{\sigma}. \qquad (10.1)$$

For $\Delta\sigma$ we obtained in § 6 the following expression [cf. Eq. (6.4)]:

$$\Delta\sigma = \frac{e\mu\beta\tau\,I_0 h}{l}(1 - e^{-kd}). \qquad (10.2)$$

For the dark conductance we can write

$$\sigma = n_0 e\mu\,\frac{hd}{l}. \qquad (10.3)$$

Substituting Eqs. (10.2) and (10.3) into Eq. (10.1), we obtain

$$v = \frac{V}{4}\frac{\beta\tau I_0}{n_0}\frac{1}{d}(1 - e^{-kd}).$$

If in place of I_0 we substitute I_0^*/hl, where I_0^* is the total light energy falling on the whole sample, then we obtain

$$v = \frac{V}{4}\frac{\beta\tau I_0^*}{n_0}\frac{1}{hld}(1 - e^{-kd}). \qquad (10.4)$$

We shall consider the sensitivity formula in the above form because the total incident light and the voltage V are frequently the quantities specified. Analysis under other conditions causes no difficulties.

The formula of Eq. (10.4) shows that the dependence of v on l and h is very simple: to increase the sensitivity these dimensions should be reduced (this concentrates the light over a smaller area).

The dependence on the sample thickness d is rather more complex. Let us consider it in greater detail. For this purpose we shall denote all the terms not containing d by A. Then

$$v = \frac{A}{d}(1 - e^{-kd});$$

the above expression shows that when d → ∞ , v → 0. When d → 0, we find, by expanding the expression in the bracket as a series, that v → Ak. Thus, the dependence of the sensitivity on the layer thickness has the form of a curve (Fig. 13) decreasing monotonically from Ak (when d → 0) to zero (when d → ∞).

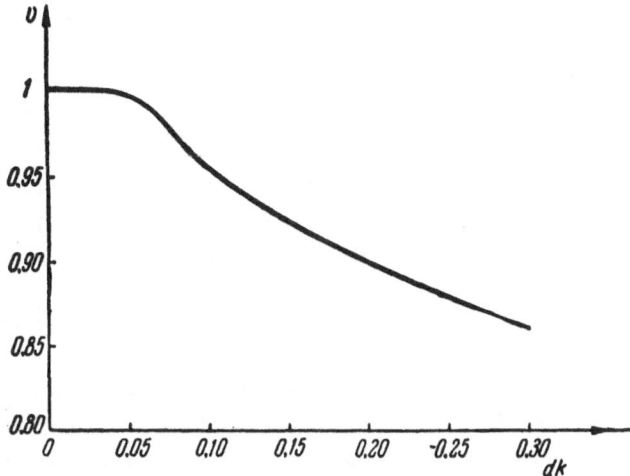

Fig. 13. Dependence of the signal on the sample thickness d in the maximum-sensitivity regime; v is measured in units of Ak.

The sensitivity is low when d is large (in spite of the fact that a larger amount of the light energy is absorbed), since the conductance of the sample is large. When d is decreased the sensitivity should fall because of the reduction of the amount of light absorbed, but the rise of the sample resistance and the resultant increase of the sensitivity prevails. Consequently the sensitivity increases on reduction of the sample thickness.

It follows that under the present conditions the photoresistor sensitivity may be increased by reducing all three dimensions: l, h, and d. *

The limit is reached when the light beam is concentrated as much as possible by modern optical methods.

§ 11.   ELIMINATION OF THE INFLUENCE OF CONTACTS

Contacts and the resultant band curvatures may affect the results of photoconductivity measurements considerably for the following reasons:

1. All the relationships given above for the calculation of the photoconductance [Eqs. (8.2), (8.4), (8.10), etc.] were derived on the assumption of complete homogeneity of the sample along its length.

Obviously the use of these relationships will lead to errors in the presence of contact inhomogeneities.

One should bear in mind that the dimensions and other parameters of the contact regions depend, in general, on the electric field, intensity

---

*It is worthwhile to reduce d until it becomes much less than the depth of penetration of light 1/k (cf. Fig. 13).

of illumination, etc. Finally, any inhomogeneities, in particular, contact ones, may affect measurements by influencing the density, diffusion, and drift of nonequilibrium carriers. *

2. The band curvature at the contact regions, changes of the relative position of the Fermi and other energy levels, and consequent changes of the electron (hole) population of these levels, may alter the principal photoconductivity parameters: the lifetime τ and the phenomenological yield.

The effect of contacts on measurements of the volume photoconductivity can be eliminated in two ways.

The first one is obvious: we can increase the sample length between the electrodes until the effect of the contact regions becomes negligibly small. Alternatively we can keep the contact layers outside the illuminated area by using the "probe" method for measuring the photoconductivity; this will be discussed later in greater detail.

A. Criterion for the Elimination of the Influence of Neutral Contacts

Obviously the first method referred to above may be used more successfully in the case of antiblocking contacts than in the case of blocking ones. In fact since the contact region resistance is in series with the sample resistance (the total resistance is governed by the region with the higher resistance), the presence of blocking layers may be neutralized only by a very great increase of the sample length.

By way of example we shall establish a criterion for eliminating the influence of contacts for a sample with unipolar (electron) conductivity and neutral contacts. The carrier density in a semiconductor at its contact with a metal is strictly prescribed and cannot be altered by illumination, a field, etc. In the case of a neutral contact this density is equal to the equilibrium density n_0 and consequently before illumination the conductivity σ_0 is the same along the whole sample. On illumination the conductivity in the central portion far from the contacts increases and becomes $\sigma_0 + \Delta\sigma$ (Fig. 14). Obviously in a certain region near the contact, the thickness of the region being l_c, the conductivity decreases continuously from $\sigma_0 + \Delta\sigma$ to σ_0 at the contact itself.

Consequently a sample with neutral contacts, homogeneous before illumination, becomes inhomogeneous during illumination, and its high-resistivity layers at the contacts remain; these layers may be called the "photoblocking layers." For an approximate estimate of the influence of the photoblocking layers on the measured photoconductivity we shall assume that over a length l_c from the contacts the conductivity is σ_0 and

---

* The phenomena of injection, extraction, etc. are involved here (cf. Chap. IV, § 19, and Chaps. XII and XIII).

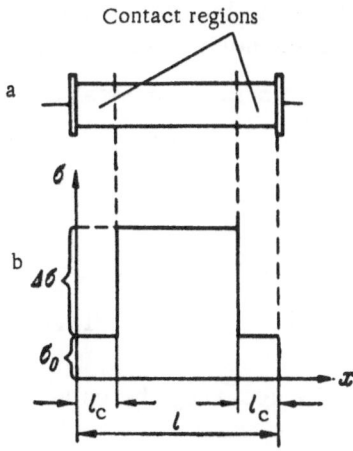

Contact regions

Fig. 14. a) Sample; b) schematic representation of the distribution of the conductivity along the sample during illumination.

then it increases suddenly to $\sigma_0 + \Delta\sigma$ (cf. Fig. 14).

We shall introduce the following notation:

$\sigma_0^* = \frac{s}{l}\sigma_0$ is the conductance of the whole sample in darkness (s is the transverse cross section of the sample);

$\sigma_i^* = \frac{s}{l}(\sigma_0 + \Delta\sigma)$ is the conductance of the whole sample during illumination without allowance for the photoblocking contact layers;

$\sigma_i^{*\prime}$ is the conductance of the whole sample during illumination with allowance for the series-connected contact layers of thickness l_c;

$\Delta\sigma^*$ and $\Delta\sigma^{*\prime}$ are the corresponding values of the photoconductance of the whole sample.

Obviously the condition that the influence of the contact layers on the photoconductivity should be weak may be formulated as a condition that the values $\Delta\sigma^*$ and $\Delta\sigma^{*\prime}$ should be nearly equal. We shall write this condition in the form

$$\frac{\Delta\sigma^* - \Delta\sigma^{*\prime}}{\Delta\sigma^*} \ll 1. \qquad (11.1)$$

From simple calculations it follows that

$$\Delta\sigma^* = \frac{s}{l}\Delta\sigma, \qquad (11.2)$$

$$\Delta\sigma^{*\prime} = \frac{s}{l}\frac{\sigma_0 \Delta\sigma (l - 2l_c)}{l\sigma_0 + 2l_c \Delta\sigma}. \qquad (11.3)$$

Substituting Eqs. (11.2) and (11.3) into (11.1) we obtain the criterion for the elimination of the influence of the contact regions in the form

$$\frac{\Delta\sigma}{\sigma_0} + 1 \ll \frac{\Delta\sigma}{\sigma_0} + \frac{l}{2l_c}. \qquad (11.4)$$

In the case of low photoconductivity $\left(\frac{\Delta\sigma}{\sigma_0} \ll 1\right)$ this gives

$$\frac{l}{2l_c} \gg 1, \qquad (11.5)$$

i.e., the sample length should be much greater than the lengths of the contact regions, and to ensure, for example, 5% accuracy we must make l such that

$$l \gg 40 l_c.$$

In the case of high photoconductivity $\left(\frac{\Delta\sigma}{\sigma_0} \gg 1 \right)$ the criterion becomes much more stringent:

$$\frac{l}{2l_c} \gg \frac{\Delta\sigma}{\sigma_0}. \tag{11.6}$$

If, for example, $\Delta\sigma/\sigma_0 = 10$, then to ensure 5% accuracy the value of l should be 400 times greater than l_c.

Obviously in poorly conducting semiconductors where σ_0 is small and $\Delta\sigma/\sigma_0$ is, as a rule, high, the role of contacts may be particularly important.

It follows that in studying the volume photoconductivity of such materials it is particularly important to ensure that the contacts produce strong antiblocking band curvature (so that for any likely intensity of illumination no photoblocking layers should appear, i.e., no layers with resistance higher than the volume resistance).

The thickness of the contact layer l_c is of decisive important in the problem of the influence of contacts. We shall not deal with the details of this problem but we shall note that in the unipolar conductivity case the thickness is governed by the so-called screening length, and in the ambipolar conductivity case it is governed by the effective diffusion length (cf. Chaps. XII and XIII).

The above quantitative analysis for neutral contacts applies only to the unipolar conduction case in weak electric fields and is not concerned with the problem of the effect of band curvature on τ and β.

However, the latter circumstance should not affect greatly our criterion.

B. Probe Method for Measuring Photoconductance

We pointed out above that only contacts with sufficiently strong antiblocking properties do not distort measurements of the volume nonequilibrium conductivity. Blocking layers and even neutral contacts may strongly distort the results considerably. Their influence can be eliminated by making the samples longer but this is not always possible. For this reason quantitative results are sometimes obtained using the probe method, in which only the central part of the sample, which is sufficiently far from the contacts (Fig. 15), is illuminated. The total change of the conductivity of the whole sample is due merely to the change in the central

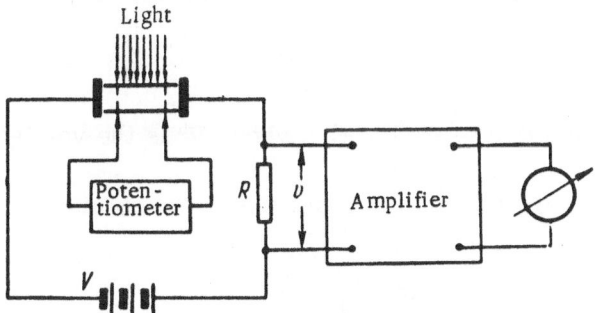

Fig. 15. Measuring circuit (probe method).

portion which is not affected by the contact fields (it is assumed that the lengths of the illuminated region and of the dark contact regions are sufficiently great so that the diffusion and drift of carriers cannot have much effect).

We shall calculate the conductance of the central portion of the sample, assuming that we know the circuit parameters V and R, the dark resistance of the whole sample r_0^* and that of the central illuminated portion r_0^0, and also the value of the alternating voltage v across the resistance R when the illumination is modulated.

In a circuit with elements connected in series the total change of the resistance is equal to the sum of the changes of resistance of the individual elements. In our case the illumination alters only the resistance of the central portion of the sample by an amount Δr^0. Consequently the total change of the resistance of the whole sample Δr^* is equal to Δr^0 and may be written in the form [cf. Eq. (8.2)]

$$\Delta r^0 = \Delta r^* = \frac{v\left(R + r_0^*\right)^2}{VR + v\left(r^* + R\right)}. \tag{11.7}$$

On the other hand, according to Eq. (8.3)

$$\Delta r^0 = \frac{r_0^{0^2} \Delta \sigma^0}{1 + r_0^0 \Delta \sigma^0}. \tag{11.8}$$

Comparing Eqs. (11.7) and (11.8), we find [2]

$$\Delta \sigma^0 = \frac{v\left(r_0^* + R\right)^2}{r_0^{0^2} R\left(V + v\right) + v\left(r_0^{0^2} r_0^* - r_0^0 r_0^{*^2} - 2r_0^0 r_0^* R - r_0^0 R^2\right)}. \tag{11.9}$$

This expression for $\Delta \sigma^0$ was obtained without neglecting any terms.

In the case of a sufficiently low photoconductivity (and consequently low v) the above formula simplifies considerably:

$$\Delta\sigma^0 = \frac{v}{VR} \frac{(r_0^* + R)^2}{r_0^{02}},$$ (11.10)

or, bearing in mind that $\dfrac{r_0^* + R}{r_0^0} = \dfrac{V}{V^0}$ where V^0 is the voltage drop across

the central portion of the sample in darkness, we obtain

$$\Delta\sigma^0 = \frac{1}{R} \frac{Vv}{V^{02}}.$$ (11.11)

Thus, in this case the photoconductance of a sample should be found using thin wire probes and measuring the voltage drop in darkness across the central portion of the sample V^0 by the compensation method (a potentiometer is shown in Fig. 15).

§12. ELIMINATION OF THE INFLUENCE OF NONUNIFORMITY OF ILLUMINATION

We have assumed above that the photoconductance is measured using a uniform light beam. However, under actual conditions the intensity of illumination of various parts of the sample and consequently the photoconductance of these parts may not be the same. Then the measured photoconductance of the whole sample $\Delta\sigma^*$ depends in a complex fashion on the nature of the Δn (or $\Delta\sigma$) distribution, and the interpretation of the results becomes difficult.

However, it is found that under certain conditions the measured photoconductance $\Delta\sigma^*$ ceases to depend on the nature of the distribution of nonequilibrium carriers in the sample and is governed only by the total number of such carriers.

Let us establish the appropriate condition. Assume that in a sample which is homogeneous before illumination the distribution of the nonequilibrium-carrier (for example, electron) density varies along the sample: $\Delta n = \Delta n(x)$ (Fig. 16) and consequently $\Delta\sigma = \Delta\sigma(x)$.

We shall find the value of the photoconductance of the sample under these conditions:

$$\Delta\sigma^* = \sigma_i^* - \sigma_0^*.$$ (12.1)

Substituting into Eq. (12.1) the quantities

$$\sigma_0^* = \frac{s}{l}\sigma_0 \quad \text{and} \quad \sigma_i^* = \frac{s}{\displaystyle\int_0^l \frac{dx}{\sigma_0 + \Delta\sigma(x)}},$$

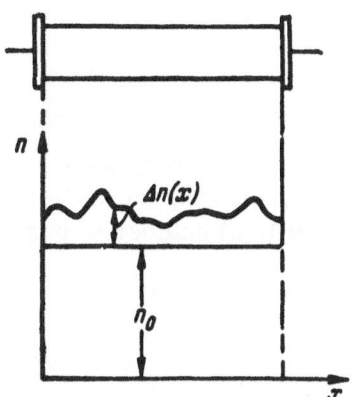

Fig. 16. Distribution of carrier density along the sample during arbitrarily nonuniform illumination.

we obtain, after simple transformations,

$$\Delta\sigma^* = \frac{s\sigma_0}{l} \frac{l - \int\limits_0^l \frac{dx}{1 + \frac{\Delta\sigma(x)}{\sigma_0}}}{\int\limits_0^l \frac{dx}{1 + \frac{\Delta\sigma(x)}{\sigma_0}}}. \qquad (12.2)$$

From Eq. (12.2) we see that the measured $\Delta\sigma^*$, as stressed above, is a complex function of $\Delta\sigma(x)$.

This expression simplifies considerably only in the case when the photoconductance is much less than the equilibrium conductance

$$\frac{\Delta\sigma(x)}{\sigma_0} \ll 1 \qquad (12.3)$$

(or $\frac{\Delta n(x)}{n_0} \ll 1$).

Since

$$\frac{1}{1 + \frac{\Delta\sigma(x)}{\sigma_0}} = 1 - \frac{\Delta\sigma(x)}{\sigma_0}, \qquad (12.4)$$

we obtain for $\Delta\sigma^*$

$$\Delta\sigma^* = \frac{s\sigma_0}{l} \frac{l - \int\limits_0^l \left(1 - \frac{\Delta\sigma(x)}{\sigma_0}\right) dx}{\int\limits_0^l \left(1 - \frac{\Delta\sigma(x)}{\sigma_0}\right) dx} = \frac{s}{l^2} \frac{\int\limits_0^l \Delta\sigma(x)\, dx}{1 - \frac{\int\limits_0^l \Delta\sigma(x)\, dx}{l\sigma_0}}. \qquad (12.5)$$

Since at any point $\Delta\sigma(x) \ll \sigma_0$, then obviously the average photoconductance is much less than the equilibrium conductance, i.e.,

$$\frac{\int\limits_0^l \Delta\sigma(x)\, dx}{l} \ll \sigma_0.$$

Consequently we can neglect the second term in the denominator of Eq. (12.5) compared with unity.

Finally, we obtain

$$\Delta\sigma^* = \frac{s}{l^2} \int_0^l \Delta\sigma(x)\,dx = \frac{s}{l^2}\,e\mu \int_0^l \Delta n(x)\,dx. \qquad (12.6)$$

Thus, if the nonequilibrium conductance (density) is at every point much less than the equilibrium conductance (density), then the measured value of the photoconductance is governed only by the total number of non-equilibrium carriers in the sample

$$\Delta N = \int_0^l \Delta n(x)\,dx$$

and is independent of the carrier distribution along the sample.

The expression (12.6) may be written alternatively:

$$\Delta\sigma^* = \frac{s}{l}\,\overline{\Delta\sigma} = \frac{s}{l}\,e\mu\,\overline{\Delta n}, \qquad (12.7)$$

where $\overline{\Delta\sigma}$ and $\overline{\Delta n}$ are the average changes of the photoconductivity and the nonequilibrium carrier density.

Chapter III

DETERMINATION OF THE PRINCIPAL PHENOMENOLOGICAL PARAMETERS β AND τ FROM STUDIES OF PHOTOCONDUCTIVITY KINETICS

In §5 of Chap.I we mentioned that the two principal phenomenological parameters – the yield β and lifetime τ – may be determined by visual inspection and analysis of the photoconductivity relaxation curves when the photoconductivity is excited by square light pulses. Apart from this general method, other methods are also used, which may be divided into two main groups:

1. Methods for the determination of τ and β by investigating of the "frequency dependence" [i.e., the dependence of the amplitude of the alternating component of the photoconductivity on the frequency of chopping (modulation)].

2. Methods for the determination of τ by compensating the phase difference between the modulated exciting light beam and the photoconductivity. Essentially the methods of both the first and second groups are suitable only for studies of linear photoconductivity.

Below, in §§13 and 14, we shall consider the basis of these two groups of methods in the case of linear photoconductivity; §15 deals with some methods for studying the nonlinear case, and §16 describes methods of modulating light beams.

§13. FREQUENCY DEPENDENCE OF PHOTOCONDUCTIVITY

The alternating component of the photoconductivity measured using chopped illumination depends not only on the chopping frequency but also on the form of modulation of the light beam intensity.

Below we shall consider "sinusoidal" and "square" modulations, which are the two most widely used.

A. Square Modulation of Light Intensity

1. Symmetric Square Light Waves. Periodic "square" light pulses of definite spectral composition may be described by three principal pa-

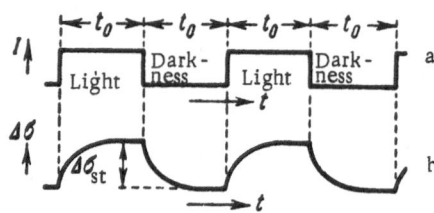

Fig. 17. a) Square light waves; b) $\Delta \sigma = f(t)$
when $t_0 \gg \tau$.

rameters: t_0^i, the duration of the light pulse; t_0^d, the duration of the "dark" interval between two pulses; and I_0, intensity of the light beam during the pulse.

Photoconductivity is frequently excited by "symmetric square light waves," for which $t_0^i = t_0^d = t_0$ (Fig. 17). Obviously if

$$t_0 \gg \tau, \tag{13.1}$$

then steady-state photoconductivity may be established during the period of illumination and this photoconductivity may decay to zero during the period of darkness (Fig. 17b). The amplitude of the alternating component of the photoconductivity $\Delta \sigma_\sim$ measured by means of the usual circuit for chopped illumination (cf. for example, Fig. 9), is equal to the steady-state photoconductivity $\Delta \sigma_{st}$.

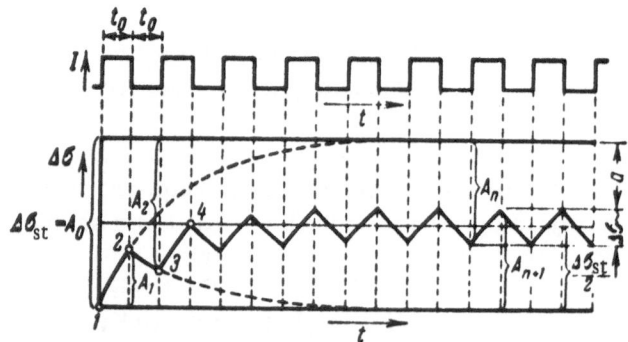

Fig. 18. Dependence of the photoconductivity on time during illumination of a semiconductor by symmetric square light waves. The case $t_0 < \tau$.

We shall now consider in detail how the conductivity of a substance varies with time during illumination with square light waves if the time t_0 does not satisfy the condition of Eq. (13.1). This corresponds to the case when steady-state photoconductivity cannot be established in the time t_0.

We shall assume that a sample previously in darkness is illuminated at a certain moment $t = 0$ with square light waves (Fig. 18). From the moment $t = 0$ until $t = t_0$ the photoconductivity increases in accordance with the law [cf. Eq. (5.2)]

$$\Delta \sigma = A_0 \left(1 - e^{-\frac{t}{\tau}} \right), \tag{13.2}$$

i.e., the photoconductivity tends to approach its steady-state value $\Delta \sigma_{st} = A_0$. However, at $t = t_0$ the illumination ceases and the photoconductivity begins to decay in accordance with the law [cf. Eq. (5.5)]

$$\Delta \sigma = A_1 e^{-\frac{t}{\tau}}. \qquad (13.3)$$

Since $A_1 < A_0$, the photoconductivity decays during the dark interval to a value smaller than that to which it rose in the preceding illumination period. * Consequently point 3 lies above point 1. The next illumination period produces a rise $\Delta \sigma$ from the value at point 3 according to the law

$$\Delta \sigma = A_2 \left(1 - e^{-\frac{t}{\tau}}\right). \qquad (13.4)$$

Since $A_2 > A_1$, point 4 is higher than point 2. The subsequent behavior of the curve is clear from Fig. 18. Finally a state should be reached when the curve giving the variation of $\Delta \sigma$ lies centrally between the upper and lower equilibrium positions. Such a

Fig. 19. Modulator disks for the observation of the transient process of Fig. 18.

state obviously occurs when $A_n \approx A_{n+1}$. The photoconductivity then has a constant component $\Delta \sigma_= = \Delta \sigma_{st}/2$ and an alternating component $\Delta \sigma_\sim$ the amplitude of which can be measured experimentally.†

We shall calculate $\Delta \sigma_\sim$ (or, more precisely, the amplitude of this quantity) by assuming that we know $\Delta \sigma_{st}$, τ, and t_0. Figure 18 shows that

$$\Delta \sigma_\sim = A_0 - 2a, \qquad (13.5)$$

---

* Exponential relaxation occurs in those cases where the rate at which a given system returns to equilibrium is proportional to the departure from this equilibrium. Because of this the rate of approach to equilibrium in equal intervals of time will increase with the departure from equilibrium, i.e., with increase of the pre-exponential multiplier.

† A transient rise process, like that shown in Fig. 18, cannot usually be observed on an oscillograph since the time required for the curve to reach its position between the lower and upper equilibrium positions is very short (of the order of τ), and since the process occurs only once the recorded oscillogram always represents the steady-state case.

One can, however, observe the development of a transient process repeating it periodically by illumination of the sample through a disk with a special profile (Fig. 19). Figure 20 gives a series of oscillograms of the variation of photoconductivity obtained with this disk.

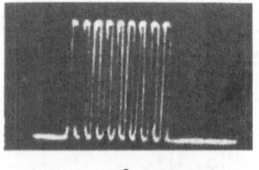

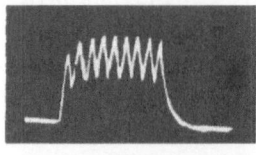

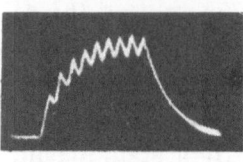

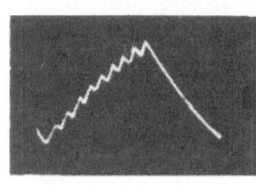

Fig. 20. Oscillograms of a transient process. a) $\tau \ll t_0$; e) $\tau \gg t_0$. The value of τ/t_0 increases from a to e.

as well as

$$a = (A_0 - a)\, e^{-\frac{t_0}{\tau}}. \tag{13.6}$$

Hence

$$a = A_0 \frac{e^{-\frac{t_0}{\tau}}}{1 + e^{-\frac{t_0}{\tau}}}. \tag{13.7}$$

Substituting Eq. (13.7) into Eq. (13.5), we obtain [7]

$$\Delta\sigma_\sim = A_0 \frac{1 - e^{-\frac{t_0}{\tau}}}{1 + e^{-\frac{t_0}{\tau}}} = A_0 \frac{e^{\frac{t_0}{2\tau}} - e^{-\frac{t_0}{2\tau}}}{e^{\frac{t_0}{2\tau}} + e^{-\frac{t_0}{2\tau}}} = A_0 \ \tanh\ \frac{t_0}{2\tau} \tag{13.8}$$

or

$$\Delta\sigma_\sim = \Delta\sigma_{st}\, \tanh \frac{t_0}{2\tau}. \tag{13.9}$$

Figure 21 shows a curve plotted using Eq. (13.9). It is seen that on increase of τ (or decrease of t_0) the experimentally measured value of $\Delta\sigma_\sim$ decreases and departs more and more from $\Delta\sigma_{st}$. On the other hand, when $t_0 \gg \tau$ we have $\Delta\sigma_\sim = \Delta\sigma_{st}$.

By investigating the dependence of $\Delta\sigma_\sim$ on t_0 experimentally we can determine the value of τ in several ways. For example, by drawing the tangent to the experimental curve at the point $t_0 = 0$ and dropping normal onto the time axis from the point of intersection of the tangent with a straight line parallel to the time axis and distant $\Delta\sigma_{st}$ from it, we obtain the value $t_0 = 2\tau$. The same value $t_0 = 2\tau$ is obtained if we drop a normal from the point of intersection of the experimental curve with a straight line parallel to the time axis and distant $0.76\Delta\sigma_{st}$ from it (tanh 1 = 0.76).

The "yield" β may be determined from the values of $\Delta\sigma_\sim$ when the pulse duration t_0 is short. Replacing, for $t_0 \ll \tau$, the hyperbolic tangent in Eq. (13.9) by its argument, we obtain

$$\Delta\sigma_\sim = \Delta\sigma_{st}\, \frac{t_0}{2\tau} = \frac{e\mu k I t_0}{2}\, \beta, \tag{13.10}$$

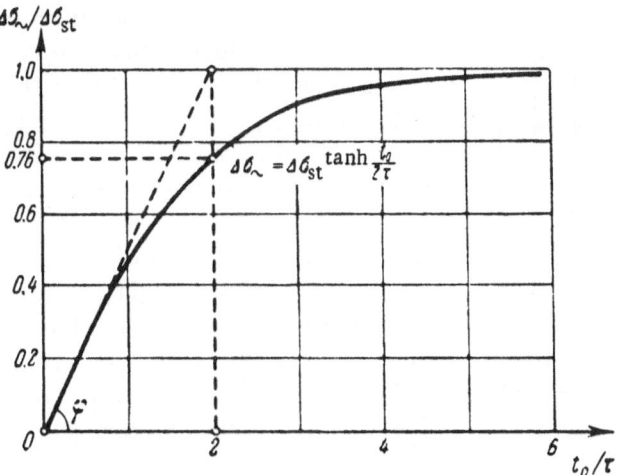

Fig. 21. Dependence of the alternating signal amplitude on
the duration of square light pulses.

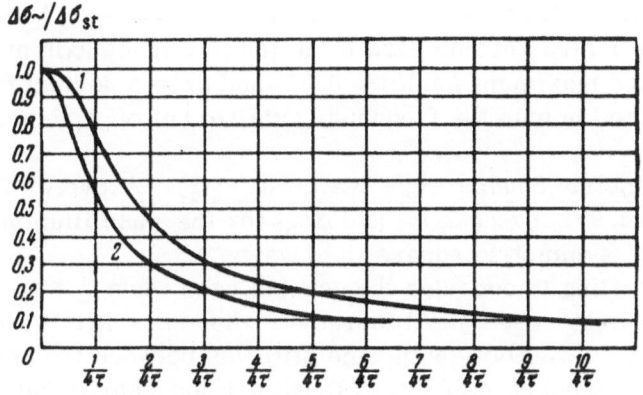

Fig. 22. "Frequency" dependence of the photoconductivity
for various modulations of the excitation intensity. Curve 1:
square modulation, $\Delta\sigma_\sim = \Delta\sigma_{st} \tanh\frac{1}{4\tau f}$; curve 2: sinusoidal

modulation $\Delta\sigma_\sim = \Delta\sigma_{st} \frac{1}{\sqrt{1+(2\pi f\tau)^2}}$.

i.e., a quantity which does not contain τ and is proportional to β.* Hence
(cf. Fig. 21)

$$\beta = \frac{2}{e\mu kI} \tan\varphi. \qquad (13.11)$$

* To determine β quantitatively we must also measure independently the mobility μ and the
absorbed light energy kI.

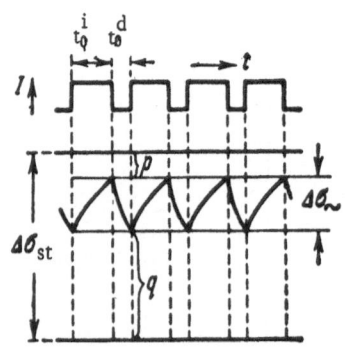

Fig. 23. $\Delta\sigma = f(t)$ during excitation with asymmetric square light waves.

The β is found from the slope of the initial part of the dependence $\Delta\sigma_\sim = f(t_0)$.

It should be stressed that this method of determining τ is suitable only for linear conductivity. The method applicable in the general case will be described below.

The experimental results are frequently presented in the form of a dependence of $\Delta\sigma_\sim$ on the frequency of light modulation (and not on the pulse duration t_0). To obtain the frequency characteristic we replace t_0 in Eq. (13.9) by $t_0 = 1/2f$, where f is the chopping frequency. Then we have

$$\Delta\sigma_\sim = \Delta\sigma_{st} \tanh \frac{1}{4\tau f}. \qquad (13.12)$$

The frequency dependence for square modulation of light is given in Fig. 22. To calculate τ from the experimental curve $\Delta\sigma_\sim = f(f)$ we can, for example, draw a straight line parallel to the frequency axis at a height of 0.76 from the maximum ($\tanh 1 = 0.76$) and drop a normal from the point of intersection onto the frequency axis, to cut off a segment equal to $1/4\tau$.

2. **Asymmetric Square Light Waves** ($t_0^i \neq t_0^d$). Above we derived the relationship (13.9) between $\Delta\sigma_\sim$ and $\Delta\sigma_{st}$ for the case illumination are equal: $t_0^i = t_0^d$ ("symmetric square light waves").

It is interesting to consider the general case when $t_0^i \neq t_0^d$ ("asymmetric square light waves") and moreover when $\tau_i \neq \tau_d$.* Then the variation of the photoconductivity with time still lies between the upper and lower equilibrium positions (Fig. 23), but not necessarily halfway between these two positions, as was the case with symmetric square waves and $\tau_i = \tau_d$.

The principal relationships for the general case, written in terms of the notation explained in Fig. 23, are

$$\Delta\sigma_{st} = p + \Delta\sigma_\sim + q. \qquad (13.13)$$

---

*In other words, the photoconductivity rise during illumination obeys the law

$$\Delta\sigma \propto \left(1 - e^{-t/\tau_i}\right),$$

and its decay after illumination obeys the law

$$\Delta\sigma \propto e^{-t/\tau_d},$$

i.e., the time constants for the rise and decay are not equal.

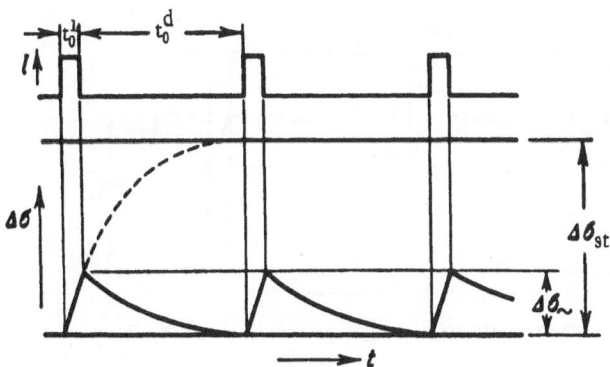

Fig. 24. Asymmetric square light waves $t_0^i \ll \tau \ll t_0^d$. The amplitude of the signal is proportional to the yield β.

$$(\Delta\sigma_\sim + p)\, e^{-t_0^i/\tau_i} = p, \qquad (13.14)$$

$$(\Delta\sigma_\sim + q)\, e^{-t_0^d/\tau_d} = q. \qquad (13.15)$$

From Eq. (13.14) we find

$$p = \frac{\Delta\sigma_\sim e^{-t_0^i/\tau_i}}{1 - e^{-t_0^i/\tau_i}}. \qquad (13.16)$$

From Eq. (13.15) we find

$$q = \frac{\Delta\sigma_\sim e^{-t_0^d/\tau_d}}{1 - e^{-t_0^d/\tau_d}}. \qquad (13.17)$$

Substituting the values of p and q into Eq. (13.13), we obtain the following expression for the relationship between $\Delta\sigma_{st}$ and $\Delta\sigma_\sim$:

$$\Delta\sigma_{st} = \Delta\sigma_\sim \left(1 + \frac{e^{-t_0^i/\tau_i}}{1 - e^{-t_0^i/\tau_i}} + \frac{e^{-t_0^d/\tau_d}}{1 - e^{-t_0^d/\tau_d}} \right). \qquad (13.18)$$

It is easily seen that when $t_0^i = t_0^d = t_0$ and $\tau_i = \tau_d = \tau$, Eq. (13.18) reduces to the earlier formula (13.9).

An analysis of Eq. (13.18) shows that using asymmetric waves to determine τ and β we need measure only two values of $\Delta\sigma_\sim$ at two different values of the ratio of t_0^i to t_0^d, instead of investigating the whole frequency characteristic.

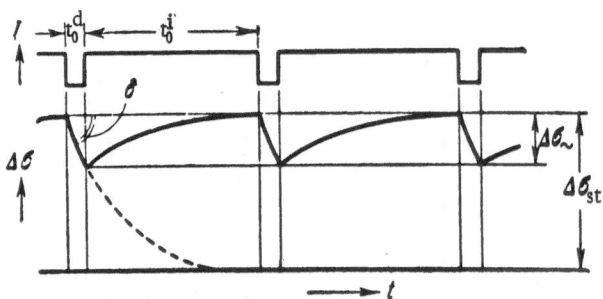

Fig. 25. Asymmetric square light waves $t_0^d \ll \tau \ll t_0^i$.

Measurement of β. If we use short light pulses $(t_0^i \ll \tau)$, separated by long dark intervals $(t_0^d \gg \tau)$,* then it follows from Eq. (13.18) that

$$\Delta\sigma_\sim = \Delta\sigma_{st} \frac{t_0^i}{\tau} = \frac{e\mu\tau\beta kI}{\tau} t_0^i = e\mu kI t_0^i \beta, \qquad (13.19)$$

i.e., the quantity $\Delta\sigma_\sim$ is proportional to the yield β and the expression does not contain τ.

This result also follows directly from a consideration of Fig. 24; here the quantity $\Delta\sigma_\sim$ is governed only by the initial linear portion of the photoconductivity rise which is not affected by recombination. We note that this method is in principle applicable both to linear and nonlinear photoconductivity and is therefore more general than that described at the beginning of the present section [cf. Eq. (13.11)].

Measurement of τ. To determine the lifetime τ we have to use long light pulses $(t_0^i \gg \tau)$ separated by short dark intervals $(t_0^d \ll \tau)$. The signal then has the form shown in Fig. 25 and it follows from Eq. (13.18) that

$$\Delta\sigma_\sim = \Delta\sigma_{st} \frac{t_0^i}{\tau}.$$

Hence

$$\tau = \frac{\Delta\sigma_{st}}{\Delta\sigma_\sim} t_0^i = \frac{\Delta\sigma_{st}}{\tan\delta}. \qquad (13.20)$$

Consequently to determine τ we must also measure $\Delta\sigma_{st}$ in addition to the quantities $\Delta\sigma_\sim$ and t_0^d found directly by experiment.

The basis of the method of determining τ from the slope of the initial stage of the photoconductivity decay is quite obvious: in this method we essentially measure the instantaneous value of the lifetime in the steady state (cf. §5 in Chap. I)

---

*The nature of the photoconductivity variation during illumination with such pulses is shown in Fig. 24.

$$\tau_{st} = -\frac{\Delta \sigma_{st}}{\left(\frac{d\sigma}{dt}\right)_{t=0}} = -\frac{\Delta \sigma_{st}}{\tan \delta}. \tag{13.21}$$

Consequently this method may also be used in the case of nonlinear photoconductivity.

B.   Sinusoidal Modulation of Light Intensity

Let the intensity of light vary sinusoidally with time (Fig. 26a)

$$I = I_{amp}(1 - \cos \omega t), \tag{13.22}$$

where $\omega = 2\pi f$ is the angular frequency.

It is easily seen that the intensity of light then has a constant component I_{amp} and an alternating component whose amplitude is I_{amp}. The time dependence of the carrier density Δn (and consequently of the photoconductivity $\Delta \sigma$) in the linear case is obtained from the solution of Eq. (5.1) in Chap. I, which, allowing for Eq. (13.22), is written in the form

$$\frac{d}{dt} \Delta n = \beta k I_{amp}(1 - \cos \omega t) - \frac{\Delta n}{\tau}. \tag{13.23}$$

The solution of Eq. (13.23) has the following form:

$$\Delta n = \beta k I_{amp}\tau + \frac{\beta k I_{amp}\tau}{1 + \omega^2 \tau^2} (\tau \omega \sin \omega t + \cos \omega t) + Ce^{-t/\tau}, \tag{13.24}$$

where C is a constant of integration.

If we assume that n = 0 at t = 0, then Δn is given by

$$\Delta n = \beta k I_{amp}\tau \left[1 - \frac{2 + (\tau \omega)^2}{1 + (\tau \omega)^2} e^{-t/\tau} \right]$$

$$+ \frac{\beta k I_{amp}\tau}{1 + (\tau \omega)^2} (\tau \omega \sin \omega t + \cos \omega t). \tag{13.25}$$

Figure 26b gives the complete curve of the variation of Δn with time. The transient conditions at the beginning of the curve, governed by the first (exponential) term in Eq. (13.25), are analogous to the transient regime in the case of square modulation (Figs. 18 and 20). The first term in Eq. (13.25) tends to a constant value $\beta k I_{amp}\tau$ under steady-state conditions (when $t \to \infty$); this constant value does not affect the results of measurements by means of an ac amplifier.

Thus the experimentally measured quantity $\Delta n_{\sim}$ is governed by the second term. After a simple transformation of this term we find [1]

$$\Delta n_{\sim} = \frac{\beta k I_{amp}\tau}{\sqrt{1 + (\tau \omega)^2}} \cos [\omega t - \tan^{-1} (\tau \omega)]. \tag{13.26}$$

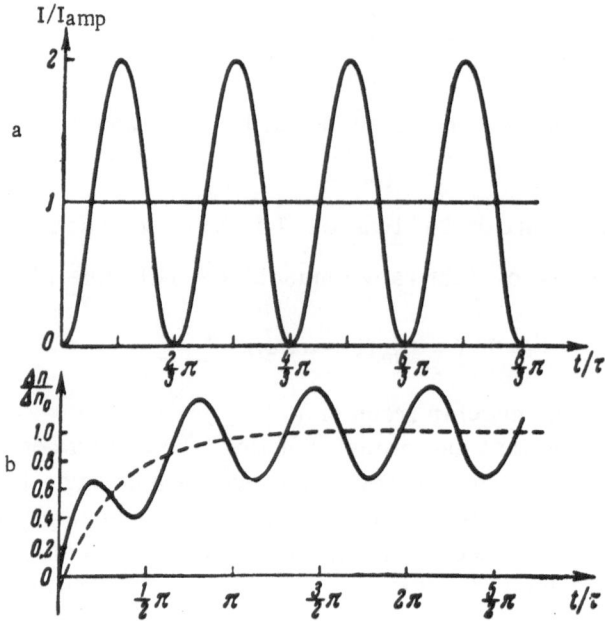

Fig. 26. Sinusoidal modulation of light. a) Dependence of the light intensity on time; b) example of the time dependence of photoconductivity.

It follows from Eq. (13.26) that under steady-state conditions the relationship between the modulation frequency ω and the lifetime τ governs not only the maximum (peak) value of the alternating component (for which we shall keep the earlier notation $\Delta n_\sim$),

$$\Delta n_\sim = 2 \frac{\beta k I_{amp}\tau}{\sqrt{1+(\tau\omega)^2}},\qquad(13.27)$$

but also the phase difference between the excitation and photoconductivity, given by the quantity

$$\varphi = \tan^{-1}(\tau\omega).\qquad(13.28)$$

The problem of using the phase difference to determine τ will be considered in §14 and here we shall restrict ourselves to a discussion of the frequency dependence of Δn.

Equation (13.27) is illustrated by the lower curve in Fig. 22. To determine τ from the experimental curve $\Delta n_\sim = f(f)$ we can, for example, draw a straight line parallel to the time axis at a distance of 0.71 units from the maximum until it intersects the experimental curve and drop a perpendicular to the time axis to cut off a segment equal to $1/2\pi\tau$.

Analogously with the square modulation case the yield β can be determined from the values of $\Delta n_\sim$ at a sufficiently high frequency f. In fact when $f \gg 2\pi\tau$ we find from Eq. (13.27)

$$\Delta n_\sim = \frac{kI_{amp}\beta}{\pi f},\qquad\qquad(13.29)$$

i.e., the experimentally measured quantity is proportional to the yield.

§ 14. DETERMINATION OF THE LIFETIME BY THE PHASE-SHIFT COMPENSATION METHOD

A. Sinusoidal Modulation

As pointed out above [cf. Eqs. (13.26) and (13.28)], in the case of sinusoidal modulation we have a phase shift φ between the exciting signal and the alternating component of the photoconductivity, depending on the relationship between τ and the modulation frequency f:

$$\varphi = \tan^{-1}(2\pi f\tau).\qquad\qquad(14.1)$$

It follows that by measuring φ at a given f we can determine τ. The simplest method of measuring the phase shift φ is the compensation method [1, 4, 5, 8, 22, 30]. The basis of the method is shown in Fig. 27. A sinusoidally modulated light beam is split into two parts by means of a mirror Mi: one part falls on the sample Sa and the other on a rapid-response receiver (in some cases we can use for this purpose a vacuum photocell or a photomultiplier PM). Obviously the photoconductivity of the sample and the saturation photocurrent in the photocell circuit differ by the phase angle φ. Sinusoidal signals from two load resistances R_1 and R are applied to two mutually perpendicular plates of an oscillograph which then shows a Lissajou curve (an ellipse).

By varying the signal phase in the photocell circuit by means of a calibrated phase shifter we can reduce to zero the phase difference between the signals reaching the oscillograph. Then the ellipse degenerates into a sloping straight line and the phase shifter scale gives the phase shift φ.

It follows from Eq. (14.1) that the phase shift φ may vary from zero for $\tau \ll 1/2\pi f$ to $\pi/2$ for $\tau \gg 1/2\pi f$.

To compensate the phase shift within these limits it is simplest to use an RC-circuit with variable R and C. Such a phase shifter in the photocell circuit (Fig. 27) may consist of a load resistance R shunted by a capacitor bank C. It is easily seen that then we do not need to determine the value of the phase shift φ since at any frequency f the value of RC corresponding to compensation (i.e., when the ellipse becomes a straight line) is equal to the lifetime τ.

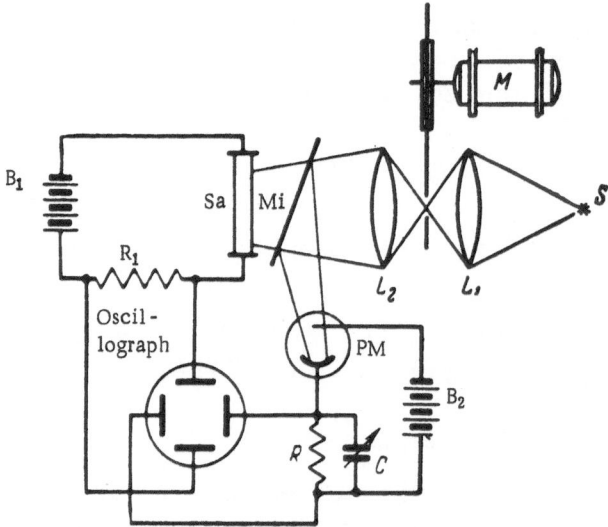

Fig. 27. Basic circuit for the determination of τ by the phase-shift compensation method. S is a source of light, L_1 and L_2 are lenses, M is a motor which rotates a disk, B_1 and B_2 are batteries. [Sa is the sample].

In fact the variation of the potential in the RC circuit may be written in the form

$$dv = \frac{dQ}{C} = \frac{i_C \, dt}{C},\tag{14.2}$$

where dQ is the variation of the charge in the capacitor C and i_C is the charging current. Since $i_C = i - i_R = i - (v/R)$, where i is the saturation current of the rapid-response photocell, which in the case of sinusoidal modulation is given by [cf. Eq. (13.22)]

$$i = i_{amp}(1 - \cos \omega t),\tag{14.3}$$

we obtain

$$\frac{dv}{dt} = \frac{i_{amp}}{C}(1 - \cos \omega t) - \frac{v}{RC}.\tag{14.4}$$

Comparison of Eqs. (14.4) and (13.23) shows that when RC = τ the variation of v and Δn with time [and consequently the variation with time of the signal taken from the resistor in the sample circuit (Fig. 27)] are given by the same differential equations and (because the initial conditions are identical) are described by the same curves.

Since the oscillograph screen always shows a straight line when two identical signals are applied to its mutually perpendicular plates, irrespective of the frequencies of these signals, the appearance of a straight line indicates that RC = τ. We shall now find the parameters of the ellipse on the screen in the case when RC $\neq \tau$. This is needed in sub-section C below for the analysis of the sensitivity of the phase-shift compensation method.

The signals recorded by the two mutually perpendicular plates of the oscillograph may be written in the form *

$$x = x_0 \cos(\omega t - \varphi_1),$$
$$y = y_0 \cos(\omega t - \varphi_2). \qquad (14.6)$$

Eliminating time from Eqs. (14.6), we find that the ellipse on the oscillograph screen is given by

$$\frac{x^2}{x_0^2} + \frac{y^2}{y_0^2} - \frac{2xy}{x_0 y_0} \cos(\varphi_2 - \varphi_1) = \sin^2(\varphi_2 - \varphi_1). \qquad (14.7)$$

It is easily seen that in the case of full compensation, i.e., when $\varphi_1 = \varphi_2$ [which corresponds to RC $= \tau$, in accordance with Eq. (14.5)], the ellipse degenerates into a straight line

$$\frac{x}{x_0} = \frac{y}{y_0}. \qquad (14.8)$$

B.   Square Modulation

We have shown above that the method of compensating the phase shift between the excitation and photoconductivity which arises in the case of sinusoidal modulation may be used to determine the lifetime τ.

It is easily proved that this null method of "rectifying the ellipse" may be used not only for sinusoidal excitation but also for any other form. The only essential condition is the complete identity of the exciting signals incident on the test sample and on the sensitivie element (in our case a vacuum photocell) in the compensating circuit.† The photoconductivity lag in the linear case and the lag due to the RC combination in the fast-response photocell circuit both distort the form of any light signal in the same way if RC $= \tau$.

The differential equations for the signals taken from the load resistances in the main circuit $v_1(t)$ and in the compensating circuit $v_2(t)$ are

---

* We shall denote the phase shift between the light beam and the test sample signal by φ_2, and between the light beam and the compensating signal in the RC circuit by φ_1; consequently

$$\varphi_2 = \tan^{-1} \omega \tau,$$
$$\varphi_1 = \tan^{-1} \omega RC. \qquad (14.5)$$

† In the circuit of Fig. 27 this identity is obtained by splitting the exciting light beam into two parts by means of a semitransparent mirror.

$$\frac{dv_1}{dt} = A_1 \Phi(t) - \frac{v_1}{\tau},$$ (14.9)

$$\frac{dv_2}{dt} = A_2 \Phi(t) - \frac{v_2}{RC}.$$ (14.10)

Here $\Phi(t)$ is the variation of the intensity of excitation with time (arbitrary in this case); A_1 and A_2 are scale multipliers which do not affect the forms of the curves.

Obviously when $RC = \tau$ and the initial conditions for both linear equations are the same, their solutions are the same to within a scale factor (defined as the ratio of A_1 to A_2). By applying the signals $v_1(t)$ and $v_2(t)$ to the mutually perpendicular plates of an oscillograph we obtain a straight line when $RC = \tau$ and a more complex curve when $RC \neq \tau$.

Thus the null method for compensation of phase shifts, using an RC circuit, may, in principle, be used for any form of excitation.

In practice the sinusoidal and square modulation methods are the most widely used, the latter having been employed for the first time in the phase-shift compensation method by N. A. Tolstoi and P. P. Feofilov [3, 6]. The wide use of this method is due to the fact that square modulation gives a clearer picture of the relaxation processes on the oscillograph screen. In this way one can establish the characteristic features of the relaxation process and, in particular, find deviations from the linear case (exponential relaxation when square excitation pulses are used).

We shall deal below in greater detail with the use of square modulation and exponential display for nonlinear processes (in connection with the "partial time" method), but here we shall restrict ourselves to linear processes only.

We showed earlier that when a compensation circuit (Fig. 27) is used and $RC = \tau$ the oscillograph screen shows a straight line irrespective of the form of excitation (including square excitation pulses). We shall now find the form of the curve for square pulse excitation and $RC \neq \tau$. This is required later, in the analysis of the sensitivity of the method. Assume that the two sets of oscillograph plates receive two synchronous signals in the form of rising and decaying exponentials, which have different time constants (Fig. 28). We shall, for example, assume that the test signal applied to the Y plates has the form:

rise curve

$$y = y_0 \left(1 - e^{-\frac{t}{\tau}}\right),$$ (14.11)

decay curve

$$y = y_0 e^{-\frac{t}{\tau}}.$$ (14.12)

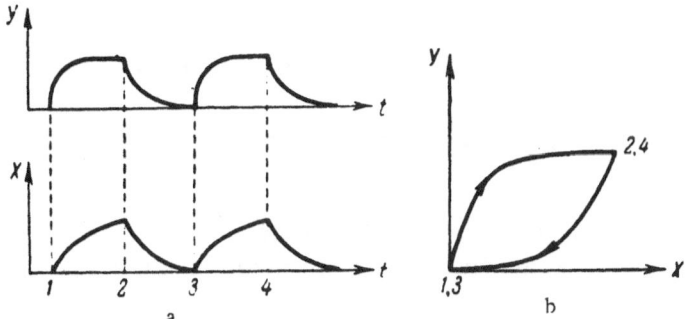

Fig. 28. a) Synchronous signals with different time constants (RC ≠ τ); b) loop on the oscillograph screen for RC ≠ τ.

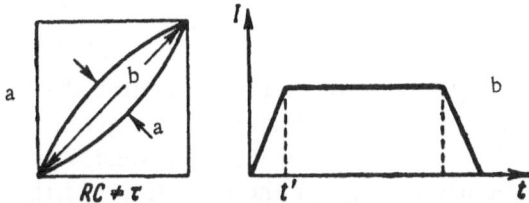

Fig. 29. a) Characteristic dimensions of a loop on the oscillograph screen for RC ≠ τ; b) square pulse with finite rise and decay fronts.

For the compensating signal along the X axis we then have

$$x = x_0 \left(1 - e^{-\frac{t}{RC}} \right), \tag{14.13}$$

$$x = x_0 e^{-\frac{t}{RC}}. \tag{14.14}$$

The resultant curve, traced out by the luminous spot on the oscillograph screen, is, in general, a loop of the type shown in Fig. 28b. Figure 28a shows that from the moment 1 the luminous spot begins to move rapidly upward (along the Y axis) and slowly toward the right (along the X axis), describing a curve sharply veering toward the X axis. However, the Y signal then reaches saturation and ceases to vary, while the displacement along the X axis still continues. The slope of the curve then decreases and point 2 represents saturation along both axes.

Reasoning similarly for the decaying branches of the exponentials in Fig. 28a, we obtain the second ("return") part of the loop shown in Fig. 28b between points 2 and 1. Thus, when RC ≠ τ the screen shows a closed loop.

Eliminating time from Eqs. (14.11) and (14.13), as well as from Eqs. (14.12) and (14.14), we find that the forward and return branches of the loop are parabolas:

forward branch $\quad \dfrac{y}{y_0} = 1 - \left(1 - \dfrac{x}{x_0}\right)^{\frac{RC}{\tau}},$ $\qquad\qquad$ (14.15)

return branch $\quad \dfrac{y}{y_0} = \left(\dfrac{x}{x_0}\right)^{\frac{RC}{\tau}}.$ $\qquad\qquad$ (14.16)

It is easily seen that in the case of full compensation, i.e., when $RC = \tau$, both parabolas degenerate into coincident straight lines

$$\frac{y}{y_0} = \frac{x}{x_0}.$$ $\qquad\qquad$ (14.17)

This condition, known as taumeter* tuning, is found visually.

C. Sensitivity of the Phase-Shift Compensation Method

1. Square Modulation. Obviously the greater the "width" a of the loop on the oscillograph screen (Fig. 29a) for a given "detuning" (i.e., for a given difference between τ and RC), the easier it is to detect when $\tau \neq RC$ and consequently the higher the sensitivity of the method. The ratio $\varepsilon = a/b$ may be used as the quantitative characteristic of the sensitivity. Obviously each value of "detuning" has a corresponding value of ε.

The detuning will be denoted by δ and it can be represented by the ratio of the difference $(\tau - RC)$ to $\tau_{av} = (\tau + RC)/2$, which is the average of the times τ and RC, i.e.,

$$\delta = 2\,\frac{\tau - RC}{\tau + RC}.$$ $\qquad\qquad$ (14.18)

The sensitivity ε is constant for a given detuning if the square pulses are "ideal" with vertically rising fronts and ends. In practice the pulses always have finite rise and decay times,† i.e., the "square" pulses are really "trapezoidal" (Fig. 29b). Although, in principle, such a distortion of the pulse shape does not prevent measurement of τ (which, as stressed above, can be carried out for any shape of the signal), it affects the sensitivity of the method considerably.

We shall consider in greater detail the sensitivity of the taumeter method allowing for the departure of the square pulses from ideal shape (later we shall also take into account the delay effect of the amplifying devices employed to strengthen the signals fed to the oscillograph) [9].

---

* The RC unit in the compensating circuit, used to determine the lifetime τ, is frequently called the "taumeter," and the whole method using square pulse excitation, proposed by N. A. Tolstoi and P. P. Feofilov, is known as the "taumeter method."

† The rise and decay times are sometimes called "cutoff times."

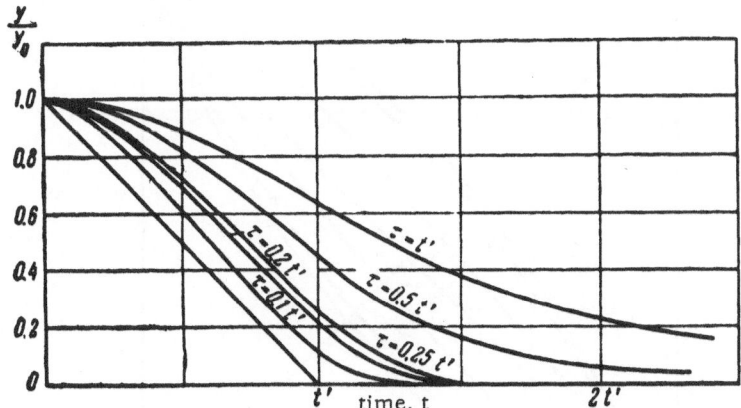

Fig. 30. Relaxation curves for the linear decay of excitation.

We shall assume that the excitation pulses acting on the test sample and on the taumeter circuit are completely synchronous and that, if they are distorted by the final cutoff, the distortion is the same in both cases. The excitation pulse has the trapezoidal shape shown in Fig. 29b. This pulse excites a test sample with the intrinsic time constant τ.

We shall assume that the pulse duration is sufficient for the process investigated to reach its steady state, and we shall consider the process beginning at a time t = 0, when the excitation begins to decrease linearly with time (Fig. 30).

A linear relaxation process obeys the equation

$$\frac{dy}{dt} = I - \frac{y}{\tau}.$$
(14.19)

where y is the investigated quantity, t is the time, τ is the time constant of the process and I represents the intensity of excitation.

Obviously the variation of the investigated quantity with time is given by different analytic expressions in the regions from 0 to t', where the excitation decreases linearly with time, and from t' to ∞, where the excitation is constant and equal to zero.

For the region from t = 0 to t = t' we have

$$\frac{dy}{dt} = I_0 \left(1 - \frac{t}{t'}\right) - \frac{y}{\tau}.$$
(14.20)

where I_0 is the maximum value of the excitation intensity. If $y = y_0$ ($y_0 = I_0 \tau$) when t = 0, then the solution of the above equation has the form

$$y = y_0 \left(1 - \frac{t}{t'}\right) + y_0 \frac{\tau}{t'} \left(1 - e^{-\frac{t}{\tau}}\right).$$
(14.21)

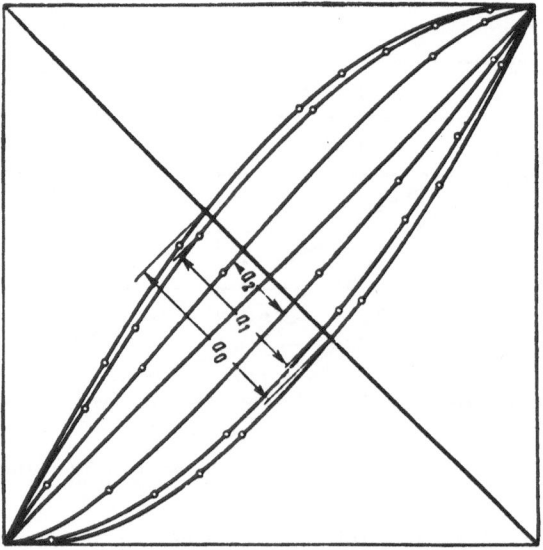

Fig. 31. Loops for different ratios of τ to t' when
$\delta = 67\%$: a_0) $\tau_{av}/t' = \infty$; a_1) $\tau_{av}/t' = 0.75$; a_2)
$\tau_{av}/t' = 0.15$.

For the region from t = t' to t = ∞, we have

$$\frac{dy}{dt} = -\frac{y}{\tau}. \tag{14.22}$$

From Eq. (14.21) we find that when t = t'

$$y = y_0 \frac{\tau}{t'} \left(1 - e^{-\frac{t'}{\tau}}\right). \tag{14.23}$$

Using the initial condition given by Eq. (14.23) we solve Eq. (14.22) to
find

$$y = y_0 \frac{\tau}{t'} \left(1 - e^{-\frac{t'}{\tau}}\right) e^{-\frac{t-t'}{\tau}}. \tag{14.24}$$

It is easily proved that at t = t' the derivatives of Eq. (14.21) and (14.24)
are equal and consequently the decay curve is not only continuous but its
first derivative is also continuous. A series of such curves is given in
Fig. 30 for various ratios of t' to τ. The straight line in that figure represents the linear decay of the excitation pulse.

   If the mutually perpendicular plates of an oscillograph are excited by
synchronized signals which vary with time in accordance with Eqs. (14.23)
and (14.24) represented by the curves in Fig. 30, but have different τ *
───────────────
* For one of these curves τ is a measured quantity, and for the other τ represents RC, which
is the time constant of the taumeter.

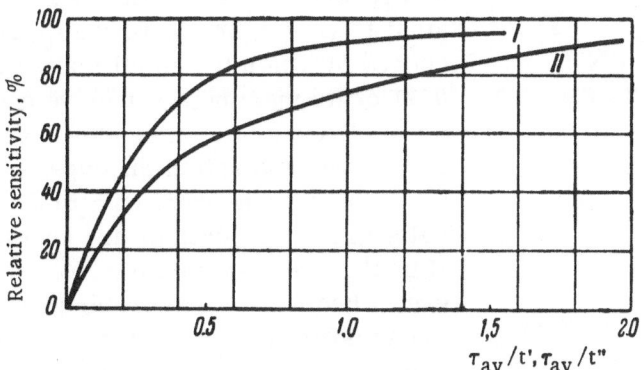

Fig. 32. Sensitivity of the "taumeter" method. I) Dependence
on the cutoff (rise or decay) time; II) dependence on the delay
by amplifying devices.

(obviously if both values are equal a straight line is obtained on the screen),
then the oscillograph screen shows a curve which differs from a straight
line depending on the degree of detuning. The forward and return branches
of the curve (corresponding to the rise and decay of excitation) form a
closed loop on the screen.

Three such loops are shown in Fig. 31 corresponding to various
ratios of t' and $\tau_{av} = (\tau + RC)/2$, but having the same relative detuning:
$\delta = (\tau - RC)/\tau_{av} = 67\%$. The loops were plotted graphically by eliminating time from the curves of Fig. 30. Figure 31 shows that the larger
the value of the cutoff time t' compared with τ_{av}, the narrower the loop
on the oscillograph screen and therefore the lower the sensitivity $\varepsilon = a/b$.

To represent the sensitivity we shall now use the ratio of ε to the
maximum possible value of ε (for a given detuning) which is obtained using
an ideal pulse ($t' = 0$, $\tau_{av}/t' = \infty$). This ratio is equal to a/a_0, where a_0
represents the width of the loop for $\tau_{av}/t' \to \infty$ (cf. Fig. 31).

Figure 32 gives the sensitivity as a function of the ratio τ_{av}/t' (curve
I). This figure shows that when $\tau_{av} = t'$ the sensitivity of the method is
still within the region of the maximum value corresponding to the condition
$t' \ll \tau_{av}$. When τ_{av} is five times smaller than t' the sensitivity is 48%
of its maximum value, and when τ_{av} is ten times smaller than t' the
sensitivity drops to 25%.

We have just considered the influence on the sensitivity of the deviations of square pulses from the ideal shape. Another quantity which is
important in practice is related to the delay due to the devices usually
necessary for amplifying the signals applied to the oscillograph plates.
After passing through an amplifier a signal is strengthened (this is unimportant for our purpose) and distorted because the frequency response
of the amplifier drops at high frequencies. Considering an amplifier as

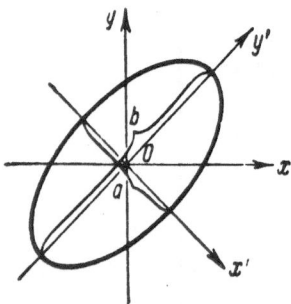

Fig. 33. Ellipse for RC $\neq$ τ.

an approximate RC system with a time constant t'' (t'' = $1/2\pi f_{th}$, where f_{th} is the frequency at which the amplification factor is 0.71 of its maximum value) we can proceed as in the case of a pulse with finite rise or decay time and obtain the dependence of the sensitivity on the ratio of τ to t''. The results of the calculations are shown by curve II in Fig. 32. From the curve we see that even when t'' is considerably greater than τ, the sensitivity still retains a relatively high value.

Thus, in the case of purely linear processes we can use the taumeter method for pulse rise and decay times t' and amplifier time constants t'' considerably longer than the intrinsic time constant of the process without greatly altering the sensitivity.

2. Sinusoidal Modulation. The sensitivity in the case of sinusoidal modulation is defined, as for square modulation, by the ratio of the width of the loop on the oscillograph screen to its length, i.e., by the ratio of the minor and major axes of the ellipse (Fig. 33)

$$\varepsilon = \frac{a}{b}. \tag{14.25}$$

The value of ε for the ellipse given by Eq. (14.7) may be calculated by transforming the coordinates so that Eq. (14.7) becomes (in new coordinates x', y')

$$\frac{x'^2}{a^2} + \frac{y'^2}{b^2} = 1. \tag{14.26}$$

Without going into details of the calculation we can give the final expression for the sensitivity

$$\varepsilon = \frac{x_0}{y_0} \tan \frac{\varphi_2 - \varphi_1}{2}. \tag{14.27}$$

Here x_0, y_0, φ_2, and φ_1 have the same meaning as in Eqs. (14.6) and (14.5).

We shall plot the dependence of ε on the modulation period T = $2\pi/\omega$ when τ = 2RC, i.e., for a detuning given by

$$\delta = \frac{\tau - RC}{\tau_{av}} = 2\frac{\tau - RC}{\tau + RC} = 0.67 = 67\%.$$

Replacing, in Eq. (14.5), τ and RC by τ_{av}, and ω by $2\pi/T$, we find

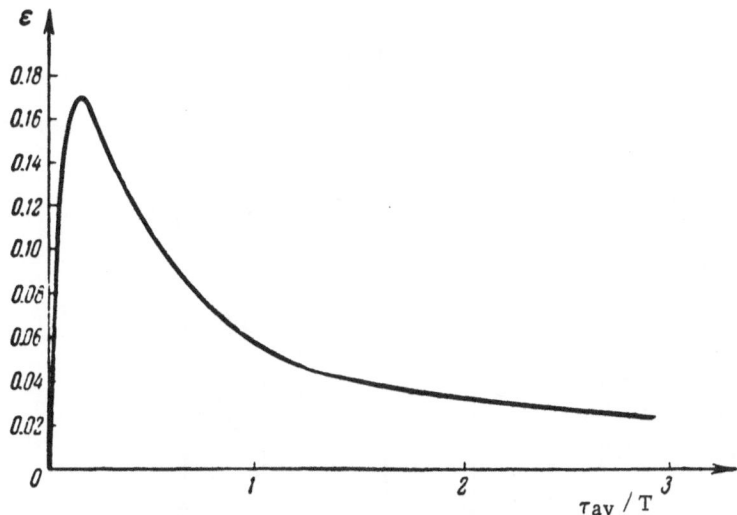

Fig. 34. Sensitivity of the phase-shift compensation method in the case of
sinusoidal modulation, $\varepsilon = f(\tau_{av}/T)$.

$$\left.\begin{array}{l} \varphi_2 = \tan^{-1} \dfrac{8\pi}{3} \dfrac{\tau_{av}}{T}, \\[2mm] \varphi_1 = \tan^{-1} \dfrac{4\pi}{3} \dfrac{\tau_{av}}{T} \end{array}\right\} \qquad (14.28)$$

and consequently

$$\varepsilon = \frac{x_0}{y_0} \tan \frac{1}{2}\left(\tan^{-1} \frac{8\pi}{3} \frac{\tau_{av}}{T} - \tan^{-1} \frac{4\pi}{3} \frac{\tau_{av}}{T}\right). \qquad (14.29)$$

Figure 34 gives the dependence of ε on τ_{av}/T for $x_0 = y_0$. In contrast to the monotonic curves in the case of square modulation (Fig. 32), this dependence has a maximum at $\tau_{av}/T = 0.169$ (i.e., when $\omega = 1.06/\tau_{av}$).* The absence of a maximum in the case of the curves in Fig. 32 and its presence in Fig. 34 are due to the fact that in the former case we assumed, in investigating the influence of the duration of the rise time of a square pulse on the sensitivity, that the pulse itself is of infinite duration and the relaxation process always reaches the steady state during the pulse; whereas in the case of sinusoidal modulation the "front" and "total duration of the pulse" are closely related: a reduction of the "front" is accompanied by a reduction of the "total duration of the pulse."

The sensitivity in Fig. 34 rises initially with increase of the frequency because of a reduction in the duration of the sinusoidal pulse "front," but having reached a maximum the sensitivity decreases because although the "front" still continues to decrease, approaching the ideal maximum steepness, the "total duration of the pulse" decreases.

* When detuning approaches zero, $\delta \to 0$, a maximum is reached at $\omega = 1/\tau$.

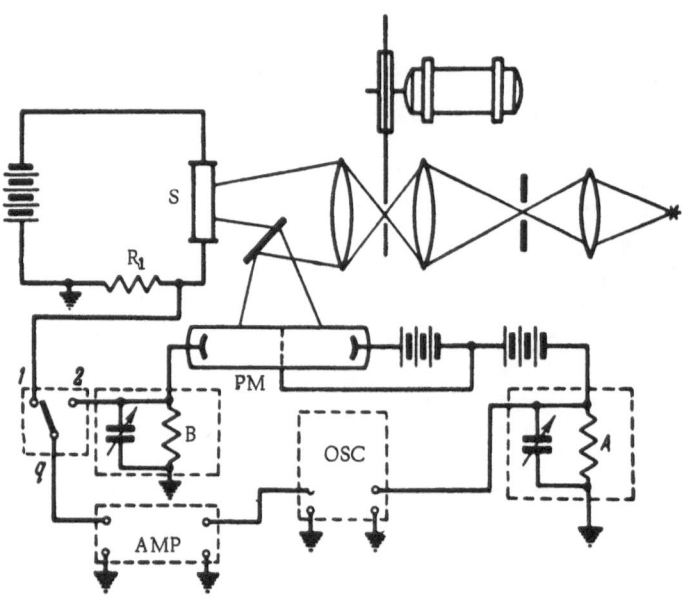

Fig. 35. Substitution method. Basic circuit. S is the sample and
AMP is the amplifier.

Then, to a gradually increasing extent, the process fails to reach, during
the "pulse," its maximum corresponding to the amplitude of the excitation intensity.

Thus, in contrast to the square modulation case when (provided the
total duration of the pulse is sufficiently long) a reduction of the rise
time increases the sensitivity, there is an optimum condition for measurements with $\omega = 1.06/\tau_{av}$ (or $f = 0.169/2\pi\tau_{av}$). *

It is worth noting that the absolute maximum values of ε are similar
for sinusoidal and square modulations. However, sinusoidal modulation
allows us to use narrow-band amplifiers and in practice gives higher sensitivity. On the other hand, square modulation shows more clearly the
deviations from simple linear relaxation which can be observed visually
by examining the form of the curves on the oscillograph screen.

3. Substitution Method. We have assumed above that when the taumeter method is used not only are the excitation pulses incident on the sample and the photocell identical [this is achieved quite simply by splitting
the light beam with a semitransparent mirror (Fig. 27)] but also the amplifiers in both channels (X and Y) are completely identical, i.e., even
if they introduce any distortion this distortion is the same for both chan-

---

* If in the case of square modulation the relationship between the pulse duration and its decay
and rise times is definite, the value of ε is maximum at some value of ω (or over a
finite range of ω values).

nels. It is not always possible to achieve this in practice. For this
reason the "substitution method" is sometimes used (Fig. 35) [9].

The test signal in the sample circuit is fed from a resistor R_1 to a
wide-band amplifier (the switch q is in the position 1) and after amplification is applied to the vertical plates of an oscillograph. To obtain
a scanning exponential, part of the light beam is deflected by a mirror
to a photocell or, better, to a photomultiplier PM. Two taumeters A
and B are connected in the photomultiplier circuit. The use of two taumeters is the characteristic feature of the substitution method: using the
taumeter A, connected to the horizontal plates of the oscillograph, a
straight line is obtained when the test signal (with the switch q in the
position 1) is applied to the vertical plates. Then the switch q is placed
in the position 2 and again a straight line is obtained on the screen by
means of the taumeter B. *

Since the signal of the taumeter B has passed through the same amplifying channel as the test signal and their identity has been proved by
successive "rectification" by means of the same signal from the taumeter
A, it is obvious that the value of the lifetime τ can be read off from the
taumeter B. The substitution method obviates the necessity of making
sure that the amplifying channels of the horizontal and vertical plates are
identical.

§15. INVESTIGATION OF RELAXATION IN THE NONLINEAR CASE

In the linear case the relaxation curves of nonequilibrium conductivity, like the excitation parameters, are represented by a single constant parameter of the substance – the lifetime τ. (In the special case
of square pulse excitation the relaxation curves have the form of ascending and descending exponentials with a time constant equal to τ.)

In the linear case the value of τ may be found relatively simply
either from the frequency dependence or by the phase-shift compensation method.

The problem of investigating nonequilibrium conductivity relaxation
and the parameters representing this process are much more complex in
the nonlinear case. As pointed out in §5 of Chap. I, the nonlinear process
is represented by an infinite set of instantaneous lifetimes depending, in
general, on the excitation intensity, time, etc.

Obviously the frequency dependence or the phase-shift compensation
methods, in the forms described above for the linear case, are inapplicable here. It is best then to analyze the relaxation curves directly in the

---

* In those cases where complete "rectification" is not obtained because of nonlinear distortions
introduced by an amplifier, it is sufficient to obtain the same figures on the screen in these two
measurements.

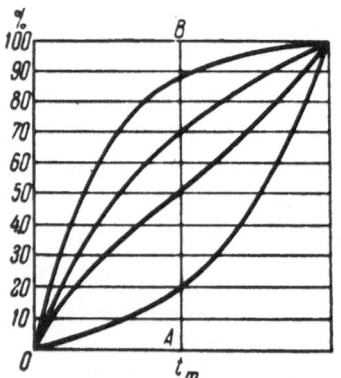

Fig. 36. Partial time method. Relaxation curve in the case of exponential scanning and different time constants.

case of square modulation, these curves being observed visually on the oscillograph screen or obtained in the form of the dependence $\Delta\sigma = f(t)$ by some other method. Square modulation is obviously most convenient, since the relaxation then occurs under conditions of constant excitation intensity (I = const for the rise curve, I = const = 0 for the decay curve) and the nature of the relaxation curves is governed by the parameters of the relaxing system itself.

The relaxation curves may be photographed directly from the oscillograph screen or plotted from measurements obtained, for example, by the "partial time" method [6]. In this method the oscillograph screen is fitted with a scale grid as shown in Fig. 36. Using the taumeter circuit (Fig. 27) we can observe the rise or decay curves using variable exponential scanning. If we position the ends of the investigated branch of the loop (ascending or descending) in opposite corners of the rectangle (Fig. 36), then obviously the point of intersection of the loop with the vertical line AB, which divides the rectangle into half, corresponds to the time t_m which is necessary for the scanning exponential to travel halfway along the X axis. This time can easily be determined. If, for example, x increases according to the law

$$x = x_0\left(1 - e^{-\frac{t}{RC}}\right),\qquad(15.1)$$

and decreases according to

$$x = x_0 e^{-\frac{t}{RC}},\qquad(15.2)$$

then, in both cases the value x = $x_0/2$ corresponds to t = t_m = RC ln 2. The time t_m may be varied by varying R or C of the taumeter circuit. This makes the investigated branch of the loop intersect the line AB at varying heights. Selecting t_m so that the intersection occurs at a height corresponding to 90%, 80%, etc. * we can obtain a series of values of the time t, which correspond to equally spaced values of the investigated quantity (for example, $\Delta\sigma$), and then plot the relaxation curve $\Delta\sigma = f(t)$ on any scale.

---

* It is assumed that complete relaxation curves (up to saturation) are being investigated.

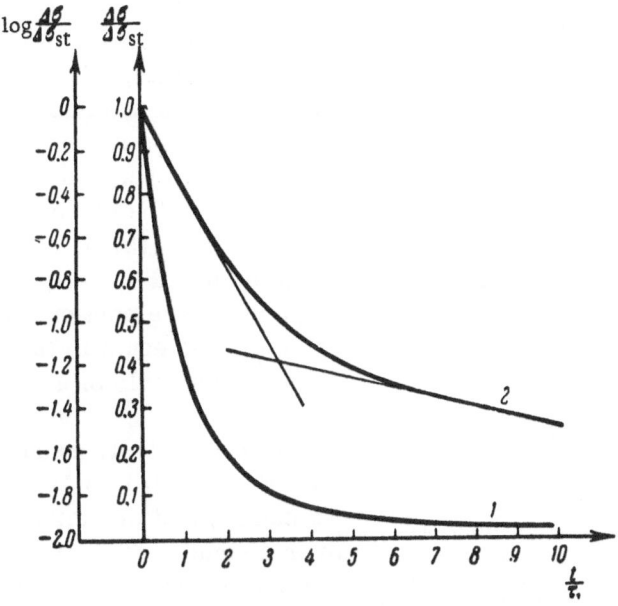

Fig. 37. Relaxation curves on ordinary (1) and semilogarithmic (2) scales.

In the method of "partial times" (i.e., times corresponding to equally
spaced values of the investigated quantity $\Delta\sigma$), it is possible to study
both fast (initial) and slow (final) stages of the relaxation process with the
same accuracy by using variable exponential scanning.

The dependences $\Delta\sigma = f(t)$ obtained in this way or in the form of the
usual oscillograms (linear scanning), have to be analyzed in order to
determine the mechanism of nonequilibrium conduction. For this purpose
it is convenient to use certain special coordinates.

If we expect considerable exponential portions in a complex relaxation curve (i.e., if we expect that over a large region the relaxation is
characterized by a constant lifetime), it is convenient to plot the results
in the coordinates $\ln(\Delta\sigma/\Delta\sigma_{st}) = f(t)$ for the decay curve or $\ln[1-(\Delta\sigma/\Delta\sigma_{st})]$
$= f(t)$ for the rise curve. Then the exponential parts are represented by
straight lines with slopes which determine τ.

Assume, for example, that the photoconductivity decays in accordance
with the law

$$\Delta\sigma = A_1 e^{-\frac{t}{\tau_1}} + A_2 e^{-\frac{t}{\tau_2}}, \qquad (15.3)$$

where

$$A_1 \gg A_2 \text{ and } \tau_1 \ll \tau_2. \qquad (15.4)$$

Then, on the usual linear scale, the relaxation curve has the form of
curve 1 in Fig. 37. This curve consists of two regions which merge

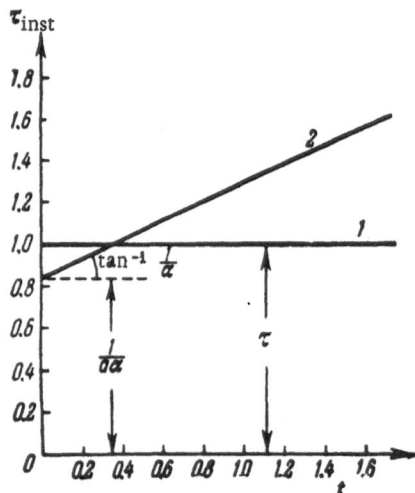

Fig. 38.   Dependence of τ_{inst} on time: 1) Exponential relaxation law; 2) hyperbolic law.

smoothly, but it is difficult to obtain detailed quantitative information from them.   Replotting the same dependence on the scale $\ln(\Delta\sigma)=f(t)$ (curve 2) shows more clearly that there are two rectilinear portions representing the two exponential terms of Eq. (15.3).   The values of τ can be found directly from the slopes of these portions (the absolute value of τ is equal to the cotangent of the slope angle of a rectilinear portion).

It should be stressed that such a simple result is obtained only when we can neglect one term in Eq. (15.3) over a considerable time interval, i.e., when we have the conditions given by Eq. (15.4).

In general it is desirable to carry out a more detailed analysis of the experimental curves for the purpose of plotting the instantaneous lifetimes as a function of time: $\tau_{inst} = f(t)$.

In §5 of Chap. I, where we defined and analyzed the concept of "the instantaneous value of the lifetime," we described in essentials the method for determining this lifetime experimentally; this reduces to the following: having obtained an oscillogram (or having plotted a curve by the partial-time or some other method) which gives the dependence of the photoconductivity on time, we determine at each point the derivative (from the slope of the tangent) and the value of the photoconductivity and then we find τ_{inst} from the following formula for the descending branch*

$$\tau_{inst} = -\frac{\Delta\sigma}{\dfrac{d}{dt}(\Delta\sigma)} \qquad (15.5)$$

When τ_{inst} is determined for any point of the ascending branch, for which

$$\tau_{inst} = \frac{\Delta\sigma}{\beta kI - \dfrac{d}{dt}(\Delta\sigma)}, \qquad (15.6)$$

we must also find separately the slope of the tangent at the moment illumination begins (t = 0), which is equal to βkI.   The dependence $\tau_{inst} = f(t)$

_____
* Electrical differentiation [10] may be used to determine the derivative $d(\Delta\sigma)/dt$.

for the exponential type of relaxation is represented by the horizontal straight line τ = const (Fig. 38, line 1).

This dependence is also relatively simple in several more complex relaxation curves.

Thus, when the photoconductivity decay can be represented by a hyperbola of the type *

$$\Delta\sigma = \frac{\Delta\sigma_{st}}{(1 + at)^{\alpha}} \tag{15.7}$$

the dependence $\tau_{inst} = f(t)$ is also a straight line (line 2 in Fig. 38):

$$\tau_{inst} = \frac{1}{a\alpha} + \frac{1}{\alpha}t. \tag{15.8}†$$

From the slope of this line we can find the power exponent α in the denominator of Eq. (15.7) and then, having found the intercept on the ordinate, we can determine a.

If the experimental dependence $\tau_{inst} = f(t)$ is not a straight line but has quite long linear portions, it means that in these portions the relaxation is exponential if the linear parts are horizontal, or hyperbolic if they are not.

In conclusion, we note that even in studies of nonlinear kinetics we can frequently apply some special relatively simple experimental method in order to reduce the analysis to that used for linear processes. The essence of such a special method is the simultaneous illumination of the sample with a modulated (sinusoidal or square pulse) beam and a beam whose intensity is constant with time. The intensity of the constant illumination is usually greater than the alternating intensity so that the relaxation of the alternating component of the photoconductivity is linear, i.e., it is represented by a single time constant for the whole process, the time constant depending in general on the intensity of the constant illumination. The stronger constant illumination governs the electron (hole) population of all the energy levels. The weaker alternating component cannot change this population greatly and therefore relaxation occurs under conditions of practically constant energy-level population; thus the time constant does not vary.

We shall illustrate this for the case of quadratic recombination. The transport equation for this case is [cf. Eq. (5.6) in Chap. I]

---

* In the case of nonlinear photoresistors the relaxation decay is frequently best represented by a hyperbola of the type in Eq. (15.7).

† We obtain Eq. (15.8) from $\tau_{inst} = -\dfrac{\Delta\sigma}{\dfrac{d}{dt}(\Delta\sigma)}$, where $\Delta\sigma$ is taken from Eq. (15.7).

$$\frac{dn}{dt} = \beta kI - \gamma n^2. \tag{15.9}$$

As shown in §5 of Chap. I, the rise and decay curves in the case of quadratic recombination and square modulation are, respectively, a hyperbolic tangent and a hyperbola. The instantaneous lifetime along these curves is not constant, the relaxation is nonlinear, and the methods described in §§13 and 14 are not applicable.

Assume, however, that the sample is illuminated with a light beam with intensity given by

$$I = I_c + \Delta I, \tag{15.10}$$

where I_c is the intensity of the constant illumination and ΔI is the intensity of the alternating component. We shall assume that

$$I_c \gg \Delta I. \tag{15.11}$$

The carrier density may also be represented in the form

$$n = n_c + \Delta n, \tag{15.12}$$

where n_c is the carrier density produced by the action of the constant illumination I_c, and Δn is the density due to ΔI.

Then, substituting Eq. (15.10) into (15.9) and bearing in mind that

$$\frac{dn_c}{dt} = \beta kI_c - \gamma n_c^2 = 0, \tag{15.13}$$

we have

$$\frac{d\Delta n}{dt} = \beta k\Delta I - 2\gamma n_c \Delta n - \gamma \Delta n^2. \tag{15.14}$$

From Eq. (15.11) we may also assume that $\Delta n \ll n_c$. Consequently, neglecting the last term in Eq. (15.14), we obtain

$$\frac{d}{dt} \Delta n = \beta k\Delta I - \frac{\Delta n}{\tau}, \tag{15.15}$$

where $\tau = 1/2\gamma n_c$ is a constant quantity governed by I_c. Equation (15.15) is linear and consequently the relaxation of Δn may be investigated by the methods suitable for linear relaxation and described in §§13 and 14. By varying the intensity of the constant illumination I_c and consequently of n_c we can vary τ and thus establish the dependence of τ on n_c, which may help in drawing important conclusions on the nature of the process. *

* One should bear in mind that τ measured in this way is not in general equal to the steady-state lifetime τ_c when only one type of illumination is present. For example, in the case considered $\tau_c = 1/\gamma n_c$, while the measured value is $\tau = 1/2\gamma n_c$.

§16. SOME METHODS OF MODULATING
THE LIGHT-BEAM INTENSITY*

The experimental study of relaxation processes resulting from the action of light requires the use of optical excitation mostly in the form of square light pulses or in the form of a sinusoidally modulated light beam (§§13, 14, and 15).

It is desirable that the characteristic time constants of the modulated light (for example, the rise time of a square pulse) should be smaller or at least comparable with the time constants of the process being studied. In various cases the study of the inertia of photoelectric phenomena in semiconducting materials and devices involves time constants extending over a very wide range and therefore it is necessary to use sinusoidal modulation, of frequency from several cycles per second to tens of mega-cycles, or pulse modulation, having rise times down to several nanoseconds, various pulse durations, and pulse repetition rates.

In the case of fast processes the characteristics of the light modula-tion become decisive since in the majority of cases it is the parameters of the modulated light that govern the limits of applicability of various methods for measuring the time constants. Therefore the light modulator is an important part of the measuring apparatus and other devices.

We shall describe briefly the most widely used methods of modulat-ing the light beam.

To obtain square light pulses, a mechanical modulator consisting of a rotating disk with a cutaway sector is frequently used: it periodically interrupts the light beam incident at right angles to the disk surface (Fig. 39) (cf., for example, [7], [12], [15] as well as [11], [27]). The light pulses produced may be represented by three time constants (Fig. 40): (1) pulse duration t_0^i, (2) duration of the interval between pulses t_0^d, or the off-duty factor $\theta = (t_0^i + t_0^d)/t_0^i$, (3) pulse rise and decay times t'. The times t_0^i and t_0^d are governed by the angular velocity of rotation of the disk ω and by the values of the sector angles φ_i and $2\pi - \varphi_i$,

$$t_0^i = \frac{\varphi_i}{\omega}, \qquad t_0^d = \frac{2\pi - \varphi_i}{\omega},$$

the time t' is governed by the linear velocity of rotation of the edge of the disk ωr and the cross-sectional width of the light beam a_0 in the plane of the disk

$$t' = \frac{a_0}{\omega r}.$$

---

* This section was written by L. G. Paritskii.

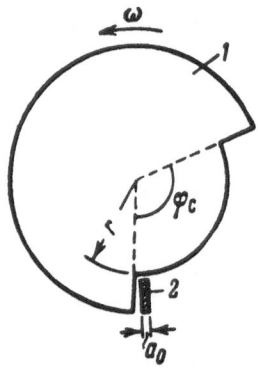

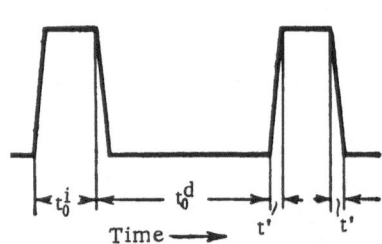

Fig. 39. 1) Modulator
disk with a cutaway
sector; 2) cross section of
the light beam in the
plane of the disk.

Fig. 40. t_0^i is the duration of a light
pulse; t_0^d is the duration of the
interval between pulses; t' is the
rise or decay time.

When an electric motor is used to rotate a disk of radius r = 15 cm at
a rate of 6500 rev/min, the pulse rise time t' is 10^{-5} sec for a_0 = 1 mm.
Increase of the angular velocity and radius of the disk is limited by air
resistance to the motion. This may be overcome by placing the disk in
an enclosure pumped down to fore-vacuum pressure [28]. After this,
further reduction of the pulse rise time is limited only by the strength
of the disk material. By rotating a disk of r = 30 cm at 12,000 rev/min
in vacuum it was possible to obtain t' ≈ 10^{-6} sec for a_0 = 0.5 mm.

If the radius of the disk r and the width of the light beam a_0 are fixed,
then the quantities t_0^i, t_0^d, and t' are not independent. By fixing one of
them (for example, t_0^i) we restrict the range of variation of t', and if t'
and t_0^i are fixed this gives only one possible value of t_0^d (or the off-duty
factor θ). This circumstance makes it impossible to obtain pulses with
large intervals t_0^d between them (while still keeping the rise front steep)
and consequently it is difficult to study complex relaxation phenomena involving both fast and slow processes. *

Pulses with steep fronts, separated by large intervals of time, may
be obtained with a two-disk modulator [12], the basis of which is shown
schematically in Fig. 41. Light from a source S falls on a test object T

---

* Such processes are frequently observed in the photoconductivity of semiconductors. For example, to investigate the initial stages of relaxation in CdS it is necessary to use light pulses
with relatively steep fronts of ~10^{-5} sec. Then, however, in the presence of slow components
in the relaxation process in CdS (i.e., components represented by times of the order of seconds),
it is necessary to ensure that the spacing between the pulses is sufficiently great so that
a new light pulse begins after the end of the relaxation process connected with the previous
pulse, in order to give the system a chance to return to the unexcited state.

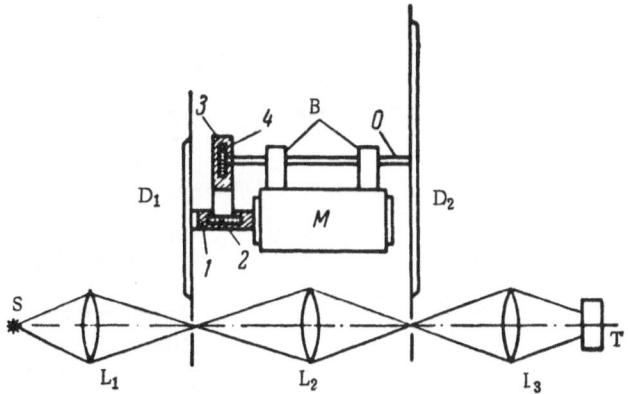

Fig. 41. Two-disk modulator. L_1, L_2, L_3 are lenses; B denotes
the bearings of the controlling disk shaft; 1, 2, 3, 4 are the gears.

after passing through apertures (cutaway sectors) in two disks D_1 and
D_2, which are rotated by a motor M. The disk D_1 is fixed directly on
the motor shaft and the disk D_2 is fixed on a shaft O which is connected to
the motor shaft by a set of gears with a transmission ratio $n = \omega / \Omega$ (ω
and Ω are the angular velocities of the disks D_1 and D_2 respectively). Obviously the light falls on the test object only when its path is not interrupted by either of the disks. Only some of the pulses formed by the
faster disk D_1 reach the test object T: these are pulses which are passed
by the slower disk D_2, whose angular velocity may be varied. Consequently the formation of a pulse and its time constants t' and t_0^i are determined
by the disk D_1 but their repetition frequency, i.e., the duty factor, is
controlled by the disk D_2.

Normally the modulator can only work when certain quantitative relationships between the parameters of the forming and controlling disks are
satisfied. Figure 42 shows both disks of the modulator and Fig. 43 shows
four typical successive positions of a cut in the controlling disk with respect to the light beam. The dimensions of the cuts and the transmission
ratio were selected so that during the time taken by the controlling disk to
move from position 1 to position 2 and also from position 3 to position 4
the forming disk interrupted the light beam, while during the time taken
by the controlling disk to move from position 2 to position 3 the cut in the
forming disk lets through the light beam which will then reach the object.
The ratio of the total period to the pulse duration (the off-duty factor) increases in this case by a factor n compared with the single-disk modulator.
In an actual modulator in which the ratio of the angular velocities of the
forming and controlling disks was n = 160, the front rise time was
$t' \approx 10^{-5}$ sec and the pulse was from 2×10^{-2} to 2×10^{-5} sec duration;
the dark interval reached 2.13 sec, i.e., the off-duty factor varied from
87 to 1.6×10^5.

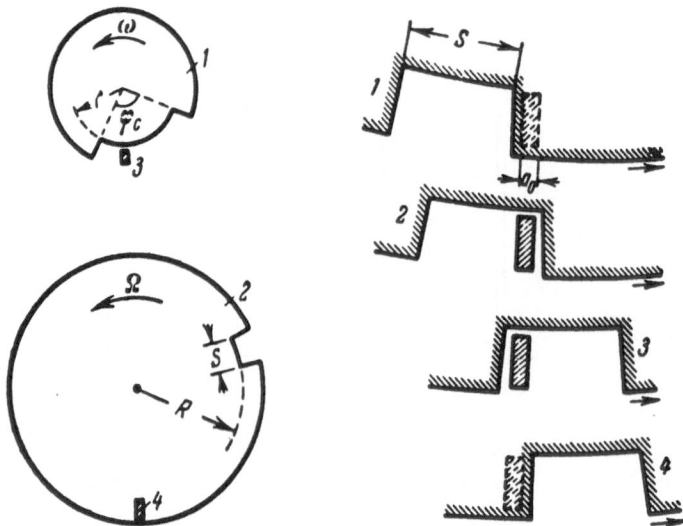

Fig. 42. 1) Pulse forming disk;
2) controlling disk; 3, 4) cross
sections of the light beam in
the disk planes.

Fig. 43. Successive positions of the
light-beam cross section and the
cutaway   sector in the controlling
disk.

A simpler method of obtaining single square pulses is the combined
use of a rotating disk and a mechanical shutter, as shown schematically
in Fig. 44 [31]. Light from a source S reaches a test object T only when
the cut in the disk D is in the path of the beam and the shutter SH is open.
By selecting the shutter exposure t_{SH} to be equal to the duration of the
dark interval between two consecutive pulses we may reach a state of
very high probability that during the time that the shutter is open only one
square pulse formed by the disk D will reach the test object (cf. Fig. 45).
This probability is given by the ratio $w = t_{SH}/(t_0^d + t_0^i)$, and under usual
conditions, for example when $t_{SH} = 7.3$ msec, $t_0^d = 200$ μsec, represents
97% which is quite satisfactory.*

Of the mechanical devices we should also mention the widely used
mirror modulators. These are based on the principle of the optical lever.
One such mirror modulator with a rotating mirror is shown in Fig. 46.
Light from a source S is passed through an optical system L to a rotating
mirror M and at a considerable distance R from the latter it forms a
sharp image of the slit SL in the form of a uniformly illuminated rectangle. During rotation of the mirror this image passes over a slit of

---

*It should be noted that the appearance of an imperfect light pulse, i.e., a pulse which reaches
the object T when the shutter SH is not yet completely open or completely closed, will be seen
at once because a double pulse and not a single one is then obtained.

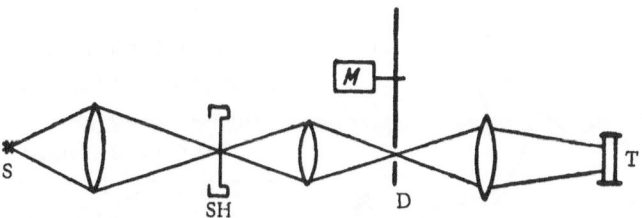

Fig. 44. Modulator combining a disk with a shutter.

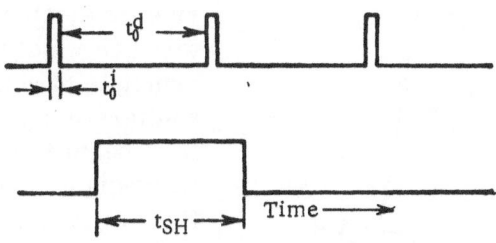

Fig. 45. Relative positions in time of the light transmission by the disk and the shutter of Fig. 44.

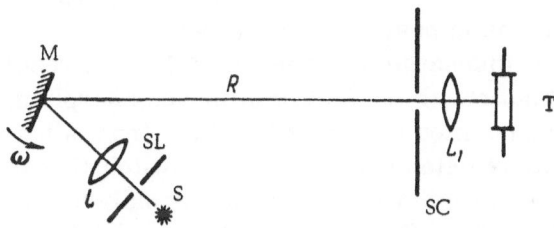

Fig. 46. Simplest mirror modulator.

width d in a screen SC, forming a square pulse at the test object, the pulse rise time of which is given by the linear velocity v at which the image moves in the plane of the screen:

$$t' = \frac{d}{v} = \frac{d}{2\omega R}.$$

When the mirror is rotated at a rate of 25,000 rev/min, R = 1.5 m, d = 1.5 cm, the modulator gives periodic square pulses with a pulse rise time of 10^{-7} sec.

Figure 47 shows the basic principle of a more complex mirror modulator with a rotating mirror prism [17], which uses the increase of the angular velocity of a light beam on reflection from a rotating mirror. A light beam reflected in turn from each face of a hexagonal prism doubles its angular velocity at each reflection. Such a modulator can be used to produce light pulses of nanosecond duration (cf. also [29]).

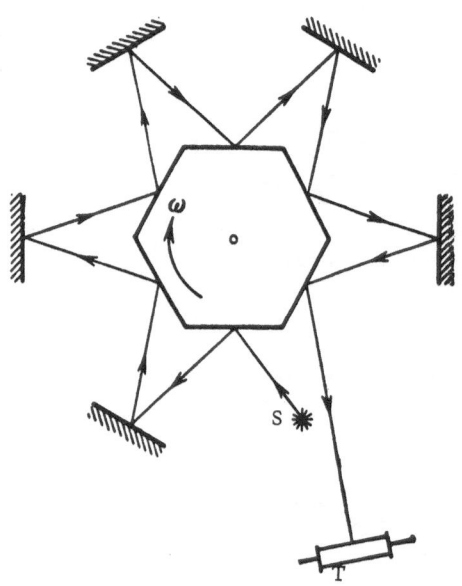

Fig. 47. A modulator with a mirror prism. S is the source of light, T is the test object.

To obtain single light pulses of relatively long duration and arbitrarily large off-duty factor, and also to produce a square "excitation step" with a short pulse rise time, electrodynamic shutters are widely used [16]. The principle of action of one of these shutters is shown in Fig. 48. A light beam is interrupted by a lightweight screen A which is fixed by a hollow steel support rod SR to the end of an audio-frequency coil CO (taken from an electrodynamic loudspeaker); the coil is located in the narrow air gap of a permanent magnet M.

When the switch K is in position 1, a strong current pulse passes through the coil due to the discharge of a capacitor C_1; as a result the moving part of the modulator with the screen A moves sharply upwards and is kept in the raised position by a small current from a battery E passing through a limiting resistance R. When the switch K is placed in position 2 the coil with the screen A moves equally rapidly downward under the action of a current pulse of opposite direction, due to the discharge of a capacitor C_2. The rise or decay times for a modulator of this kind are in the range 10^{-4}-10^{-5} sec.

Of modulators using the magneto-optical Faraday effect (rotation of the plane of polarization of light in certain substances under the action of a magnetic field) and the electro-optical effects, we shall consider only a device which is widely used in the form of the so-called Kerr cell [2, 19, 22, 30]. The modulator is shown schematically in Fig. 49. Light from a source S passes first through a polarizer P (for example a Nicol prism) so that it becomes linearly polarized, and then passes through a Kerr cell which is a parallel-plate electric capacitor filled with a liquid of high electro-optical constant (usually nitrobenzene). The light then passes through an analyzer A and reaches the test object T. In the absence of electrical stress nitrobenzene is optically isotropic and if the polarizing prisms are crossed at an angle of 90°, the system is opaque. On application of an electric field, usually directed at an angle of 45° to the plane of polarization of the light, the nitrobenzene becomes aniso-

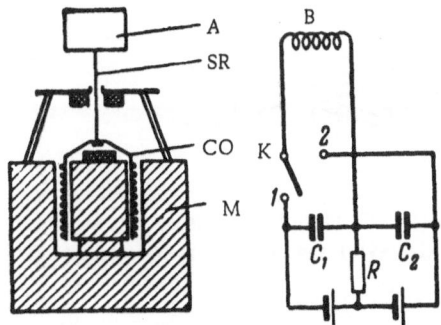

Fig. 48. Electrodynamic shutter (cross section) on the left, and the electric circuit of the coil CO.

tropic, acquiring the properties of a uniaxial crystal; because of birefringence the light is found to be elliptically polarized and passes through the analyzer to the object at an intensity approximately proportional to the electric field. The Kerr effect time constant does not exceed 10^{-9} sec, which makes it possible to modulate the light beam at a very high frequency. The cell consists of a cuvette of prismatic or cylindrical shape with parallel entry and exit windows. The electrodes are either parallel or inclined. In some cases multi-electrode cells are used. The cell works best at a voltage E_m cor-

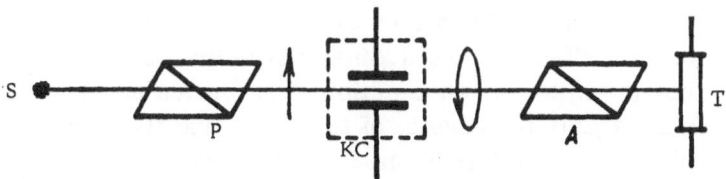

Fig. 49. Modulator based on the use of a Kerr cell. P is the polarizer, KC is the Kerr cell and A is the analyzer.

responding to the first maximum (Fig. 50), which usually requires an alternating voltage of the order of 3 kV. This alternating voltage is frequently obtained by using a dc bias of ~2 kV and an alternating modulating signal of ~1 kV. It should be borne in mind that in the absence of an alternating voltage the Nicol prisms cannot be crossed so as to obtain complete darkness, and therefore a fluctuating light signal reaches the test object even in the absence of constant illumination, which is sometimes undesirable. The apparatus described in the literature for photoelectric studies by means of Kerr cells includes modulators with an alternating sinusoidal light-beam frequency of several tens of megacycles, and, in the case of pulse modulation, with fronts of 10^{-7}–2×10^{-8} sec.

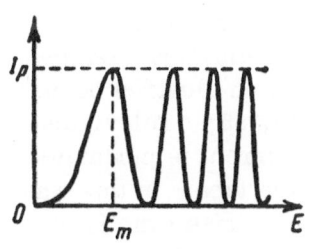

Fig. 50. Dependence of the magnitude of the Kerr effect I_p in nitrobenzene on the electric field E.

Sinusoidal modulation of a light beam may also be produced by diffraction modulators [14, 20], which have relatively high transmission and, unlike Kerr cells, require only a low-

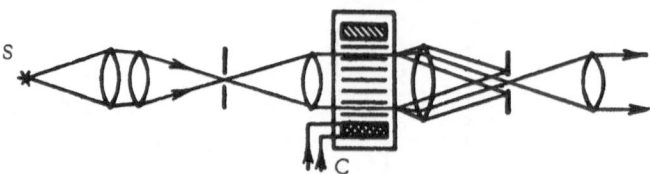

Fig. 51. Diffraction modulator using ultrasonic vibrations. S is the light source, C is a cuvette filled with xylol and containing a vibrator.

voltage high-frequency supply. The working principle of such modulators is based on the diffraction of light on passing through a liquid in which there is a field of standing ultrasonic waves. A modulator of this type is shown schematically in Fig. 51. A piezoelectric vibrator is placed at the bottom of a closed cuvette fitted with two windows, and filled with a liquid (usually xylol). This vibrator may be a barium titanate plate tuned to a certain mechanical vibration frequency. An identical plate is fixed to the top of the cuvette. Vibrations are excited in the plates by the high-frequency source. Standing longitudinal waves are established in the liquid, the compression and rarefaction at antinodes being so strong that they alter considerably the density of the liquid and its refractive index. A regular volume "grating" is formed which appears and disappears at double the frequency of the plate vibrations. Since the wavelength of ultrasound is comparable with the wavelength of light, the light is diffracted by this "grating"; the light at the principal ("zeroth") maximum is not modulated, so the test object is usually illuminated with light deflected in the direction of the first diffraction maximum. In modulators of this type the vibration frequency is between 5 and 20 Mc.

Recently light sources which give a beam varying with time have become popular as modulators. Among such modulators we shall consider three types: (1) a luminescent screen radiating under the action of an electron beam, (2) a spark discharge in air, (3) pulse discharge lamps.

The luminescence of a screen under the action of an electron beam may be conveniently arranged for fast modulation [13, 25]. Television tubes are usually employed for this purpose. The current is modulated by passing a triggering voltage pulse to the grid of the tube. In this way a very bright square light pulse of about 10^{-7} sec pulse rise time is obtained. Deflecting plates may be used to modulate the light by directing the electron beam to a certain point on the screen (which produces luminescence at this point).

A spark discharge is one of the commonest forms of electric discharge in gases. In air under normal conditions (atmospheric pressure, normal humidity and room temperature) a discharge occurs when the electric field is above 31 kV. The spark discharge begins by forming a discharge channel (in a time of the order of 10^{-8} sec), and then the chan-

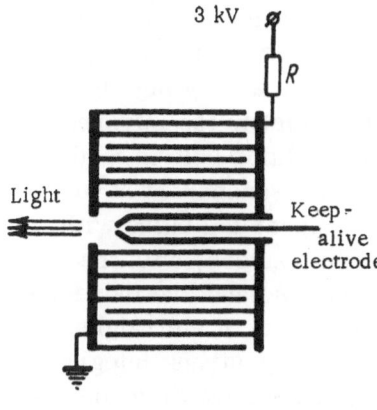

3 kV

R

Light

Keep-
alive
electrode

Fig. 52. Spark source of light.

nel increases in size; the discharge current becomes stronger and the voltage, which after formation of the channel amounts to hundreds of volts, falls and the discharge stops. Since most of the energy is evolved in the main part of the channel in a very short time (~10^{-9} sec), the temperature in the discharge plasma, according to spectrometric measurements, may reach 10,000°K. Then the degree of ionization of the gas is close to 100%. The discharge is accompanied by strong light emission. In the case of photoelectric investigations the spark discharge must satisfy the following requirements to serve as a light source: the light pulse should have the shortest possible duration; the rise of the light intensity during the pulse should be as rapid as possible, and the intensity should be high; the intensity should be constant during the pulse; there should be the smallest possible scatter in the position of the spark channel in the discharge gap; and the discharge frequency should be controllable. These requirements are to a great extent satisfied by a device the principle of which is shown in Fig. 52 [23]. The light source is an electric capacitor of 0.2 μF capacitance charged to 3 kV, and having the dielectric in the form of polystyrene film. One of the electrodes is in the form of a disk with an aperture of 1 mm diameter; the point of the second electrode is close to the aperture, so that the discharge occurs between a plane and a point and therefore the spark channel has an approximately constant position. The test object is illuminated by light emitted along the spark channel and this reduces to minimum the dependence of the intensity of light on the position of the breakdown. The luminous region is very small (d $\approx$ 1 mm). This has great advantages in focusing.

Special attention has been paid to reducing the capacitor inductance and also reducing the resistance in series with the discharge gap. To achieve this the capacitor has the an "induction-free" winding and the shortest possible distances between electrodes of large cross sections; this ensures a high light intensity and short duration of the complete discharge (0.2 μsec). This source may work under self-discharge conditions at a frequency determined by the time constant of the supply circuit, as well as under controllable conditions; in the latter case the discharge frequency can be varied, using gas ionization in the discharge gap, by means of a low-power high-voltage discharge between the plate and a special auxiliary electrode. In other devices without a capacitor and in some using long lines, spark discharges of 10^{-8} sec duration have been obtained (see, for example, [26]).

In pulse-discharge lamps the emission (is used) of a discharge in an inert-gas atmosphere. The discharge usually produces emission in a relatively large volume of gas. The emission reaches a very high intensity and has a continuous spectrum from the ultraviolet to the infrared (to use the extreme wavelengths of the spectrum the lamp bulb is made of quartz or fitted with windows of material which is transparent at these wavelengths). The lamps are made in the form of spherical bulbs with closely spaced electrodes (they then approach point sources, with the luminous volume only a few millimeters in dimensions, and are therefore convenient in focusing) or with one electrode in the form of a straight, bent or spiral tube (in the latter case they produce very strong integral light flux, cf. [24, 21]). In many types of lamp a keep-alive electrode is used to control the discharge.

The duration of a light pulse from a discharge lamp depends on the electronic circuit (in complete analogy to the supply circuit of a spark discharge in air) and the lamp construction: it may vary from a fraction of a microsecond to several milliseconds. The rise time of the light pulse varies from 10^{-8} sec to several microseconds for various types of lamp. The available pulse repetition frequency is from single pulses (~ 0.1 cps) to 10 kc (in a stroboscopic pulse lamps).

The advantage of pulse-discharge lamps in photoelectric investigations [18] is their high intensity of light, which makes it possible to carry out measurements (especially those involving variation of wavelength) with low amplification of the electric signals, and also to establish, even in low-resistivity semiconductors, conditions for high excitation level. Because of the short duration of the pulses or their short pulse rise time, we can investigate by means of pulse-discharge lamps various types of relaxation process and, in particular, we can measure short relaxation time constants. The phase-shift compensation method may then by used.

This short review shows that the selection of a modulator which satisfies the requirements of photoelectric measurements in respect of the range of wavelengths, intensity of light, and time parameters of the signal, should be based on the actual requirements of the experiment.

In the near future considerable progress may be expected in fast light modulators in view of the latest achievements in semiconductor electronic (lasers, recombination radiation of injected carriers, etc.).

Chapter IV

PROCESSES OF GENERATION OF NONEQUILIBRIUM CARRIERS

Nonequilibrium carriers may be generated as the result of internal ionization in a semiconductor on absorption of radiation, or by the disturbance of the equilibrium distribution by a strong field, injection, extraction, etc. (when current passes through an inhomogeneous system, for example through a p-n junction).

We shall consider briefly the characteristic features of some of these processes.

§17. INTERNAL PHOTOEFFECT
(EXCITATION DUE TO ABSORPTION OF LIGHT)

We noted in §1 of Chap. I that if I quanta pass per unit time through unit area of a substance at right angles to the direction of propagation of light, then kI quanta are absorbed per unit volume, where k is the absorption coefficient. This leads to the generation, during the same time in the same unit volume, of Δn' electrons and Δp' holes, where

$$\Delta n' = \beta_n k I, \qquad (17.1)$$

$$\Delta p' = \beta_p k I. \qquad (17.2)$$

Here the coefficients β_n and β_p represent, respectively, the electron and hole quantum yields, i.e., they give the number of electrons (holes) produced by a single absorbed quantum.

It should be pointed out that since each absorbed quantum necessarily excites one electron, then the quantum yield of a single microscopic absorption process is identically equal to unity. The coefficients β_n and β_p have a meaning only because absorption processes which produce free electrons and holes may coexist with "competing" absorption mechanisms which are unaccompanied by electron transitions, or are accompanied by

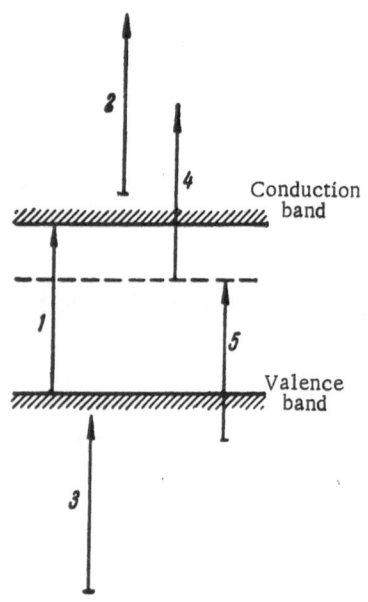

Fig. 53. Possible transitions in the
fundamental absorption region.

such transitions as do not produce free carriers. *

It should also be mentioned that these competing absorption mechanisms are not necessarily due to some foreign occlusions in the material (impurities, defects, grain boundaries in polycrystals, etc.). In principle even a perfect single crystal – absolutely free of impurities – may exhibit such absorption. For example, absorption in the fundamental band may produce electron transitions from the valence band to the conduction band (Fig. 53, transition 1), but some absorbed quanta may be used for transitions within the allowed bands (transitions 2 and 3 in Fig. 53). In the presence of local impurity centers we can also have transitions 4 and 5, each of which leads to the generation of carriers of one sign only. The transitions 2, 3, 4, and 5, which compete with transition 1, may obviously reduce β_n and β_p, the quantum yields of free electron and hole generation, to below unity.

Below we shall consider the intrinsic internal photoeffect by assuming that the competing absorption mechanisms are unimportant. Then the absorption of a quantum from the fundamental band produces a pair of free carriers: an electron in the conduction band and a hole in the valence band. Consequently $\beta_n = \beta_p = 1$ and Δn and Δp are governed only by the parameter k, so that the theory of the internal photoeffect reduces essentially to the theory of the absorption of light. †

The obvious necessary condition for the absorption of a photon and formation of an electron-hole pair is the condition $h\nu \geqslant \Delta\mathscr{E}$ ‡ , where $\Delta\mathscr{E}$ is the width of the forbidden band of the semiconductor. However, in the case of optical transitions the concept of forbidden band width is quite complicated. For one thing this width may, in principle, be greater for

---

* Sometimes β_n and β_p are understood to represent the "effective" or "phenomenological" quantum yield, which may be related not only to the direct phototransitions under the action of phonons, but to the subsequent fast (and therefore unobservable) processes of of charge redistribution between the energy levels in a semiconductor (cf. Chap. VI).

† By the "internal photoeffect" we understand the process of internal ionization. The subsequent processes of diffusion, drift, and recombination represent the phenomenon of "photoconductivity."

‡ Strictly speaking this condition is valid only when the role of phonons in the absorption processes is ignored (see below).

optical transitions than for thermal ones. This follows from the Franck-
Condon principle and is explained in general as follows.

The transition of an electron, following photon absorption, from one
state to another which has a higher energy is accompanied in general by
a change of the wave function of the electron. This may alter the strength
or nature of the interaction of the electron (or the center to which it be-
longs) with the immediate surroundings in the crystal lattice, which in
turn should produce changes in these surroundings (due, for example, to
displacements of the ions which form the ionic crystal lattice).

If such changes proceed slowly compared with electron transitions,
then obviously as a result of a rapid electron phototransition the crystal
near the point where the photon was absorbed becomes excited and then
returns to its normal state during the "rearrangement time." Part of the
energy used for the phototransition is then evolved as heat.

Thus the energy necessary for the phototransition is found to be
greater than the energy gap between the initial and final states, and therefore it is greater than the thermal ionization energy. This is important
in the case of ionic crystals in which the optical permittivity (governed
only by the fast electron polarization which is rapid) is considerably
smaller than the static permittivity (governed by the total polarization
including displacements of atoms or ions). In the case of covalent crystals
both permittivities are practically identical and consequently the thermal
and optical ionization energies have very similar values.

Another reason can also be given for the difference between the values
of the thermal and optical ionization energies of covalent crystals. Considering phototransitions we were interested only in the question of the
relationship between the quantity $h\nu$ and the forbidden bandwidth $\Delta \mathcal{E}$, i.e.,
we started from the necessity of satisfying the law of conservation of energy. Let us now consider the restrictions imposed on phototransitions
by the law of conservation of momentum. When an electron is transferred
from the valence band to the conduction band the difference between its
momentum in the final and initial states $\mathbf{p} - \mathbf{p}'$ should be equal to the momentum of the absorbed photon $\frac{h\nu}{c} \mathbf{s}$

$$\mathbf{p} - \mathbf{p}' = \frac{h\nu}{c} \mathbf{s}. \qquad (17.3)$$

Here $\mathbf{s}$ is a unit vector in the direction of motion of the photon. *

However, since the momentum of an optical photon is negligibly
small†, the expression (17.3) can be quite satisfactorily replaced by

---

* In a crystal lattice the quantity corresponding to the momentum of a free electron is called
quasi-momentum. From now on we shall use this term.

† Thus, for a photon corresponding to $\lambda = 2\,\mu$ the momentum is $1.87 \times 10^{-29}\ \mathrm{g \cdot cm \cdot sec^{-1}}$.
When the momentum of an electron is changed by this amount its energy changes only by
3.5×10^{-7} eV.

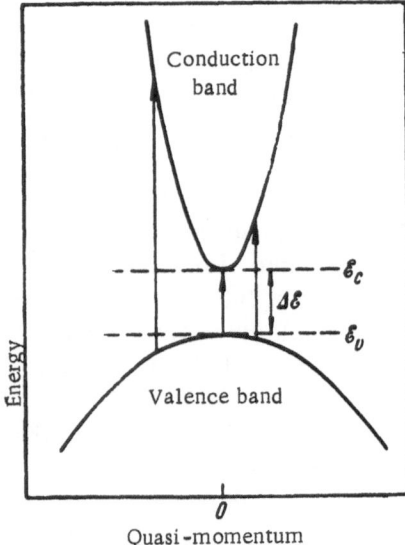

Fig. 54. Dependence of the energy on the quasimomentum for the valence and conduction bands based on a model of a semiconductor with a "simple energy gap."

$$p - p' = 0, \qquad (17.4)$$

i.e., only those transitions are allowed for which the magnitude and direction of the quasi-momentum of the electron undergoing the transition remain fixed.

Figure 54 shows the well-known dependence of the energy on the quasi-momentum for the valence and conduction bands in the simplest case where the lower edge of the conduction band and the upper edge of the valence band correspond to zero quasi-momentum, and the energy of band electrons and holes depends quadratically on the quasi-momentum.

Allowed transitions (i.e., transitions which do not alter quasi-momentum) correspond to "vertical" (or "direct") transitions. Examples of such transitions are shown by vertical arrows in Fig. 54. This figure shows that the minimum photon energy $h\nu_{min}$ which is still capable of transferring and electron from the valence to the conduction band (transition of an electron with quasi-momentum equal to zero) is exactly equal to the forbidden bandwidth $\Delta\mathcal{E}$, and consequently the thermal and optical ionization energies are in this case identical. *

However, strictly speaking in a crystal lattice we have not the law of conservation of momentum but the law of conservation of the total quantum number $\mathbf{k}$. Only in the simplest case of the band structure shown in Fig. 54 can the wave number $\mathbf{k}$ be related directly to the momentum (or more precisely the quasi-momentum) by $\mathbf{p} = \hbar\mathbf{k}$. In substances for which the dependence of the energy on the wave number is known (the number of such substances is still small) [6], this dependence is not as simple as in Fig. 54.

Figure 55 shows the energy band structures for germanium and silicon obtained from experiments on cyclotron resonance and from optical measurements [14]. Figure 55 shows that the band structure is very complex and differs considerably from that shown in Fig. 54. Thus it is evident from Fig. 55 that the valence bands of germanium and silicon consist of

---

* This is correct if the considerations relating to the Franck-Condon principle are unimportant.

Table 1

	Ge	Si	AlSb	GaSb	InAs	InSb
$\Delta\mathcal{E}$, eV	0.65	1.09	1.65	0.69	0.35	0.18
$h\nu_{min}$	0.62	1.05	1.5	0.7	0.35	0.18

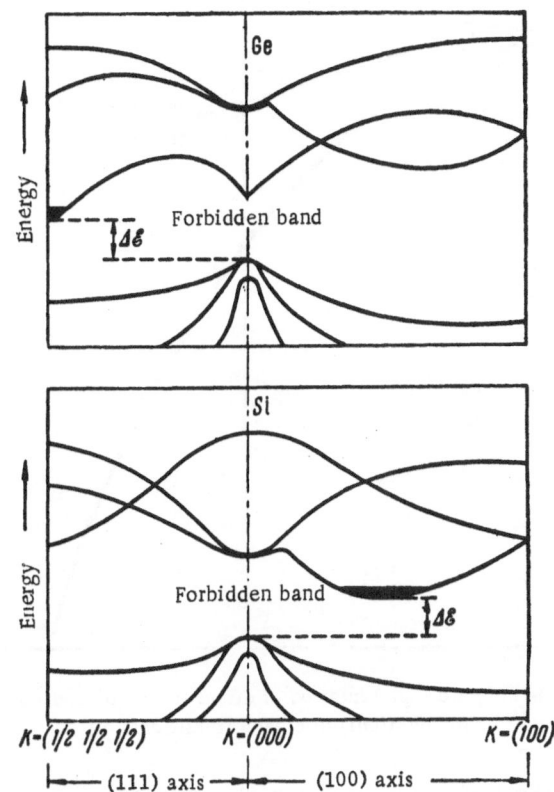

Fig. 55. Schematic representation of the energy bands
in germanium and silicon along the (100) and (111)
axes. The states normally occupied by electrons at
room temperature are shown in black.

three sub-bands (two of which are degenerate when $\mathbf{k} = 0$) and moreover
(which is particularly important) different wave numbers $\mathbf{k}$ correspond to
the bottom of the conduction band and the ceiling of the valence band. (In
the simple case shown in Fig. 54 the bottom and the ceiling have the same
value of the quasi-momentum $\hbar\mathbf{k} = 0$.)

Obviously if we consider the allowed (i.e., "vertical") optical transitions, then the minimum energy of a photon $h\nu_{min}$ still capable of transferring an electron from the valence to the conduction band is now in general greater than the minimum distance between the bands $\Delta\mathcal{E}$ (i.e., greater

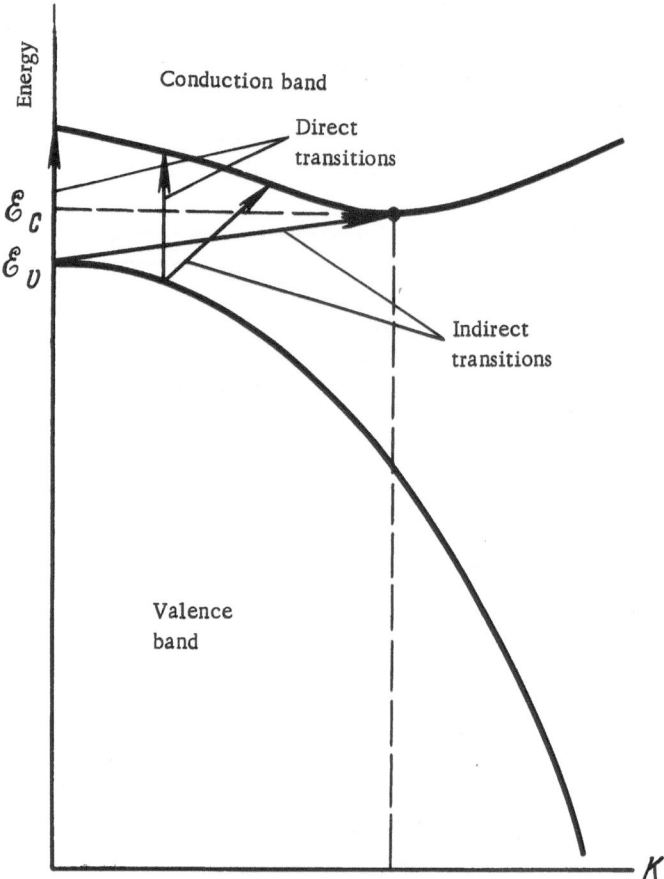

Fig. 56. Schematic representation of direct (vertical) and indirect
(nonvertical) interband transitions.

than the width of the forbidden band which determines the thermal ionization energy; cf. Fig. 55). Consequently the optical ionization energy
should apparently be greater than the thermal energy.

However, for several substances (including germanium and silicon)
it has been found that the thermal and optical ionization energies are
quite close. Data for some of these substances at room temperature are
given in Table 1.

To explain this fact it is necessary to assume that in addition to the
vertical optical transitions referred to above we also have (with high probability) "nonvertical" (or "indirect") transitions in which, apart from the
energy, the wave vector **k** also changes (Fig. 56).

Since, as shown above, such transitions are impossible when only
"two bodies" (the photon and the electron) interact, because this would
violate the law of conservation of the wave number, it is necessary to
assume that there is a high probability of three-body processes involving

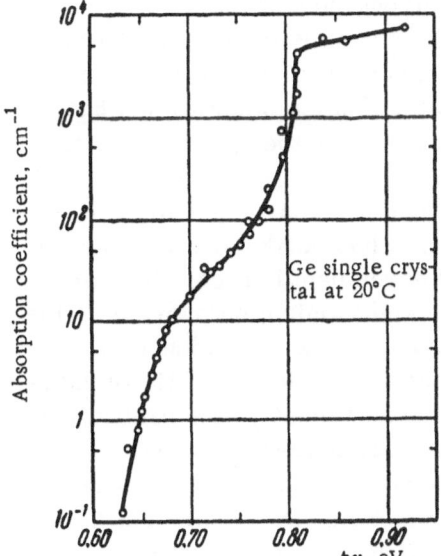

Fig. 57. Dependence of the absorption coefficient of germanium on the energy of incident quanta.

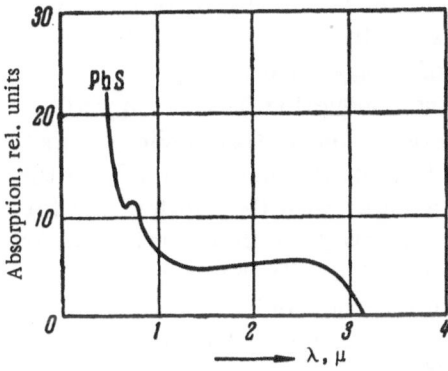

Fig. 58. Fundamental absorption spectrum of PbS.

a photon, an electron, and a lattice vibration quantum, i.e., the phonon. As a result of this interaction the electron acquires the main part of the photon energy and changes its wave number at the expense of the phonon, i.e.. by interacting with the lattice.

Figure 56 gives examples of direct and indirect phototransitions. A transition between the ceiling of the valence band and the bottom of the conduction band is indirect when it requires an energy $h\nu = \mathscr{E}_c - \mathscr{E}_{v}$.

Since the probability of indirect transitions (for which the three-body interaction must be postulated) should, in general, be lower than the probability of direct transitions (for which the two-body interaction is sufficient), the absorption spectra should exhibit a fairly rapid increase of absorption on going toward photons of higher energies capable of producing direct transitions. A careful study of the fundamental absorption bands of germanium, silicon, and some other materials (cf. review [6]) confirms this expectation. Figure 57 shows the absorption spectrum in the fundamental band of germanium. Considerable absorption begins from the photon energy $h\nu \approx 0.65$ eV. This energy agrees well with the value of the forbidden bandwidth of germanium at room temperature and consequently the absorption is connected with indirect transitions. On increase of $h\nu$ the absorption rises but the really strong rise (the curve in Fig. 57 is plotted on logarithmic scale) begins with $h\nu \approx 0.81$ eV. This increase is related to the appearance of direct transitions at $h\nu > 0.81$ eV.

Figure 58 shows a similar curve for lead sulfide. The absorption in the wavelength range from 3 to ~1 μ corresponds to indirect transitions.*

* Then the width of the forbidden band calculated from the long-wavelength edge should be, as expected, close to the thermal ionization energy.

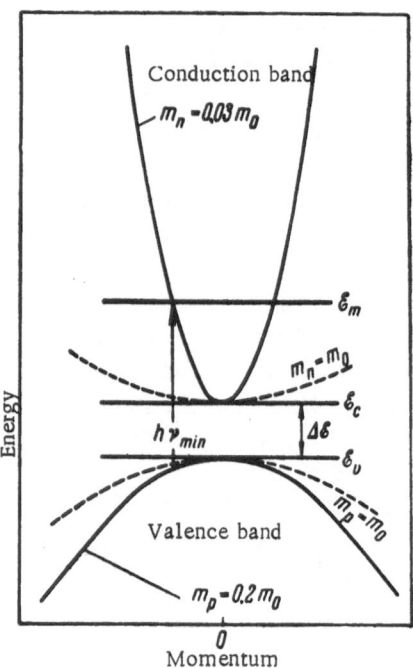

Fig. 59. Dependence of the energy on the
quasi-momentum for InSb.

The strong rise of the absorption at $\lambda \approx 0.9\,\mu$ ($h\nu \approx 1.3$ eV) is due to the appearance of direct transitions. (In later studies other values have been given for the direct-transition threshold.)

Thus for a large group of important photoconductors we observe good agreement between the thermal and optical ionization energies, this being due to the coincidence of the optical and static permittivity values and the sufficiently high probability of indirect phototransitions.

In conclusion we should note that in some cases the absorption (and, in particular, the absorption near the long-wavelength edge) may be governed not only by the form of the energy spectrum (e.g., by the width of the forbidden band) but also by the nature of the population of energy states with equilibrium electrons and holes. If, for example, in the case of a degenerate n-type semiconductor, the equilibrium electrons in the conduction band occupy all the states in a region close to the bottom of the band, then obviously the energy of the quantum $h\nu_{min}$ necessary to transfer an electron from the valence to the conduction band should be greater than the width of the forbidden band $\Delta\mathscr{E}$ by an amount equal to the effective width $\mathscr{E}_m$ of the region in the conduction band occupied by electrons, i.e., in the present case $h\nu_{min} = \Delta\mathscr{E} + \mathscr{E}_m$. In this case a variation of the equilibrium electron density in the conduction band may alter the long-wavelength edge of the fundamental absorption and of the photoeffect.

This phenomenon was detected in indium antimonide [10]. The effective electron (m_n) and hole (m_p) masses in indium antimonide, and consequently the densities of states in the conduction and valence bands (N_c and P_v), are very different: $m_n \approx 0.03m_0$, $m_p \approx 0.2m_0$. Since the density of states in the conduction band is anomalously small, then in the case of degeneracy the electrons which occupy the states at the bottom of the conduction band represent a considerable energy interval $\mathscr{E}_m$, and this leads to a shift of the long-wavelength edge toward short wavelengths.

Thus, for example, at room temperature the long-wavelength edge of pure InSb lies at $\sim 7.1\ \mu$, while the edge for material with $\sim 5\times10^{18}$cm^{-3} electrons is shifted to $3.5\ \mu$.

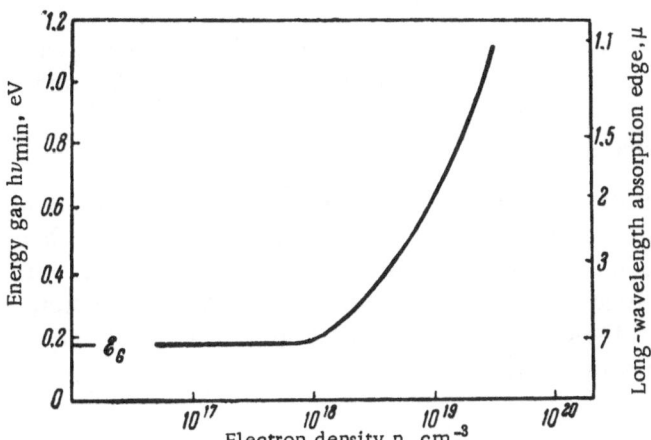

Fig. 60. Dependence of the optical energy gap $h\nu_{min}$ on the
carrier density in n-type InSb.

Figure 59 gives the dependence of the energy on the momentum for
InSb and shows a direct transition corresponding to $h\nu_{min}$. The lowest
unfilled state in the conduction band for an n-type sample with an elec-
tron density 5×10^{18} cm^{-3} is shown in the figure $(\mathscr{E}_m)$. The dashed curves
represent a semiconductor in which $m_n = m_p = m_0$.

Figure 60 gives the calculated dependence of $h\nu_{min}$ on the electron
density, which is in good agreement with the experimental data. In p-type
InSb under the same conditions the shift of the edge is slight because of
the much greater density of states in the valence band.

So far we have considered only the problem of the long-wavelength
edge, which has been studied much more thoroughly than the complete
fundamental absorption spectrum.

It would seem that the structure of the experimental absorption curve
should give important information on the density of the energy states in
the bands (because the intensity of phototransitions, deduced from the absorption at a certain wavelength, is related to the density of states in those
regions of the valence and conduction bands between which the transition
takes place). However, an analysis of the shape of the absorption spectrum for the purpose of obtaining information on the electron energy band
structure meets with considerable difficulties, among which are:

1. The probability of a phototransition is related to the density
    of states in both bands and not in one of them. Moreover
    other factors may affect the probability.
2. The absorption corresponding to direct transitions between
    definite states in the bands with the same wave numbers may
    be superimposed on the absorption due to indirect transitions
    and may thus distort the picture observed.

3. Direct investigation of the absorption in the fundamental
band is difficult because the absorption is strong and it
is necessary to use very thin samples. The results obtained are not very accurate.

It is only in the case of very great changes in the density of the energy states in the bands, associated with overlapping of the bands, that
we can expect some singularities in the absorption spectra which could
then be used to draw conclusions about the energy band structure.

The third of the complications listed above has until recently made
such studies very difficult. *

§ 18. IONIZATION BY HIGH-ENERGY QUANTA AND PARTICLES

The fundamental absorption band, which always has a well-defined
long-wavelength edge, may also in principle have a short-wavelength
edge determined by the highest photon energy for transitions between the
valence and conduction bands.

However, there are reasons for assuming that in many cases the
conduction band touches or overlaps higher allowed bands, forming a
continuous spectrum which extends to infinity. In this case the absorption due to transitions between the valence and conduction bands has superimposed on it, at sufficiently short wavelengths, the absorption due to
transitions to the conduction band from deeper bands or levels. This
complicates the phenomenon. There is a further complication at short
wavelengths. The assumption that the internal photoeffect (in the absence
of competing mechanisms) is in fact explained by the absorption theory
is valid only in a certain region close to the long-wavelength edge, because in the case of sufficiently energetic photons the phototransitions
may be accompanied by impact ionization which increases the effective
quantum yield. † If, for example, an electron transferred to an upper
band on absorption of a quantum $h\nu$ (Fig. 61a) possesses an energy $\mathscr{E}$, which
is greater than the forbidden band width $\Delta\mathscr{E}$, such an electron may transfer its energy to a valence band electron which then moves to the conduction band (transition a). The first electron, having lost its energy
(transition b) remains in the conduction band. Consequently the absorp-

---

* Recently considerable success has been achieved in the study of the energy band structure
using the absorption spectra for intraband phototransitions connected with the absorption of
light by free carriers (this absorption is studied outside the fundamental band and therefore
one does not meet with the difficulties described as the third category above).
† Obviously if the quantum yield $\beta \neq 1$, then the internal photoeffect theory does not reduce
to the absorption theory.

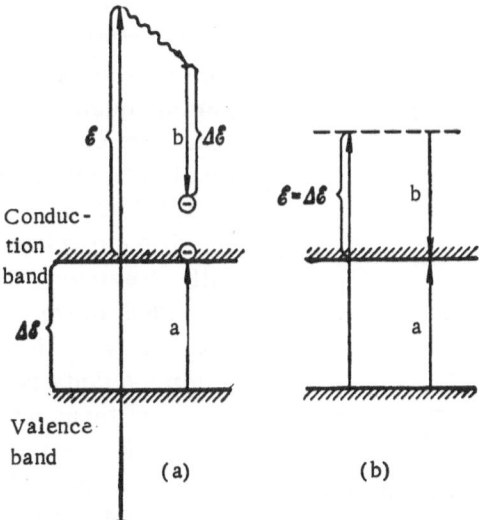

Fig. 61. Electron transitions on absorption of a
photon with energy several times greater than the
forbidden band width.

tion of a single photon leads to the liberation of two electrons and two
holes (instead of one electron-hole pair as in the usual case) due to the
secondary process of impact ionization, i.e., we are dealing with carrier
"multiplication."

If the photon energy is increased further, multiple impact ionization
processes become possible, with the electron and hole formed by the primary absorption act generating additional electron-hole pairs, the energies of these electrons and holes being sufficient to generate further
pairs by impact ionization.

When x rays and gamma quanta are absorbed, such multiplication
processes by multiple impact ionization may consist of tens or hundreds
of thousands of stages, until all the energy of the initial quantum is used
up by ionization, i.e., by the formation of free electrons and holes in the
conduction and valence bands. *

The ionization process is approximately similar when, instead of
photons, charged high-energy particles and used (β particles, α par-

---

* In the case of high-energy quanta (for example, gamma rays) the effect is more complex
than in the above description. The high-energy quanta are mainly absorbed not by single
internal-photoeffect acts but by the Compton effect, in which the energy of the quantum is
transferred in several collisions to recoil electrons. Moreover, the energy exchange is not
limited to impact ionization processes. Thus, if ionization liberates electrons from the deep
shells of the atom, characteristic x rays may appear which, when absorbed, will produce further ionization.

For our purpose, it is important that all these complex processes are rapidly completed and
end finally with the formation of nonequilibrium electrons and holes in the bands.

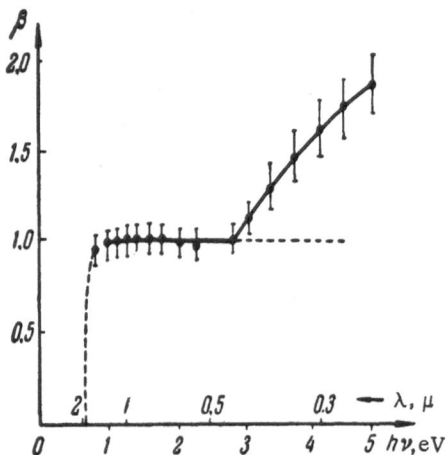

Fig. 62. Dependence of the quantum yield in germanium on the energy of incident quanta [13].

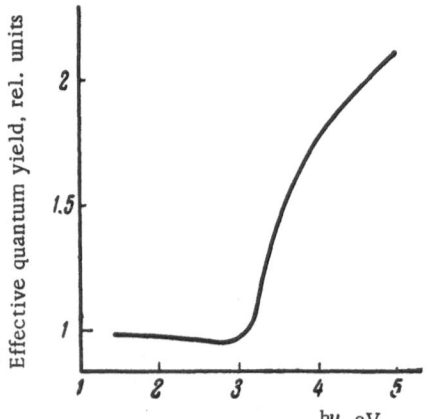

Fig. 63. Dependence of the quantum yield in silicon on the energy of incident quanta.

ticles, protons, etc.). The primary act of ionization by these particles generates, as in the case of gamma quanta, fast electrons which then liberate a large number of electrons and holes by multistage multiplication.

The experimental data are in full agreement with the explanations provided above. The effective quantum yield for ionization by gamma quanta reaches many hundreds of thousands, while for quanta which are not much greater in energy than the forbidden bandwidth, the yield is usually close to unity [3, 4].

Figure 62 shows that in germanium the quantum yield changes from $\beta = 1$ to $\beta > 1$ in the region where $h\nu$ is several times the forbidden band width [9, 13]. It is clear from Fig. 63 that for silicon the yield β becomes greater than unity, and consequently impact ionization becomes important, when $h\nu > 3$ eV [13, 15].

From the law of conservation of energy it follows that to produce impact ionization an electron or hole should have kinetic energy not less than the width of the forbidden band. The generation of such electrons and holes would seem to be possible on absorption of quanta with $h\nu = 2\Delta\mathscr{E}$ (such a process is shown in Fig. 61b). However, the experimental results show that impact ionization becomes important only at much higher values of $h\nu$. This is obviously due to the selection rules which govern photoionization as well as the impact ionization process.

The direct phototransition shown in Fig. 61b, which satisfies the law of conservation of energy, may conflict with the law of conservation of momentum, and consequently the intense generation of free electrons and holes with energies sufficient for impact ionization will occur only when

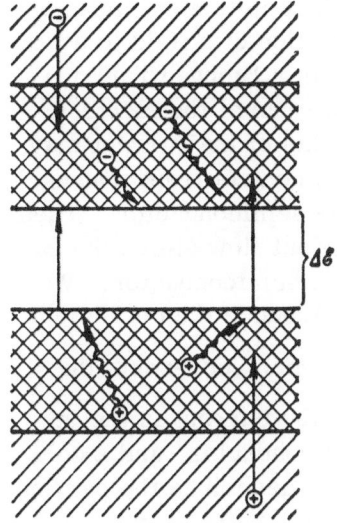

Fig. 64. Transition scheme explaining why in the case of ionization by high-energy particles and quanta the energy required to form an electron-hole pair is considerably greater than the forbidden bandwidth.

quanta with energy considerably higher than $2\Delta\mathscr{E}$ are absorbed (Fig. 61a) [11,19]. Restrictions related to the conservation of momentum in the impact ionization may also be important. This is illustrated in Fig. 64. In this figure the parts of the allowed bands near the edges, where the carrier energy is insufficient for impact ionization, are shown by double hatching. The impact ionization, indicated by straight arrows, is due to carriers which are initially outside these regions. As a consequence of the law of conservation of momentum the doubly hatched regions may be somewhat wider than the forbidden band.

When high-energy particles and quanta are used the formation of an electron-hole pair absorbs energy which is considerably greater than the width of the forbidden band. This is because the multistage process of impact ionization frequently ends with the formation of carriers which are quite far from the band edges but are still within the doubly hatched portions (Fig. 64). Such carriers do not produce further impact ionization and lose their excess energy by collisions with the lattice phonons (wavy arrows). Moreover, even the carriers which still have sufficient energy for further impact ionization may lose some of this energy on interaction with the lattice phonons. The occurrence of these processes of the conversion of energy into heat is the reason why the average energy for the formation of an electron-hole pair by ionization with high-energy particles and quanta is considerably greater than the forbidden bandwidth.

§ 19. OTHER METHODS OF "GENERATING" NONEQUILIBRIUM CARRIERS

A. Use of "Nonneutral" Contacts

The carrier density in a semiconductor may depart from its equilibrium value not only as a result of ionization by light or other radiations, but also (under certain conditions) due to the application of a voltage to the semiconductor across a contact with some other materials. Departure of the density from equilibrium is possible if the contact is not neutral.*

* If the energy bands of a semiconductor are curved at the contact with a material having a different work function, then this contact is "nonneutral."

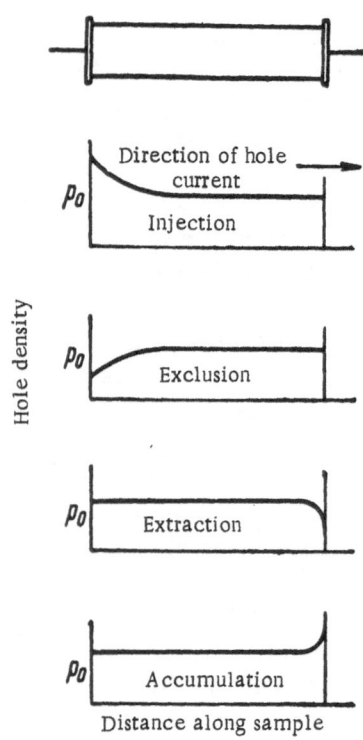

Fig. 65. Change in the hole density along a sample for different types of departure from the equilibrium density.

Enrichment or depletion of carriers occurs, on application of a field to a semiconductor with a "nonneutral" contact, because the ratio of the electron and hole components of the current may be different at the contact from that in the interior of the semiconductor. Then obviously the current flow alters the carrier density in the semiconductor. We shall therefore introduce the coefficient

$\gamma_0 = \frac{\mu_p p_0}{\mu_p p_0 + \mu_n n_0}$, which represents the

"composition" of the current in the interior of the semiconductor, and a coefficient γ_c, which describes the current at the contact [7, 8].

Obviously if $\gamma_c = \gamma_0$, i.e., the ratios of the hole to electron components of the current are the same at the contact and in the interior, a current will not alter the electron and hole densities.

The number of carriers of a given type emitted by the contact into the neighboring part of the semiconductor is equal to the number of carriers leaving that part.

The situation is different when $\gamma_c \neq \gamma_0$. Then the composition of the current at the contact is different from that in the interior (although the total current over any cross section is the same). Consequently the flow of a current alters the carrier density from its equilibrium value.

We can distinguish four possible cases of carrier density change because of two possible types of "nonneutral" contact, having blocking or antiblocking layers, and two possible polarities of the voltage applied to the system with a contact.

Let us consider these cases for an n-type semiconductor:

1. Blocking layer (for example, p-n junction):
    a. The contact is positive with respect to the interior (forward direction), $\gamma_c > \gamma_0$, and the region next to the contact is enriched with carriers; this is known as injection;
    b. The contact is negative with respect to the interior (reverse direction), $\gamma_c < \gamma_0$, and the region next to the contact is depleted of carriers; this is known as extraction.

2. Antiblocking layer (for example, n-n' junction):

a. The contact is positive, $\gamma_c < \gamma_0$, and carriers are depleted in the region next to the contact; this is known as exclusion;

b. The contact is negative, $\gamma_c > \gamma_0$, and carriers are enriched in the region next to the contact; this is known as accumulation.

The minority carrier distribution in these four cases is shown in Fig. 65. In the two upper curves the "nonneutral" contact lies on the left, and in the two lower ones it lies on the right.

Of the four possible variants of carrier density change in a semiconductor with a "nonneutral" contact, injection is most important, being widely used in semiconductor devices and as a method of generating nonequilibrium carriers.

B. Formation of Nonequilibrium Carriers by Impact Ionization

In § 18 it was stated that carriers having kinetic energy greater than the forbidden bandwidth may, in principle, generate further carriers by impact ionization.

The kinetic energy of carriers may be raised by applying an external electric field. If this field is sufficiently strong so that during a time interval of the order of the relaxation time the carriers are able to acquire from the external field sufficient energy for impact ionization, then as the result of this ionization the carrier density will increase, i.e., nonequilibrium carriers will appear.

"Intrinsic" impact ionization (related to interband transitions) requires very strong fields of the order of 10^5 V/cm (such fields are easily obtained in p-n junctions) [16]. However, in addition to the "intrinsic" effect, we can also have "impurity" impact ionization which does not require such high fields. For example, the impact ionization of impurities in the form of elements of the third and fifth groups of the periodic table occurs in germanium in fields of ~5-10 V/cm [12, 18]. In several studies carried out at liquid-helium temperatures the impact ionization of impurities in germanium was used to generate nonequilibrium carriers.

In strong fields the effect of the increase of the average kinetic energy of the carriers, known as "heating," should be allowed for.

Thus, in contrast to the photoionization case, strong fields may produce a condition in which the carrier temperature is considerably different from the lattice temperature.

Chapter V

RECOMBINATION THROUGH SIMPLE LOCAL CENTERS

§20. INTRODUCTION. LIMITATIONS IMPOSED BY THE LAWS OF CONSERVATION OF ENERGY AND MOMENTUM

The main problem in the process of the recombination of band electrons and holes is the question of the form in which the energy liberated on recombination is evolved.

If the recombination act is accompanied by the appearance of a light quantum, i.e., the energy liberated appears as luminescence, the interpretation of the recombination mechanism presents no difficulties.

The laws of conservation of energy and momentum are simultaneously satisfied in one of two ways: (1) the electron and hole have equal but opposite momenta, and the emitted light quantum has practically zero momentum but carries away all the energy liberated on recombination; (2) the recombination act involves three "bodies": an electron, a hole a phonon. In the latter case the phonon acquiring (or losing) a small portion of the liberated energy may acquire (or lose) practically all its momentum, and thus make it possible for electrons and holes with different momenta to recombine with the emission of a photon.

These two variants of radiative recombination are the converse of light absorption leading to the "direct" and "indirect" phototransitions of electrons from the valence to the conduction band. The theory of the radiative recombination of band electrons and holes was developed by Van Roosbroeck and W. Shockley (for details see Chap. VIII). By equating (under conditions of thermal equilibrium) the rate of generation of electron–hole pairs by thermal radiation and the rate of their radiative recombination, they first calculated the equilibrium radiative recombination coefficient (since the rate of generation is governed by the amount of thermal radiation energy absorbed by the semiconductor, it was found that to calculate the equilibrium radiative recombination coefficient for a particular semiconductor it was necessary to know its optical constants, in particular the absorption coefficient in the fundamental band). As the nonequilibrium distributions of electrons and holes in the bands do not differ from the equilibrium distributions, Van Roosbroeck and Shockley used the calculated coefficient to describe the radiative recombination of the nonequilibrium carriers. Without considering their calculation in detail (cf. § 37) we may point out that for pure semiconductors with the usual values of the forbidden bandwidth (1-2 eV) the intensity of radiative recombination is found to be negligible. Thus, for example, in pure germanium ($\Delta\mathscr{E} \approx 0.7$ eV) at room temperature the radiative recombination lifetime is found to be 0.75 sec. For substances with a wider forbidden band this lifetime is even longer. Since the measured lifetime τ is usually considerably shorter than the values just quoted, it is obvious that radiative recombination is relatively unimportant under real conditions,* and that there must be other mechanisms which play the dominant role. Clearly these mechanisms represent radiationless recombination by conversion of the energy evolved into heat. Radiationless recombination is observed experimentally in most cases but its interpretation meets with considerable difficulties.

The principal difficulty is connected with that fact that the energy evolved at each recombination act is of the order of the forbidden band width of a semiconductor (i.e., ~ 1-2 eV) and cannot be absorbed by a single phonon. On the other hand, any process involving simultaneous energy transfer to a large number of phonons is extremely unlikely.

Until now no complete theory describing quantitatively the mechanism of radiationless recombination, and making it possible to calculate the corresponding cross section, has been available. However, there are some indications of the possibility, in principle, of recombination with energy transfer to the lattice vibrations [24].

Mechanisms have been proposed in which the difficulty of explaining the direct transfer of energy to the lattice vibrations is avoided by allowing for possible interaction between the carriers. Thus, for example, the energy evolved on recombination of an electron with a hole may be transferred to a free carrier which then gradually transmits it to the lattice by collisions. This is known as impact recombination, because the recombination requires the collision of the recombining electron and hole with a further carrier which takes away the energy evolved (cf. § 40). A recombination mechanism has been proposed [15] in which the energy is first transferred to the collective motion of free carriers (the so-called plasma-type oscillations), as well as a mechanism involving exciton formation [16].

The above discussion of radiationless recombination applies not only to the case of a band electron and a band hole but also to the case of free-

---

* The exceptions are those semiconductors with a narrow forbidden band in which the radiative recombination rate is high and the radiative mechanism may be the dominant one. Thus, for example, in InSb (at room temperature $\Delta\mathscr{E} = 0.18$ eV) the lifetime for radiative recombination is only ~10^{-6} sec.

carrier capture by a local center (trap). In the latter case the recombination act occurs in two stages: for example, an electron is first captured
by a level of the impurity center (trap) and then the same level captures
a hole (which is equivalent to the transition of the captured electron to the
valence band).

As stressed above, it has been firmly established that impurities and
defects play a dominant role in recombination processes. The mechanism of recombination through impurity centers was proposed some time
ago and has been used by many investigators to explain their photoconductivity results (cf., for example, [1, 2]). However, only the experimental investigations of recent years, carried out mainly on substances
of the germanium (Ge, Si) and cadmium sulfide (CdS, CdSe, CdTe) groups,
have definitely established that in these substances, and probably in many
others, the recombination is governed mainly by the number and type of
impurities. Thus recombination through centers is one of the most important mechanisms which actually occurs.

§ 21. RECOMBINATION THROUGH LOCAL CENTERS (TRAPS). CAPTURE OF CARRIERS BY LOCAL CENTERS

The efficiency of recombination through an impurity center is determined primarily by the probability of the capture of an electron or a hole
by this center. However, this probability does not determine the recombination probability uniquely. For the captured electron (hole) two
alternatives are possible: it can either recombine with a free hole (electron), or it may be transferred back to the band by thermal motion (the
latter process was not considered in Chap. I).

Thus the theory of recombination through centers should deal with
two types of problems: on the one hand a consideration of the process of
the interaction of a carrier with a recombination center should allow us
to describe the process of capture and calculate the capture cross section;
on the other hand we should attempt to deduce the recombination statistics,
i.e., calculate the rate of recombination (the number of recombination
acts per unit time in unit volume) for a given number of centers with given
capture cross sections and given positions of the levels in the forbidden
band (i.e., with a given electron population and a given probability of the
return thermal transitions).

A theory of recombination statistics should include an allowance for
the electron population of the local centers (which govern the rate of capture) and for the competition of two possible processes of liberating a
captured carrier from a recombination center: the return thermal transition back to the band or the capture of a carrier from another band which
would complete the process of recombination.

As far as the mechanism of a single capture act is concerned, we shall merely refer the reader to the brief remarks made on this point in § 20.

In the statistical theory the mechanism of capture is not considered. The probability of capture is assumed to be known and represented by the capture cross section. However, one should bear in mind certain qualitative considerations, discussed below, on the ratio of the electron and hole capture cross sections for a given type of center.

Usually simple impurity centers may be present in a semiconductor lattice in two states: neutral or singly charged. Depending on the sign of the singly charged center we distinguish donors, which are singly charged positive centers, and acceptors, which are singly charged negative centers.

Some impurities give rise to centers which may be in the neutral state or in one of several charged states (singly charged, doubly charged, etc., positive or negative). Thus, for example, gold impurity atoms in germanium may be in the neutral state or be singly charged positive, or singly, doubly, and triply charged negative centers; copper atoms may be neutral, singly, doubly or triply charged negative centers, etc. Such impurity centers have several energy levels in the forbidden band (cf. § 35 in Chap. VII).

In the present chapter we shall consider the simplest impurities which may be only singly charged and which have only one level in the forbidden band.* In this case each impurity center may be represented by only two capture cross sections: the cross section for electron capture (when the level of the center is free) q_n, and the cross section for hole capture (when the level is occupied by an electron) q_p. (Such impurities will be called "simple" in contrast to multicharged ones.)

We may expect these cross sections to be unequal. Considering a donor center we see that in the neutral state it may capture a hole and in the positively charged state it may capture an electron. In the latter case the capture is clearly aided by the electrostatic attraction of the positively charted center and the negatively charged electron. Consequently one would expect the cross section for electron capture to be larger. For an acceptor center the opposite is true and the hole capture cross section should be greater than the electron capture cross section.

In real semiconductors the kinetics of nonequilibrium processes is frequently determined by recombination processes and is very complex. To explain the kinetics qualitatively one must make the assumption that the recombination processes involve several types of capture center(trap) with different cross sections and concentrations, giving rise to levels at different energies in the forbidden band.

---

*Recombination through multiply charged centers is considered in § 36 of Chap. VII.

Calculations of such complex models with several types of trap are possible only in an approximate form and only for some special cases. When the calculation is impossible one must use qualitative representations, based on some reasonable classification of traps, which can be used to estimate the role of each type of trap under various recombination conditions.

First we shall consider in detail processes in the presence of traps of one type only. The results obtained for this case are of great interest in themselves and should help in the task of providing a qualitative interpretation in more complex cases.

§22. SEMICONDUCTOR WITH ONE TYPE OF TRAP

We shall consider the transitions (Fig. 66) which can occur in a semiconductor with one type of center (trap).

We shall denote the total trap concentration by M, and the electron density at these traps by m (Fig. 66). The effective densities of states in the bands will as usual be denoted by N_C and P_V, and the free electron and hole densities by n and p. In a system with three types of state (an upper band, a lower band, and traps), six transitions (1-6) are in principle possible; they are shown in Fig. 66. These six transitions may be divided into two groups:

1. Transitions 1, 2, and 3 under the action of light or thermal motion which involve energy absorption;
2. Transitions 4, 5, and 6 which lead to capture or recombination and are accompanied by the evolution of thermal or optical energy.

Since the mechanisms of optical and thermal transitions are different and since several recombination transitions can produce heat or light, each of the transitions indicated by an arrow may in fact represent several transitions, and the complete system of possible transitions may be very complex. To interpret the actual results we usually use models with a limited number of those transitions which are most likely to occur under experimental conditions and are capable of explaining the observations.

We shall now consider the system shown in Fig. 67 which includes several of the most important and typical transitions. (The results obtained for this system may be used in qualitative analyses of more complex systems.) In the system of Fig. 67 illumination transfers electrons from the valence to the conduction band. An electron from the conduction band may be captured by traps M. The captured electron is either transferred back to the conduction band by thermal motion, or it recombines with a hole in the valence band (which is equivalent to the capture of a hole by the level M). Finally we can also have the thermal transfer

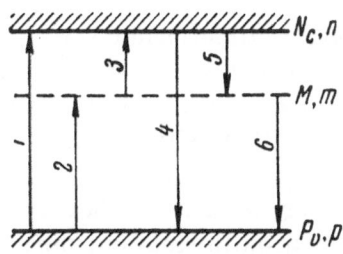

Fig. 66. Possible electron transitions in a semiconductor with one type of trap.

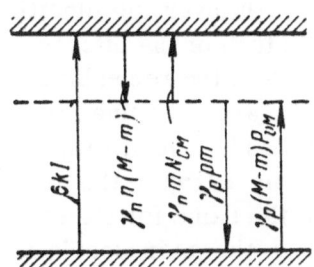

Fig. 67. Scheme showing the most important transitions in a semiconductor with a wide forbidden band in the presence of traps of one type.

of a hole from the M levels to the valence band (in other words the thermal transfer of an electron from the valence band to the levels M). Thus the system in Fig. 67 allows for only one phototransition between the bands (phototransitions in which traps participate are not very likely because of the low concentration of traps); direct recombination transitions from the conduction to the valence band are not included.

The phototransition rate between the bands may be written, as usual, in the form βkI (cf. §1 in Chap. I). We shall now find how we can express the rate of recombination (capture) transitions and of thermal transitions. Those traps which are free of electrons (their concentration is $M - m$) are able to capture electrons from the upper band; the traps occupied by electrons can capture holes from the valence band. Obviously the capture rate (the number of capture acts per unit time in unit volume) for, say, the conduction band electrons is proportional to the electron density n and to the concentration of empty traps $M - m$, i.e., the process is "bimolecular."

Introducing the coefficients of capture of electrons by free traps (γ_n) and of holes by occupied traps (γ_p), which were defined in Eqs. (4.10) and (4.11), we can express the rate of capture of electrons and holes by the traps in the following form:

$$\gamma_n n (M - m), \tag{22.1}$$

$$\gamma_p pm. \tag{22.2}$$

The treatment in §4 of Chap. I showed that the electron and hole lifetimes in the case of capture by the M centers may be written in the form

$$\tau_{cM} = \frac{1}{\gamma_n (M - m)} = \frac{1}{v_n q_n (M - m)}, \tag{22.3}$$

$$\tau_{vM} = \frac{1}{\gamma_p m} = \frac{1}{v_p q_p m}. \tag{22.4}$$

The subscripts "cM" and "vM" denote the type of transition. Obviously τ_{cM} and τ_{vM}, which depend on the variable m, are variable quantities and may depend on the intensity of illumination, time (under non-steady-state conditions), etc. Consequently the lifetimes τ_{cM} and τ_{vM} are the instantaneous lifetimes.

Thus the rate of recombination may be written either in the form given by Eqs. (22.1) and (22.2) or in the form n/τ_{cM} and p/τ_{vM}.

We shall now consider the problem of calculating the rate of thermal transitions. Since the energy distribution of the nonequilibrium carriers in the conduction and valence bands does not differ from the distribution of equilibrium carriers (in fact they are indistinguishable),* the nature of recombination processes must therefore be the same for the nonequilibrium and equilibrium carriers. It is obvious that both the average velocities and the average capture cross sections (i.e., the nature of the interaction of carriers with traps), which govern the values of the recombination coefficients γ_n and γ_p in Eqs. (22.1) and (22.2), are the same for the equilibrium and nonequilibrium carriers in the band. The average probability of thermal transition of an electron from a trap to the conduction band (or a hole to the valence band) is also the same for the equilibrium and nonequilibrium electrons (holes) because the microscopic interaction of thermal fluctuations with one electron (hole) is independent of whether the charge is an equilibrium or a nonequilibrium one. This allows us to use the coefficients (representing the average probabilities of single micro-acts, for example γ_n and γ_p) obtained from consideration of equilibrium transitions, in discussing the statistics of the transitions of nonequilibrium carriers.

We shall calculate first the relationship between the recombination coefficients (γ_n and γ_p) and the probabilities of thermal transitions from traps to a band under equilibrium conditions. These relationships will then be extended to nonequilibrium processes.

Let us consider, for example, the kinetics of transitions between the traps and the conduction band in the absence of illumination. The carrier density in the band changes (dn/dt) because of the thermal liberation of electrons from the traps and because of the capture of electrons by the empty traps. The rate of the second process may be written in the form of Eq. (22.1). As far as thermal transitions are concerned we can only say that their rate is proportional to the electron density in the traps m.

Thus

$$\frac{dn_0}{dt} = \alpha_n m_0 - \gamma_n n_0 (M - m_0), \tag{22.5}$$

where α_n is an unknown coefficient for thermal transitions to the conduction band.

---

* We shall consider the case when the bands are nondegenerate both in darkness and during illumination.

In the steady-state case $dn_0/dt = 0$ and we find from Eq. (22.5) that

$$m_0 = \frac{M}{\frac{\alpha_n}{\gamma_n n_0} + 1}. \tag{22.6}$$

On the other hand, at thermal equilibrium $m_0 = Mf$, where f is the Fermi distribution,

$$m_0 = \frac{M}{e^{\frac{-\Delta\mathscr{E}_M - F}{kT}} + 1}. \tag{22.7}$$

Comparison of Eqs. (22.6) and (22.7) gives

$$\alpha_n = \gamma_n n_0 e^{\frac{-\Delta\mathscr{E}_M - F}{kT}}. \tag{22.8}$$

Taking into account the fact that in the absence of degeneracy in the band we have $n_0 = N_c e^{\frac{F}{kT}}$, we obtain

$$\alpha_n = \gamma_n N_c e^{-\frac{\Delta\mathscr{E}_M}{kT}}. \tag{22.9}$$

Thus the thermal transition rate at thermal equilibrium is

$$\alpha_n m_0 = \gamma_n m_0 N_c e^{-\frac{\Delta\mathscr{E}_M}{kT}}. \tag{22.10}$$

In the same way we can write the thermal transition rate under non-equilibrium conditions, replacing m_0 by m which represents the total (equilibrium and nonequilibrium) concentration of charge in the traps.

The coefficient $\alpha_n = \gamma_n N_c e^{\frac{-\Delta\mathscr{E}_M}{kT}}$ retains its form under any conditions. Similarly, using the detailed balancing principle, we can easily show that the thermal transition rate of holes from the traps to the valence band can be written in the form

$$\alpha_p (M - m) = \gamma_p (M - m) P_v e^{\frac{-\Delta\mathscr{E} + \Delta\mathscr{E}_M}{kT}} \tag{22.11}$$

If we compare the form of the expressions (for example in the case of transitions between the conduction band and the traps) for the recombination rate $\gamma_n n(M - m)$ and the thermal transition rate $\gamma_n m N_c e^{-\frac{\Delta\mathscr{E}_M}{kT}}$, their similarity becomes apparent: in both cases we multiply the recombination coefficient γ_n by two densities, these being n and (M – m) in the

first case, and m and $N_c e^{-\frac{\Delta \mathcal{E}_M}{kT}}$ in the second case. This makes it possible to treat thermal transitions as "recombination" of m trapped electrons with conduction-band holes. However, the density of free states in the conduction band (holes) is not N_C but N_C reduced by the factor $e^{\frac{\Delta \mathcal{E}_M}{kT}}$ because such "recombination" requires, in contrast to the usual case, that electrons overcome an energy barrier $\Delta \mathcal{E}_M$. The quantity $N_c e^{-\frac{\Delta \mathcal{E}_M}{kT}}$ may be called the effective density of states in the conduction band, "reduced" to the trap level M.

Similarly the quantity $P_v e^{-\frac{\Delta \mathcal{E} - \Delta \mathcal{E}_M}{kT}}$ represents the effective density of free hole states (electrons) in the valence band, "reduced" to the trap level M.

For these effective densities we shall use the notation:

$$N_{cM} = N_c e^{-\frac{\Delta \mathcal{E}_M}{kT}}, \tag{22.12}$$

$$P_{vM} = P_v e^{\frac{-\Delta \mathcal{E} + \Delta \mathcal{E}_M}{kT}}. \tag{22.13}$$

Obviously each type of trap (if there are several) has its own values of the "reduced densities of states in the band."

The values of N_{cM} and P_{vM} are equal to the densities of equilibrium electrons and holes in the bands which would be obtained if the Fermi level coincided with the trap level. The quantities N_{cM} and P_{vM} will be used extensively in further discussion of recombination processes involving traps.

To describe thermal transitions we can introduce, as in the case of recombination, the concept of lifetime. Following Eq. (4.12) the lifetimes for thermal transitions from the traps to the conduction and valence bands may be written in the form:

$$\tau_{Mc} = \frac{1}{\gamma_n N_{cM}}, \tag{22.14}$$

$$\tau_{Mv} = \frac{1}{\gamma_p P_{vM}}. \tag{22.15}$$

Using Eqs. (22.1), (22.2), (22.10), and (22.11) we can write down a system of transport equations which describes transitions in the case shown in Fig. 67:

$$\frac{dn}{dt} = k\beta I - \gamma_n n\,(M - m) + \gamma_n m N_{cM}, \qquad (22.16)$$

$$\frac{dm}{dt} = \gamma_n n\,(M - m) - \gamma_n m N_{cM} - \gamma_p m p + \gamma_p P_{vM}\,(M - m), \qquad (22.17)$$

$$\frac{dp}{dt} = k\beta I - \gamma_p m p + \gamma_p (M - m)\,P_{vM}. \qquad (22.18)$$

Unfortunately, it is not possible to solve this system of three equations with three unknowns because one of the equations is the consequence of the other two. The missing equation may be obtained from the condition of conservation of the number of electrons for any transition (or from the condition of neutrality):

$$\Delta n + \Delta m = \Delta p. \qquad (22.19)$$

The solution of the system of Eqs. (22.16)-(22.19), for certain given initial conditions and for a given dependence of the illumination intensity I on time, should give a complete description of the kinetics of the processes in the system considered. However, considerable mathematical difficulties (the necessity of solving nonlinear differential equations) prevent us from obtaining the solution in the general form. Therefore, even in the case of a relatively simple system with one type of trap, a quantitative description of the processes can be obtained only for some special simple cases.

§ 23. SYSTEM WITH ONE TYPE OF TRAP. STEADY-STATE CASE

We first shall consider the "steady-state case" and calculate, in particular, the "lifetime of nonequilibrium carriers in the steady-state case" (cf. § 5 in Chap. I) for a system with one type of trap. This case, treated by Hall, Shockley, and Read [3], is very important in the interpretation of nonequilibrium processes in semiconductors.

From Eq. (22.17) for $dm/dt = 0$ we have

$$\gamma_n\,[n\,(M - m) - m N_{cM}] = \gamma_p\,[m p - (M - m)\,P_{vM}]. \qquad (23.1)$$

The left-hand part of this expression gives the "absolute rate of capture" of the conduction-band electrons by the traps (i.e., the difference between the number of captures and return transitions per unit time), and the right-hand part gives the corresponding ratio for the valence-band holes. Obviously in the steady-state case not only are these quantities equal but each of them is equal to the rate of generation of electron-hole pairs, i.e., to βkI:

$$\gamma_n \left[n \left(M - m \right) - m N_{cM} \right] = \beta k I, \tag{23.2}$$

$$\gamma_p \left[m p - \left(M - m \right) P_{vM} \right] = \beta k I. \tag{23.3}$$

To eliminate m from Eq. (23.2) or Eq. (23.3) we can use Eq. (23.1). Solving Eq. (23.1) we find that the number of trapped electrons in the steady-state case is given by

$$m = M \, \frac{\gamma_p P_{vM} + \gamma_n n}{\gamma_p \left(P_{vM} + p \right) + \gamma_n \left(N_{cM} + n \right)} \, . \tag{23.4}$$

Hence the number of trapped holes is

$$M - m = M \, \frac{\gamma_n N_{cM} + \gamma_p p}{\gamma_p \left(P_{vM} + p \right) + \gamma_n \left(N_{cM} + n \right)} \, . \tag{23.5}$$

Substituting the values of m and (M – m) from Eqs. (23.4) and (23.5) into Eq. (23.2) or Eq. (23.3), we obtain

$$M \, \frac{\gamma_n \gamma_p \left(n p - N_{cM} P_{vM} \right)}{\gamma_p \left(p + P_{vM} \right) + \gamma_n \left(n + N_{cM} \right)} = k \beta I. \tag{23.6}$$

Our main purpose is to calculate the steady-state lifetime τ. However it is difficult to do this using Eq. (23.6) in the general case. Therefore we shall consider the limiting cases of "low" and "high" trap concentrations.

A.   The Case of Low Trap Concentration

We shall assume that the number of traps M is so small that we need not allow for the change in the carrier densities in the bands when the concentration of charge in the traps is altered. Moreover we shall assume that the carrier densities in the bands n and p represent the equilibrium and nonequilibrium carriers. The equilibrium densities of electrons and holes in the bands (as with the position of the Fermi level) may be governed not only by the centers (traps) which are explicitly allowed for, but also by other impurity centers which affect the equilibrium charge density in the bands and in the traps but do not take direct part in recombination processes (the reasons why some centers cannot participate directly in recombination will be discussed later).

Denoting, as usual, the equilibrium electron and hole densities in the bands by n_0 and p_0 and the nonequilibrium carrier densities by Δn and Δp, and bearing in mind that when the number of capture centers is small the neutrality condition of Eq. (22.19) may be rewritten as

$$\Delta n = \Delta p, \tag{23.7}$$

we can write

$$\left.\begin{array}{c} n = n_0 + \Delta n, \\ p = p_0 + \Delta n. \end{array}\right\} \tag{23.8}$$

Using Eq. (5.17) the steady-state carrier lifetime τ is written as *

$$\tau = \frac{\Delta n}{k\beta I}. \tag{23.9}$$

Substituting in Eq. (23.9) the expression for $k\beta I$ from Eq. (23.6), we obtain the expression for τ:

$$\tau = \frac{n_0 + N_{cM} + \Delta n}{\gamma_p M (n_0 + p_0 + \Delta n)} + \frac{p_0 + P_{vM} + \Delta n}{\gamma_n M (n_0 + p_0 + \Delta n)}. \tag{23.10}$$

Obviously the quantities $1/\gamma_p M$ and $1/\gamma_n M$ which occur in Eq. (23.10) represent, respectively, the lifetime of holes for capture by M centers which are completely filled with electrons, and the lifetime of electrons for capture by the same centers when completely free of electrons. Denoting these quantities by τ_{p0} and τ_{n0}, we obtain

$$\tau = \tau_{p0} \frac{n_0 + N_{cM} + \Delta n}{n_0 + p_0 + \Delta n} + \tau_{n0} \frac{p_0 + P_{vM} + \Delta n}{n_0 + p_0 + \Delta n}. \tag{23.11}$$

We shall apply this general expression for the lifetime to some special cases.

1. The Case of Low Illumination Intensities (or Low Injection Level). In this case, neglecting Δn in the numerator and denominator of Eq. (23.11), we obtain

$$\tau = \tau_{p0} \frac{n_0 + N_{cM}}{n_0 + p_0} + \tau_{n0} \frac{p_0 + P_{vM}}{n_0 + p_0}. \tag{23.12}$$

From Eq. (23.12) it is clear that the quantity τ is independent of the nonequilibrium carrier density but is intrinsically related to the equilibrium carrier densities in the bands n_0 and p_0 and is also affected by the energy position of the traps $\left(N_{cM} = N_c e^{\frac{-\Delta \mathcal{E}_M}{kT}} \right)$.

Figure 68 shows the dependence of the logarithm of the lifetime, calculated using Eq. (23.12), on the position of the equilibrium Fermi level (which determines n_0 and p_0 uniquely) for a given position of the traps. (To make the case definite we shall assume that the traps lie in the upper part of the forbidden band and consequently

* This is the "relaxation" lifetime. For details see Chap. XI.

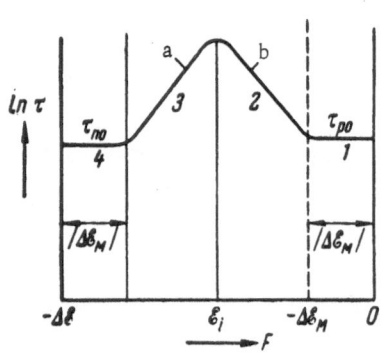

Fig. 68. Dependence of the lifetime on the Fermi level position. Region a represents the second term in Eq. (23.15) and region b represents the second term in Eq. (23.14).

$$N_{cM} \gg n_l \gg P_{vM}; \qquad (23.13)$$

the case when the traps lie in the lower part of the forbidden band does not differ in principle from the case quoted and may be derived from the latter by interchanging the roles of electrons and holes.) Using Eq. (22.13) we can replace Eq. (23.12) by two separate expressions for n-type ($n_0 \gg p_0$) and p-type ($p_0 \gg n_0$) semiconductors:

$$\tau \simeq \tau_{p0} + \tau_{p0}\frac{N_{cM}}{n_0}$$
$$= \tau_{p0} + \tau_{p0}\, e^{\frac{-\Delta\mathscr{E}_M - F}{kT}}, \qquad (23.14)$$

$$\tau \simeq \tau_{n0} + \tau_{p0}\frac{N_{cM}}{p_0} = \tau_{n0} + \tau_{p0}\frac{N_c}{P_n}\, e^{\frac{-\Delta\mathscr{E}_M + F + \Delta\mathscr{E}}{kT}}. \qquad (23.15)$$

The curve of Fig. 68 consists of four rectilinear regions (two for an n-type semiconductor and two for a p-type one). The nature of the variation of τ in these four regions may easily be explained using simple physical considerations. *

First Region. When the Fermi level is closer to the conduction band than the trap level M (Fig. 68) and consequently $n_0 \begin{cases} \gg N_{cM} \\ \gg p_0 \\ \gg P_{vM} \end{cases}$. Eq. (23.12)

gives $\tau \approx \tau_{p0}$, i.e., τ is a constant governed only by the number and properties of traps when the latter are completely filled with electrons.

In this case all the traps are filled with electrons at equilibrium and the electron density in the upper band is very high. The appearance of the nonequilibrium electrons and holes in the band initiates hole capture by the filled traps. However, such hole capture cannot greatly affect the population of the traps because, in view of the high electron density in the band, any hole captured by a trap is practically instantaneously "annihilated" by an electron from the upper band. Thus the hole lifetime (and the lifetime of an electron-hole pair) is governed by the total trap concentration (the traps being always filled), i.e., it is equal to τ_{p0}.

---

* It will always be assumed, unless otherwise stated, that τ_{p0} and τ_{n0} are quantities of the same order and that they depend weakly on temperature.

This state is obtains for all positions of the Fermi level as long as it lies at least several kT above the trap level. In the region considered the curve representing τ in Fig. 68 is horizontal.

Second Region. Assume now that the Fermi level is further from the conduction band than the trap level but closer than the level $\mathscr{E}_i$ (i.e., the electron density n_0 is still higher than the hole density p_0).

In this case, bearing in mind that $N_{cM} \gg n_0 \begin{cases} \gg p_0 \\ \gg P_{vM} \end{cases}$, we obtain from Eq. (23.12)

$$\tau = \tau_{p0} \frac{N_{cM}}{n_0} = \tau_{p0} e^{\frac{-\Delta \mathscr{E}_M - F}{kT}} . \tag{23.16}$$

The lifetime increases exponentially on lowering the Fermi level. In this region the number of electrons in the conduction band is still high and any hole captured by a trap is rapidly "annihilated" by a band electron. However, some of the traps are now free of electrons at equilibrium. The Fermi level lies below the trap level and therefore the trap population is not high and the electron density at the traps is $Me^{\frac{\Delta \mathscr{E}_M + F}{kT}}$.

Hence we can write down directly the expression for τ:

$$\tau = \frac{1}{\gamma_p Me^{\frac{\Delta \mathscr{E}_M + F}{kT}}} = \tau_{p0} e^{\frac{-\Delta \mathscr{E}_M - F}{kT}} , \tag{23.17}$$

which is identical with Eq. (23.16).

Thus the increase of the hole (and pair) lifetime on lowering the Fermi level is now governed by the reduction of the electron population of the traps and the corresponding reduction of the hole-capture probability. Subsequent electron capture is "instantaneous" and therefore does not affect the value of τ.

Third Region. When the Fermi level is in the third region the semiconductor becomes p-type, i.e., $p_0 \gg n_0$. Since in this case $N_{cM} \gg p_0$, we find from Eq. (23.12) that

$$\tau = \tau_{p0} \frac{N_{cM}}{p_0} = \frac{N_c}{P_v} \tau_{p0} e^{\frac{F + \Delta \mathscr{E} - \Delta \mathscr{E}_M}{kT}} . \tag{23.18}$$

The lifetime decreases on lowering the Fermi level. At equilibrium the traps are now almost completely empty and readily capture the nonequilibrium electrons from the band. However, a captured electron does not always recombine with a hole. Since the number of holes is not very large, the recombination process competes with the thermal transitions

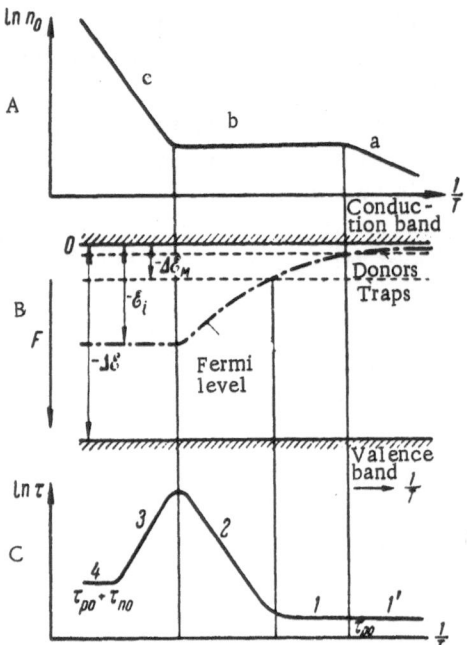

Fig. 69. Temperature dependences of: (A) the equilibrium density; (B) the Fermi level; (C) the lifetime.

of electrons from the traps back to the conduction band. Lowering the Fermi level raises the hole density, reduces the relative importance of the thermal transitions of electrons from the traps to the band, and reduces the lifetime.

Fourth Region. If the Fermi level is closer to the valence band than the trap level to the conduction band, then $p_0 \gg N_{cM}$. The semiconductor is of p-type ($p_0 \gg n_0$) and therefore Eq. (23.12) gives for the fourth region

$$\tau = \tau_{n0}. \qquad (23.18a)$$

The lifetime is now independent of the position of the Fermi level. The traps are empty and the lifetime of the nonequilibrium electrons is τ_{n0}. An electron captured by a trap recombines immediately with one of the many holes present. The return thermal transitions of electrons from the traps to the conduction band are not important.

The junctions of the rectilinear regions in the dependence of $\ln \tau$ on the Fermi level position are several kT wide and are described by the general expression (23.12).

2. Temperature Dependence of the Lifetime. Above we considered the dependence of τ on the Fermi level position at constant temperature, i.e., essentially the dependence of τ on the number and type of impurities (which do not act as recombination centers). It is of great interest to find the variation of τ with temperature in a sample of given composition [8, 17, 20]. For this purpose we shall consider an impurity (extrinsic) semiconductor of, say, n-type which exhibits all three regions of density variation when the temperature is increased from absolute zero (Fig. 69A): (a) increase of the carrier density with temperature in the impurity region; (b) a region of total ionization of the impurities and constant density; (c) a region of rapid density rise on reaching the intrinsic conduction region.

When the temperature is varied in region (a) the Fermi level lies between the donor impurity centers and the conduction band, i.e., above the levels of the traps. Consequently in this region $N_{cM} \ll n_0$ and from Eq. (23.14) we find that $\tau = \tau_{p0}$ (Fig. 69C, region 1').

On transition to region (b) the value of n_0 becomes constant (total ionization of the impurities), and $N_{cM} = N_c e^{-\frac{\Delta \mathscr{E}_M}{kT}}$ increases exponentially with temperature. However, as long as the increasing value of N_{cM} remains smaller than n_0, i.e., as long as the Fermi level does not drop below the trap, we can still neglect the ratio N_{cM}/n_0 in Eq. (23.14) as compared with unity and therefore $\tau = \tau_{p0}$ (Fig. 69C, region 1).

On further increase of temperature, n_0 remains constant and the value of N_{cM}, which is now greater than n_0, continues to increase. Therefore, neglecting unity in Eq. (23.14), we can write

$$\tau = \frac{\tau_{p0}}{n_0} N_{cM} = \frac{\tau_{p0}}{n_0} \cdot \frac{2}{h^3} (2\pi m_n k)^{3/2} T^{3/2} e^{-\frac{\Delta \mathscr{E}_M}{kT}} = \text{const } T^{3/2} e^{-\frac{\Delta \mathscr{E}_M}{kT}} \qquad (23.19)$$

It follows from the above equation that if this dependence is represented as $\ln \tau = f(1/T)$ [or, more exactly, $\ln(\tau/T^{3/2}) = f(1/T)$], we can obtain the activation energy of the traps from the slope of the resultant straight line (Fig. 69C, region 2). This method of determining $\Delta \mathscr{E}_M$ is very widely used.

If we know the equilibrium density of the majority carriers n_0 in the region of complete impurity ionization, the depth of the traps may also be estimated from the temperature of the transition from region 1 to region 2. This transition corresponds to the situation when the Fermi level coincides with the trap level and consequently we can write

$$n_0 = N_c e^{-\frac{\Delta \mathscr{E}_M}{kT_t}}, \qquad (23.20)$$

where T_t is the temperature of the transition from region 1 to region 2. Hence we find

$$\Delta \mathscr{E}_M = kT_t \ln \frac{N_c}{n_0}. \qquad (23.21)$$

The quantity $\ln(N_c/n_0)$, which occurs in Eq. (23.21), may be estimated directly from the temperature dependence of τ. It follows from Eq. (23.19) that, extrapolating the straight line 2 in Fig. 69C until it intersects the ordinate, we can find the value of $\ln(\tau_{p0} N_c/n_0)$ and hence, knowing τ_{p0} from region 1, we can find $\ln(N_c/n_0)$.

The rise of τ in region 2 is due to the reduction of the trap population on increase of temperature. In this region (cf. region 2 in Fig. 68) the number of electrons is sufficiently large for rapid "annihilation" of each nonequilibrium hole which is captured by a trap, but the trap population is small and decreases exponentially with temperature. The lifetime of the nonequilibrium holes increases and this determines the measured lifetime τ. The rise of τ with temperature continues until the intrinsic conduction region is reached. In the latter region τ first decreases exponentially (Fig. 69C, region 3) and then approaches a certain limit (region 4). Here

$$n_0 \simeq p_0 = n_i = \sqrt{N_c P_v}\, e^{-\frac{\Delta \mathscr{E}}{2kT}}, \tag{23.22}$$

and Eq. (23.12) is written in the form

$$\tau = \frac{\tau_{p0}}{2}\left(1 + \sqrt{\frac{N_c}{P_v}}\, e^{\frac{\frac{\Delta \mathscr{E}}{2} - \Delta \mathscr{E}_M}{kT}}\right) + \frac{\tau_{n0}}{2}\left(1 + \sqrt{\frac{P_v}{N_c}}\, e^{\frac{-\frac{\Delta \mathscr{E}}{2} + \Delta \mathscr{E}_M}{kT}}\right). \tag{23.23}$$

If the level of the traps is so far from the middle of the forbidden band that even in the case of intrinsic conduction there is a region where $\left|\frac{\Delta \mathscr{E}}{2} - \Delta \mathscr{E}_M\right| \gg kT$, then for that region Eq. (23.23) transforms into

$$\tau = \frac{\tau_{p0}}{2}\sqrt{\frac{N_c}{P_v}}\, e^{\frac{\frac{\Delta \mathscr{E}}{2} - \Delta \mathscr{E}_M}{kT}}. \tag{23.24}$$

Thus the lifetime in the region considered (region 3 in Fig. 69C) decreases with temperature mainly because of the sharp reduction of the carrier density in the bands, and it follows from Eq. (23.24) that the slope of region 3 is governed by the energy $\left(\frac{\Delta \mathscr{E}}{2} - \Delta \mathscr{E}_M\right)$. Consequently the sum of the slopes of regions 2 and 3 should amount to half the width of the forbidden band, $\frac{\Delta \mathscr{E}}{2}$. On increase of temperature, when we reach

$$kT \gg \left|\frac{\Delta \mathscr{E}}{2} - \Delta \mathscr{E}_M\right|,$$

the quantity τ tends to an approximately constant value (region 4):

$$\tau = \frac{\tau_{p0}}{2}\left(1 + \sqrt{\frac{N_c}{P_v}}\right) + \frac{\tau_{n0}}{2}\left(1 + \sqrt{\frac{P_v}{N_c}}\right). \tag{23.25}$$

When $N_c \approx P_v$ Eq. (23.25) transforms into

$$\tau = \tau_{p0} + \tau_{n0} = \frac{1}{M}\left(\frac{\gamma_n + \gamma_p}{\gamma_n \gamma_p}\right). \tag{23.26}$$

Thus at sufficiently high temperatures the value of τ is governed only by the number and properties of traps.

In Fig. 69C we have ignored some details of the portions of the curves at the junctions between the various regions. More accurate calculated dependences of $\log(\tau/\tau_{p0})$ on $1/T$ for three different trap levels in germanium are given in Fig. 70. In these calculations the effective masses were assumed to be $m_e^* = 0.25m_0$, $m_h^* = 0.4m_0$; the donor concentration was taken to be $N_d = 5 \times 10^{13}$ cm^{-3}; we also assumed that $10\tau_{p0} = \tau_{n0}$. To simplify the calculations we assumed an average value of the forbidden bandwidth, equal to 0.70 eV. The curves in region 4 have a small minimum, which can be shown to disappear when $m_e^* = m_h^*$ and $\tau_{p0} = \tau_{n0}$.

We note that the closer the traps are to the middle of the band the earlier region 3 goes over into region 4. When $\Delta\mathscr{E}_M = \Delta\mathscr{E}_i$ regions 3 and 2 are absent and region 1 goes over directly into region 4.† Figure 70 shows that at moderate temperatures (regions 2 and 3) the closer the traps are to the edge of the band the longer is the lifetime, other conditions being equal. Consequently in the presence of several types of trap only the deep traps affect the recombination at moderate temperatures. The shallower the traps, the greater the range of temperatures which can be called moderate where the recombination is weak and τ is large (Fig. 70). Therefore in some cases (when deep traps are present in a semiconductor) we can ignore the influence of "shallow" traps on recombination processes, though these levels are important in equilibrium conductivity.‡

We have considered above the case when the traps lie in the same part of the band as the Fermi level (in the impurity region). It is found that in other cases the temperature dependence of the lifetime on a logarithmic scale has the same form of four rectilinear regions, and the second region can be used (as before) to determine the energy separation of the trap levels from the edge of the nearest band.

Assume that, for example, in an n-type semiconductor the trap levels are in the lower half of the forbidden band at a distance $\Delta\mathscr{E}_M$ from the conduction band (and consequently at a distance $\Delta\mathscr{E} - \Delta\mathscr{E}_M$ from the valence band); then, following the treatment given above, we can distinguish four temperature regions.

<u>Region 1.</u> This is the region of very low temperatures for which

$|F| \ll |\Delta\mathscr{E} - \Delta\mathscr{E}_M|$ and, consequently $n_0 \begin{cases} \gg p_0 \\ \gg N_{cM} \\ \gg P_{vM} \end{cases}$. For this region we obtain from Eq. (23.12)

---

† Region 4 may be realized only for traps near the middle of the forbidden band. For shallow traps (cf. for example, curve for $\Delta\mathscr{E}_M$ = 0.11 eV in Fig. 70), this region corresponds to temperatures above the melting point of germanium.

‡ All the problems in the present section are discussed on the assumption that the equilibrium conductivity in the impurity region and, consequently, the values of n_0 and p_0 are governed by impurities which give rise to very shallow levels (for germanium and silicon the elements of groups III and V act as much impurities).

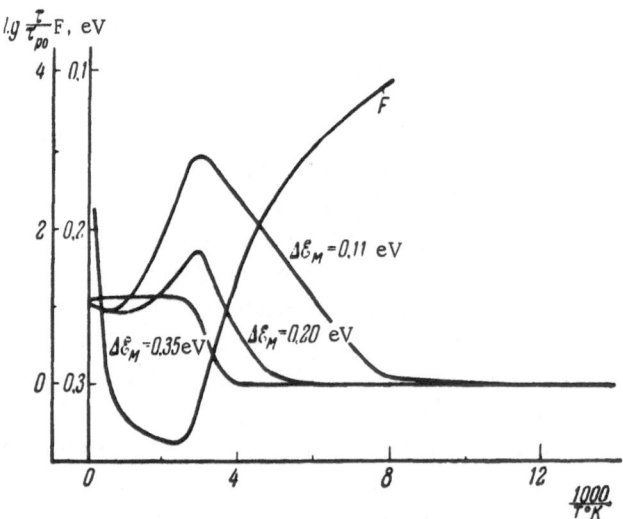

Fig. 70. Calculated temperature dependences of the lifetime
τ and the Fermi level F for germanium.

$$\tau = \tau_{p0}. \tag{23.27}$$

Region 2. This is the region of low temperatures for which

$$|\Delta\mathscr{E} - \Delta\mathscr{E}_M| \ll |F| \ll \mathscr{E}_i \text{ and, consequently, } n_0 \begin{cases} \gg p_0 \\ \gg N_{cM}. \\ \ll P_{vM} \end{cases} \text{ Here we find from}$$

Eq. (23.12) that

$$\tau = \tau_{n0}\,\frac{P_{vM}}{n_0} = \frac{\tau_{n0}}{n_0}\cdot\frac{2}{h^3}\,(2\pi mkT)^{3/2}\,e^{-\frac{\Delta\mathscr{E} - \Delta\mathscr{E}_M}{kT}} \tag{23.28}$$

The above expression differs from the corresponding expression (23.19) for traps lying in the upper half of the forbidden band by the replacement of τ_{p0} by τ_{n0}, and also by the fact that the index of the exponential function contains $(\Delta\mathscr{E} - \Delta\mathscr{E}_M)$ which is the distance from the edge of the nearest (valence) band. The quantity $\Delta\mathscr{E} - \Delta\mathscr{E}_M$ can be determined from the experimental temperature dependence of τ in this region.

The reason for the rise of τ in this region is different from that given earlier for the case of traps in the upper half of the forbidden band. In contrast to the earlier case, the traps are now completely filled with electrons and they readily capture nonequilibrium holes. However, the number of electrons in the conduction band is insufficient for immediate recombination with each captured hole, and therefore the process of recombination of captured holes has to compete with the thermal transitions back to the valence band (the latter becoming more important with

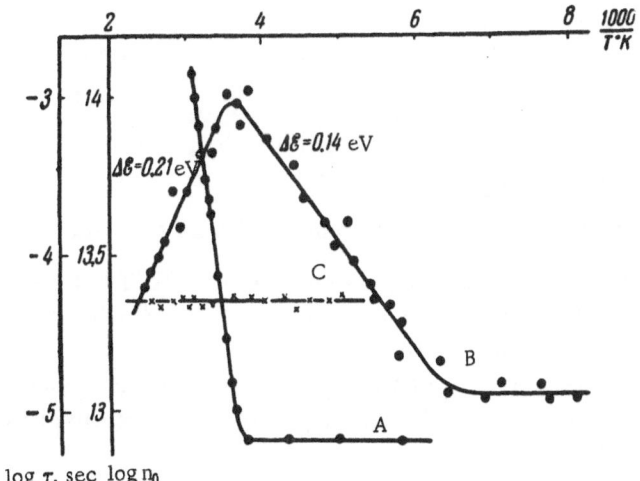

Fig. 71. Temperature dependence of equilibrium carrier density (A) and of the lifetime for low (B) and high (C) injection
levels in germanium.

increase of temperature). Thus the temperature dependence of τ in region 2 may be used to determine the energy separation of the traps from
the edge of the nearest allowed band. Temperature measurements by
themselves do not enable us to determine whether this allowed band is
conduction or valence.

Intrinsic Conduction Region. This region, as in the case of traps in
the upper half of the forbidden band, may be divided into two: region 3,
where $\left| \Delta\mathscr{E}_M - \frac{\Delta\mathscr{E}}{2} \right| \gg kT$ and, from Eq. (23.23),

$$\tau \simeq \frac{\tau_{n0}}{2} \sqrt{\frac{P_v}{N_c}} \, e^{\frac{\Delta\mathscr{E}_M - \frac{\Delta\mathscr{E}}{2}}{kT}}, \tag{23.29}$$

i.e., τ decreases with temperature, and region 4, where
$kT \gg \left| \Delta\mathscr{E}_M - \frac{\Delta\mathscr{E}}{2} \right|$ and, from Eq. (23.23),

$$\tau = \frac{\tau_{p0}}{2} \left(1 + \sqrt{\frac{N_c}{P_v}} \right) + \frac{\tau_{n0}}{2} \left(1 + \sqrt{\frac{P_v}{N_c}} \right). \tag{23.30}$$

The expression (23.30) is identical with Eq. (23.25); in this region τ is
independent of temperature if τ_{p0} and τ_{n0} are independent of temperature.
The complete temperature dependence of τ is represented qualitatively
by the curves of Fig. 69C.

The above discussion of the temperature dependence of the lifetime indicates that there are several quantitative relationships with may be checked experimentally. Among them are the following:

1. The sum of the activation energies, determined from region 2 and 3 of the temperature dependence $\ln \tau = f(1/T)$, should be

   equal to half the forbidden band width at absolute zero: $\frac{\Delta \mathcal{E}^0}{2}$;

2. The maximum of the dependence $\ln \tau = f(1/T)$, which occurs at the junction of regions 2 and 3, should correspond to a change to intrinsic conduction;

3. The temperature T_j of the junction of regions 1 and 2 is related to the trap levels $(\Delta \mathcal{E}_M)$ but by the relationship [cf. Eq. (23.21)]

$$T_j = \frac{\Delta \mathcal{E}_M}{k \ln \frac{N_c}{n_0}} .$$ (23.31)

We shall give two examples of experimental temperature dependences of τ for a sample of germanium (Fig. 71) [20]. Curve B in Fig. 71 shows clearly regions 1, 2, and 3 corresponding to the theoretical curve. The sum of the slopes of regions 2 and 3 is 0.35 eV, which is close to half the forbidden bandwidth. The junction of regions 2 and 3 occurs where the impurity conduction is replaced by intrinsic conduction.

3. The Case of High Illumination Intensities (or High Injection Level). If the illumination intensity is so high that the nonequilibrium electron density Δn (or the nonequilibrium hole density $\Delta p = \Delta n$) is much higher than all the other densities $(n_0, p_0, N_{cM}, P_{vM})$ which occur in Eq. (23.11), we then obtain *

$$\tau_\infty = \tau_{p0} + \tau_{n0} = \frac{1}{M} \left(\frac{\gamma_n + \gamma_p}{\gamma_n \gamma_p} \right).$$ (23.32)

Consequently, in contrast to the case of low illumination intensities, where τ was governed primarily by the equilibrium densities $(n_0, p_0, N_{cM}, P_{vM})$ and consequently depended strongly on temperature and sample composition, † at high illumination intensities τ is governed only by the number and properties of traps. The expression (23.32) may be obtained directly from very simple considerations. Under conditions of high densities of free electrons and holes the rate of capture is so great

---

* We shall use the notation τ_∞ to represent the value of τ at high excitation levels.

† Such impurities govern the equilibrium densities but do not take active part in the recombination.

that thermal transitions may be neglected altogether. Consequently the system of transitions (Fig. 67) becomes much simpler. An electron transferred by light to the upper band is captured by an empty trap and then it invariably recombines with a free hole (if thermal transitions are indeed absent).

Under steady-state conditions the lifetime of an electron is governed by the steady-state density of free holes $(M - m)$ and the recombination coefficient γ_n:

$$\tau_n = \frac{1}{\gamma_n (M - m)} .$$
(23.33)

The hole lifetime, which is equal to the electron lifetime, is written in the form

$$\tau_p = \frac{1}{\gamma_p m} .$$
(23.34)

Equating Eqs. (23.33) and (23.34), we obtain

$$m = M \frac{\gamma_n}{\gamma_n + \gamma_p} .$$
(23.35)

The expression (23.35) gives the steady-state population of traps during strong illumination. Substituting Eq. (23.35) into Eq. (23.33) or Eq. (23.34), we obtain

$$\tau_\infty = \tau_n = \tau_p = \frac{1}{M} \frac{\gamma_n + \gamma_p}{\gamma_n \gamma_p} ,$$
(23.36)

which is identical with Eq. (23.32).

Similar conditions (weak influence of thermal transitions) apply also at low illumination intensities at high temperature [cf. Eq. (23.26)] when the capture rate is much greater than the thermal transition rate because the density of free (but by now equilibrium) carriers in the bands increases very strongly with temperature.

In contrast to the case of direct recombination of band electrons and holes, when the recombination rate (at high illumination intensities) is proportional to the square of the carrier density (cf. § 5 in Chap. I), the recombination rate — in the case of recombination through traps — is directly proportional to the density and the quantity τ is independent of the carrier density.

The case considered here is very frequently and easily realized in pure low-conductivity semiconductors and insulators with deep traps in which the equilibrium densities of free carriers (n_0, p_0) as well as the effective reduced densities of states (N_{cM}, P_{vM}) are small and even at moderate illumination intensities the value of Δn may be considerably greater than these quantities.

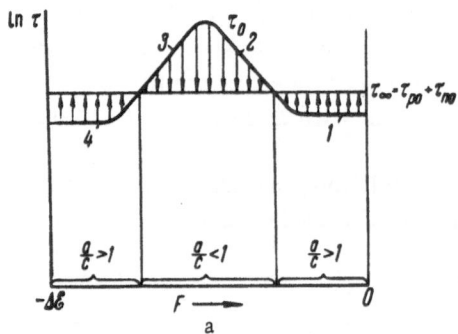

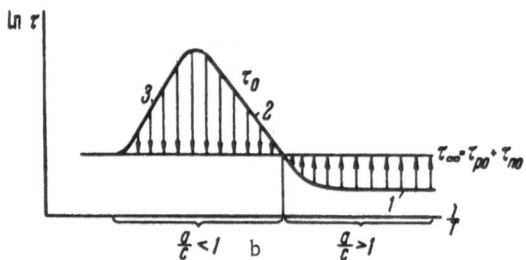

Fig. 72. Dependence of the lifetime on the Fermi level position (a) and temperature (b) for low and high illumination intensities.

In conclusion we note that in the case just considered the temperature dependence of τ, governed [cf. Eqs. (4.2) and (4.6)] by the temperature dependence of thermal velocities (v_n and v_p) and capture cross sections (q_n and q_p), may be relatively weak. Therefore the temperature dependence of τ_∞ is also weak (cf. curve C in Fig. 71).

4. The Case of Arbitrary Illumination Intensities (or Arbitrary Injection Level). We have considered in detail two extreme cases: low and high illumination intensities. In the general case we should use Eq. (23.11) which can be rewritten in a slightly different form [3]:

$$\tau = \tau_0 \frac{1 + a\,\Delta n}{1 + c\,\Delta n}. \tag{23.37}$$

Here τ_0 is the lifetime for low illumination intensities, given by Eq. (23.12); a and c are coefficients which are independent of the illumination intensity:

$$a = \frac{\tau_{p0} + \tau_{n0}}{\tau_{p0}(n_0 + N_{cM}) + \tau_{n0}(p_0 + p_{vM})}, \tag{23.38}$$

$$c = \frac{1}{n_0 + p_0}. \tag{23.39}$$

It follows from Eq. (23.37) that the steady-state lifetime varies mono-tonically with increase of Δn. The limits of such variation have been ob-tained earlier: they are τ_0 at low intensities and $\tau_\infty = \tau_{n0} + \tau_{p0}$ at high intensities.

Figure 72 gives curves for the dependence of τ_0 on the Fermi level position (a) and temperature (b). Moreover the straight line $\tau_\infty = \tau_{p0} + \tau_{n0}$ gives the dependence for τ at high intensities. The areas hatched by arrows represent all possible values of τ at any illumination intensity. The arrows indicate the direction of variation of τ with increase of the illumination intensity at given positions of the Fermi level (Fig. 72a) or at given temperatures (Fig. 72b).

It is found that τ does not always decrease with increase of the il-lumination intensity: it may also rise, as for example at "low" temper-atures (Fig. 72b).

At moderate temperatures (regions 2 and 3) τ decreases with in-crease of the illumination intensity. At high temperatures (region 4) τ may either increase or decrease depending on the ratio of N_C to P_V [cf. Eqs. (23.25)]. When $N_C = P_V$ the value of τ is independent of the illumi-nation intensity (this is precisely the case shown in Fig. 72b).

The actual nature of the dependence of τ on the illumination intensity is given analytically by the relationship between a and c in Eq. (23.37). When $(a/c) > 1$ the value of τ rises with the intensity, otherwise it de-creases. As the temperature increases the variation of this ratio (Fig. 72b) corresponds approximately to the transition from region 1 to region 2 (when $\tau_{p0} \approx \tau_{n0}$ and the traps are not too deep).

Simultaneous measurements of $\tau_\infty = \tau_{p0} + \tau_{n0}$ and τ_0 at low temper-atures (region 1), where the latter is equal to either τ_{p0} or τ_{n0}, should give the relationship between τ_{p0} and τ_{n0} and thus establish the nature of the trap (donor or acceptor) [10].

5. Dependence of the Steady-State Nonequilibrium Conductivity on the Excitation Intensity. We have established earlier the dependence of τ_{st} on Δn_{st}. This can easily be used to find the dependence of Δn_{st} on the excitation intensity [17]. In fact [cf. Eq. (5.19)]

$$\Delta n_{st} = \beta k I \tau_{st}. \tag{23.40}$$

Consequently at low intensities when $\tau_{st} = \tau_0$, i.e., when τ_{st} is inde-pendent of the illumination intensity, we have

$$\Delta n_{st} = \beta k \tau_0 I = \text{const} I. \tag{23.41}$$

At high intensities, when [cf. Eq. (23.32)] $\tau_{st} = \tau_\infty = \tau_{p0} + \tau_{n0}$ is again independent of the intensity,

$$\Delta n_{st} = \beta k (\tau_{p0} + \tau_{n0}) I = \text{const} I. \tag{23.42}$$

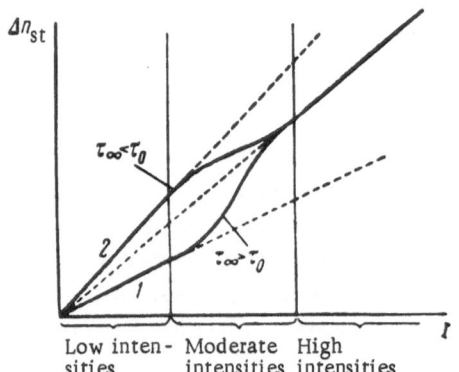

Fig. 73. Dependence of the steady-state density on the excitation intensity for $a > c$ (curve 1) and $a < c$ (curve 2).

In this region the dependence of Δn_{st} on the intensity is again linear but the slope, as indicated by comparison of Eqs. (23.41) and (23.42), is different from the low-intensity case. Since the quantities τ_0 and τ_∞, which govern these slopes, may be related in different ways [for example, $\tau_0 < \tau_\infty$ at low temperatures and $\tau_0 > \tau_\infty$ at "moderate" temperatures (cf. Fig. 72)], Fig. 73 gives two variants of the curve $\Delta n_{st} = f(I)$.

When $\tau_\infty > \tau_0$ (low temperatures), the dependence on the intensity has a superlinear region. The nature of the dependence in the transition region between the two straight lines may be found by analyzing the general expression for $\Delta n_{st} = f(I)$. This expression is obtained by substituting into Eq. (23.40) the value of τ_{st} from Eq. (23.37) and solving the resultant equation for Δn_{st}. Then we have

$$\Delta n_{st} = \frac{1}{2c}\left[a\beta k\tau_0 I - 1 + \sqrt{(a\beta k\tau_0 I - 1)^2 + 4c\beta k\tau_0 I}\,\right]. \tag{23.43}$$

It is easily seen that the expressions (23.41) and (23.42) for low and high illumination intensities follow directly from Eq. (23.43).

The theoretical dependence of the nonequilibrium conductivity on the intensity of excitation, obtained above, has two linear portions joined by a transition region, which is in good agreement with experiment, for example, for germanium [23].

Figure 74 gives on double logarithmic scale the dependence of Δn_{st} on the illumination intensity. The continuous curves are theoretical: a) for $\tau_0/\tau_\infty = 4$; b) for $\tau_0/\tau_\infty = 1/4$. The circles represent experimental values for a germanium sample with $\tau_0/\tau_\infty = 4$.

B. The Case of Arbitrary Trap Concentration

We have assumed so far that the trap concentration M is so small that the trap charge may be neglected in the neutrality condition (22.19).

However the case of high trap concentration, when all the terms must be included in the neutrality equation, is also very important. This case is frequently realized in semiconductors with moderate conductivity but with a large number of deep traps at which a considerable charge may be accumulated.

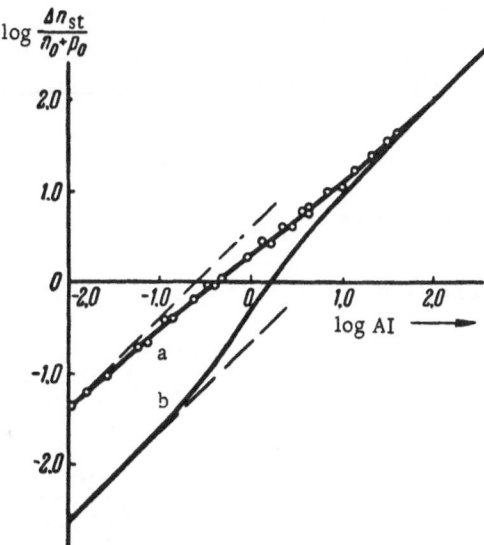

Fig. 74. Dependence of Δn_{st} on the value of I for germanium.

In the case of low trap concentration it follows from the neutrality equation (22.19) that $\Delta n = \Delta p$ and consequently the steady-state electron and hole lifetimes are equal:

$$\tau = \frac{\Delta n}{k\beta I} = \frac{\Delta p}{k\beta I}. \tag{23.44}$$

In the general case $\Delta n \neq \Delta p$ and, consequently, the electron and hole lifetimes are also unequal:

$$\left.\begin{aligned}
\tau_n &= \frac{\Delta n}{k\beta I}, \\
\tau_p &= \frac{\Delta p}{k\beta I}.
\end{aligned}\right\} \tag{23.45}$$

Our problem now is to calculate these lifetimes.

1. Low Excitation Level. Let us assume that

$$\left.\begin{aligned}
n &= n_0 + \Delta n, \\
p &= p_0 + \Delta p, \\
m &= m_0 + \Delta m,
\end{aligned}\right\} \tag{23.46}$$

and that

$$\left.\begin{aligned}
\Delta n &\ll n_0, \\
\Delta p &\ll p_0, \\
\Delta m &\ll m_0, \\
\Delta m &\ll (M - m_0).
\end{aligned}\right\} \tag{23.47}$$

Then, substituting Eq. (23.46) into the condition for the steady state of the carrier density in the conduction band (23.2) and in the valence band (23.3), we obtain, after neglecting higher orders of small quantities

$$\gamma_n \left[\Delta n (M - m_0) - \Delta m (N_{cM} + n_0) \right] = \beta k I, \tag{23.48}$$

$$\gamma_p \left[\Delta p \, m_0 + (p_0 + P_{vM}) \Delta m \right] = \beta k I. \tag{23.49}$$

Using the neutrality condition

$$\Delta p - \Delta n = \Delta m, \tag{23.50}$$

and writing m_0 in the form

$$m_0 = \frac{M}{e^{\frac{-\Delta \mathscr{E}_M - F}{kT}} + 1} = \frac{M}{\frac{N_{cM}}{n_0} + 1} = M - \frac{M}{\frac{P_{vM}}{p_0} + 1}, \tag{23.51}$$

we obtain, by simultaneous solution of Eqs. (23.48)-(23.51), the following expressions for τ_p and τ_n:

$$\tau_p = \frac{\tau_{n0} (P_0 + P_{vM}) + \tau_{p0} \left[n_0 + N_{cM} + M \left(1 + \frac{n_0}{N_{cM}} \right)^{-1} \right]}{p_0 + n_0 + M \left(1 + \frac{n_0}{N_{cM}} \right)^{-1} \left(1 + \frac{N_{cM}}{n_0} \right)^{-1}}, \tag{23.52}$$

$$\tau_n = \frac{\tau_{p0} (n_0 + N_{cM}) + \tau_{n0} \left[p_0 + P_{vM} + M \left(1 + \frac{p_0}{P_{vM}} \right)^{-1} \right]}{p_0 + n_0 + M \left(1 + \frac{p_0}{P_{vM}} \right)^{-1} \left(1 + \frac{P_{vM}}{p_0} \right)^{-1}}. \tag{23.52a}$$

It is easily seen that the above general expressions reduce to Eq. (23.12) in the case of low M. In the opposite case of very high M we have

$$\tau_p = \tau_{p0} \left(1 + \frac{N_{cM}}{n_0} \right) = \frac{1}{\gamma_p m_0}, \tag{23.53}$$

$$\tau_n = \tau_{n0} \left(1 + \frac{P_{vM}}{p_0} \right) = \frac{1}{\gamma_n (M - m_0)}. \tag{23.54}$$

In the latter case the lifetimes are simply given by the lifetimes for electron capture by empty traps and hole capture by traps filled with electrons. Thermal transitions whose rate is small compared with the capture rate can be neglected.

The dependences of τ_p and τ_n on the Fermi level position are given for this case in Figs. 75a and 75b.

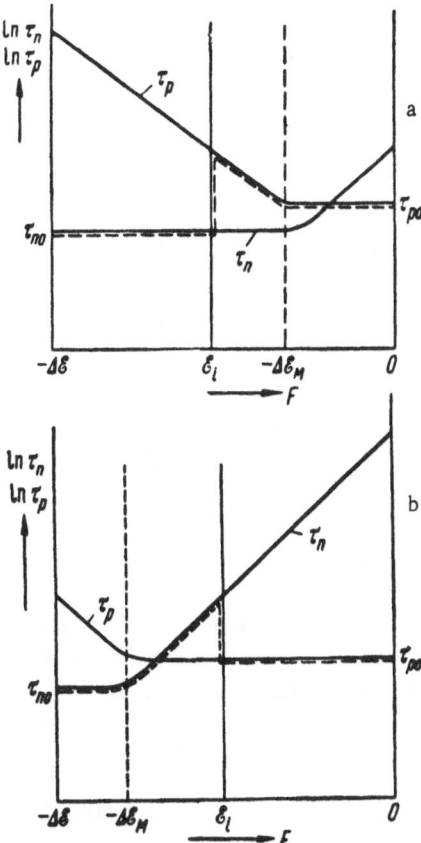

Fig. 75. Dependence of the electron and hole lifetimes on the Fermi level position: a) Traps in the upper half of the forbidden band; b) traps in the lower half.

In plotting the curve of Fig. 75a it was assumed that the traps lie in the upper half of the forbidden band ($|\Delta \mathscr{E}_M| < \mathscr{E}_i$). Figure 75b represents the case of traps in the lower half of the band. The two figures show that the nature of the dependences of τ_p and τ_n on the Fermi level position changes sharply at the point representing the coincidence of the Fermi level with the trap levels: $F = -\Delta \mathscr{E}_M$.

The values of τ_p and τ_n are not, in general, equal but they may become equal at a certain position of the Fermi level, i.e., at the point of intersection of the curves τ_n and τ_p where $\gamma_p m_0 = \gamma_n (M - m_0)$. If the lifetimes of the nonequilibrium electrons and holes are very different then the nonequilibrium conductivity is dominated by the carriers whose lifetime is longer. It is clear from Fig. 75 that if τ_{p0} and τ_{n0} are not very different, then the nonequilibrium conductivity is dominated by electrons if the Fermi level lies above the trap levels, and by holes if the Fermi level lies below the traps. The same conclusion follows from qualitative considerations: in the first case the traps are mainly filled, so hole capture is more intense than electron capture, and $\tau_n > \tau_p$. In the second case the traps are mainly empty, so electron capture predominates over hole capture, and $\tau_p > \tau_n$.

In some cases nonequilibrium processes are investigated by measuring the lifetime of the minority carriers; this applies when the behavior of the nonequilibrium majority carriers does not affect the results of the measurements because of the "selectivity" of the method (cf. § 58). The measured quantity (the lifetime of the nonequilibrium minority carriers) then depends on the Fermi level as shown by the dashed curves in Fig. 75.

2. Arbitrary Excitation Level. In poorly conducting semiconductors one frequently encounters conditions in which the trap concentration is very high compared with the carrier densities $n = n_0 + \Delta n$ and $p = p_0 + \Delta p$, and, in contrast to the case analyzed above [cf. Eqs. (23.53) and (23.54)],

Table 2

Low trap concentration	$\left\{\begin{array}{l} M \ll \Delta n(\Delta p) \ll n_0(p_0) \\ \Delta n(\Delta p) \ll M \ll n_0(p_0) \\ M \ll n_0(p_0) \ll \Delta n(\Delta p) \\ n_0(p_0) \ll M \ll \Delta n(\Delta p) \end{array}\right.$	$\left.\begin{array}{l} \text{Low excitation} \\ \quad \text{level} \\ \text{High excitation} \\ \quad \text{level} \end{array}\right\}$	(23.12) (23.32)
High trap concentration	$\left\{\begin{array}{l} \Delta n(\Delta p) \ll n_0(p_0) \ll M \\ \\ n_0(p_0) \ll \Delta n(\Delta p) \ll M \end{array}\right.$	Low excitation level High excitation level	(23.53) (23.54) (23.60) (23.62)

Note. The terminology used in the above table implies comparison of the trap concentration with the total (equilibrium and nonequilibrium) carrier densities. Consequently the case of high trap concentration at high excitation level means that the trap concentration is greater than the carrier density.

the excitation level may be arbitrary, and in particular it may be high (i.e., $\Delta n \gg n_0$; $\Delta p \gg p_0$).

We shall consider this case by assuming that not only the total trap concentration M but also the difference between the concentrations of the free and electron-occupied traps is much greater than n and p, i.e.,

$$M - m_0 \gg \left\{ \begin{array}{l} n_0 + \Delta n, \\ p_0 + \Delta p, \end{array} \right. \qquad m_0 \gg \left\{ \begin{array}{l} n_0 + \Delta n, \\ p_0 + \Delta p \end{array} \right. \tag{23.55}$$

Then using $\Delta m = \Delta p - \Delta n$ (from the condition of neutrality), we can add the following inequalities to Eq. (23.55):

$$\left. \begin{array}{l} M - m_0 \gg \Delta m, \\ m_0 \gg \Delta m \end{array} \right\} \tag{23.56}$$

In other words under these conditions the illumination does not greatly alter the equilibrium electron and hole densities in the traps.

The conditions for the steady state, Eqs. (23.2) and (23.3), are then rewritten in the form

$$\gamma_n \left[(n_0 + \Delta n)(M - m_0) - m_0 N_{cM} \right] = k\beta I. \tag{23.57}$$

$$\gamma_p \left[(p_0 + \Delta p) m_0 - (M - m_0) P_{vM} \right] = k\beta I. \tag{23.58}$$

From Eqs. (23.57) and (23.58) and $n_0(M-m_0) = m_0 N_{cM}$ and $p_0 m_0 = (M-m_0) P_{vM}$ we obtain directly

$$\Delta n = \frac{\beta k I}{\gamma_n (M - m_0)} \tag{23.59}$$

$$\tau_n = \frac{\Delta n}{\beta k I} = \frac{1}{\gamma_n (M - m_0)} \,, \tag{23.60}$$

$$\Delta p = \frac{\beta k I}{\gamma_p m_0} \,, \tag{23.61}$$

$$\tau_p = \frac{\Delta p}{\beta k I} = \frac{1}{\gamma_p m_0} \,. \tag{23.62}$$

Eqs. (23.60) and (23.62) are identical with (23.52) and (23.54).

It follows that at very high trap concentrations and arbitrary excitation levels* the electron and hole lifetimes are independent and unequal, and that they are governed only by the rate of capture of the respective carriers by the traps, as indicated by Eqs. (23.60) and (23.62).

C. Conclusions

We have considered a large number of extreme cases of recombination through traps of one type and have obtained quantitative relationships. All the different special cases can be represented by inequalities relating the trap concentration M, the equilibrium carrier densities n_0 and p_0, and the nonequilibrium carrier densities Δn and Δp.

Let us write down these inequalities (Table 2). It is easily seen that six limiting cases are possible (the extreme right-hand column in the table gives the numbers of the formulas for calculating the lifetime).

These limiting cases have been considered above and general expressions have been obtained for the transition regions between some of these cases.

§ 24. THE CASE OF SEVERAL TYPES OF TRAP

If a semiconductor has several types of trap which differ in their energy positions in the forbidden band and in their capture cross sections, then, because these traps can exchange electrons only with the conduction and valence bands (transitions between traps do not occur because they are separated in space), it would seem that the effects of traps of various types on recombination processes should be simply additive. This would mean that the effective steady-state lifetime τ could be found from the expression

$$\frac{1}{\tau} = \sum_k \frac{1}{\tau_k} \,, \tag{24.1}$$

where k signifies a particular type of trap.

*See the note in Table 2.

However, Kalashnikov [5] showed that in the case of weak excitation, if the concentration of even one type of trap is not small compared with the equilibrium carrier densities n_0 or p_0, the mutual influence of traps may be considerable. This influence is due to the fact that in the case considered the neutrality condition has the form

$$\Delta n - \Delta p = \sum_k \Delta m_k, \qquad (24.2)$$

and consequently the relationship between Δn and Δp may be strongly dependent on the presence of traps of various types. Thus, for example, the presence of a large number of traps which do not take part in the re-combination may nevertheless affect the recombination through traps of another type by changing the relationship between Δn and Δp (for details see § 30). The effects of various types of trap are additive only in the case when the number of traps of all types is so small that the neutrality condition can be expressed in the form $\Delta n = \Delta p$; the lifetime can then be calculated using Eq. (24.1).

§ 25. RELAXATION OF NONEQUILIBRIUM CONDUCTIVITY

So far we have considered steady-state conditions and calculated the steady-state lifetime of nonequilibrium carriers.

A knowledge of the steady-state lifetime is, in general, insufficient in considerations of the relaxation curves for nonequilibrium conductivity. For example, the nonlinear dependence $\Delta n_{st} = f(I)$, which is character-istic of low trap concentrations, indicates that the relaxation curves of the nonequilibrium conductivity excited by square pulses are nonexponen-tial. Hence, it is obvious that a complete description of the nonequilibri-um conductivity requires a knowledge not only of the steady-state charac-teristics such as τ_{st}, but also of the additional characteristics which can be found by solving the non-steady-state problem. One such characteristic is the instantaneous value of the lifetime τ_{inst}, which can be used to de-scribe not only the steady-state values but also the relaxation curves of the nonequilibrium conductivity.

We shall now consider the nature of relaxation in several important cases.

A. Low Trap Concentration

We shall find the dependence of the nonequilibrium carrier density on time for a semiconductor with low trap concentration [9, 23]. For this purpose it is necessary to solve the transport equations (22.16)-(22.18).

As before, we shall write the neutrality condition in the form

$$\Delta p = \Delta n + \Delta m \simeq \Delta n.$$

Assuming that the total change of the electron density in the traps Δm is negligibly small compared with Δn and Δp, and therefore $\Delta n \approx \Delta p$, it is seen that during the main part of a monotonic relaxation process (which is our present interest) the derivatives dn/dt and dp/dt are practically equal. *   Then equating the right-hand parts of Eqs. (22.16) and (22.18), we find m, and having substituted it into Eq. (22.16), we obtain

$$\frac{dn}{dt} = k\beta I + M \frac{\gamma_n \gamma_p (N_{cM} P_{vM} - np)}{\gamma_n (N_{cM} + n) + \gamma_p (P_{vM} + p)}, \qquad (25.1)$$

and using the relationships $n = n_0 + \Delta n$, $p = p_0 + \Delta n$, $N_{cM} P_{vM} = n_0 p_0$, we have finally

$$\frac{d\Delta n}{dt} + M \frac{\gamma_n \gamma_p (\Delta n + n_0 + p_0) \Delta n}{\gamma_p (\Delta n + p_0 + P_{vM}) + \gamma_n (\Delta n + n_0 + N_{cM})} = k\beta I. \qquad (25.2)$$

In Eq. (25.2) the variables are separable and, for example, the photoconductivity decay curve ($k\beta I = 0$), for the initial condition $\Delta n = \Delta n_{st}$ when $t = 0$, is found to be

$$\frac{t}{\tau_\infty} = \left(1 - \frac{\tau_0}{\tau_\infty}\right) \ln \frac{\Delta n_{st} + n_0 + p_0}{\Delta n + n_0 + p_0} - \frac{\tau_0}{\tau_\infty} \ln \frac{\Delta n}{\Delta n_{st}}. \qquad (25.3)$$

Using $\tau_{inst} = -\Delta n / (d\Delta n /dt)$, we obtain

$$\begin{aligned}
\tau_{inst} &= \frac{\gamma_p (\Delta n + p_0 + P_{vM}) + \gamma_n (\Delta n + n_0 + N_{cM})}{M \gamma_n \gamma_p (\Delta n + n_0 + p_0)} \\
&= \tau_{n0} \frac{\Delta n + p_0 + P_{vM}}{\Delta n + n_0 + p_0} + \tau_{p0} \frac{\Delta n + n_0 + N_{cM}}{\Delta n + n_0 + p_0}.
\end{aligned} \qquad (25.4)$$

It is easily seen that by analogy with the steady-state case Eq. (25.4) may be rewritten in the form [cf. Eq. (23.27)]

$$\tau_{inst} = \tau_0 \frac{1 + a \Delta n}{1 + c \Delta n}. \qquad (25.5)$$

Since density of the nonequilibrium carriers Δn changes during relaxation, τ_{inst} changes also. The expression (23.37) given in §23 for τ_{st} is the limiting case of Eq. (25.5).

Thus in the case of low trap concentration the lifetime is governed only by the nonequilibrium carrier density, and the instantaneous lifetime at some value of Δn_{inst} is equal to the steady-state lifetime at the

---

* The equality $dn/dt = dp/dt$ can also be derived by a rigorous quantitative method, but we shall not deal with this here.

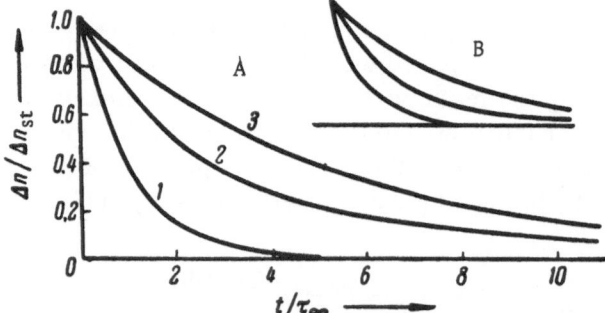

Fig. 76. A) Theoretical curves: 1) $\exp(-t/\tau_\infty)$; 2) relaxation curve plotted using Eq. (25.3); 3) $\exp(-t/\tau_0)$. B) Corresponding experimental curves (oscillograms).

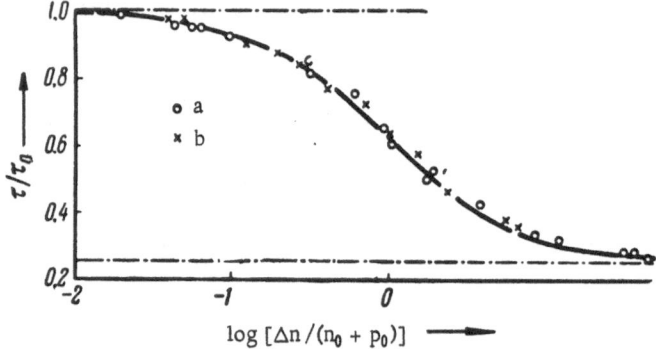

Fig. 77. The continuous curve shows the theoretical dependence $\tau_{inst} = f(\Delta n)$. a) Experimental points for $\tau_{st} = f(\Delta n_{st})$; b) experimental points for $\tau_{inst} = f(\Delta n)$.

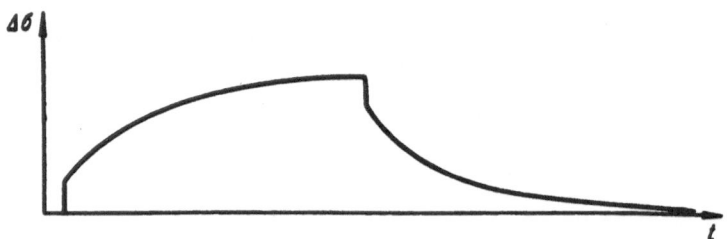

Fig. 78. Relaxation curves for high trap concentration.

steady-state density given by $\Delta n_{st} = \Delta n_{inst}$. Since for $\Delta n \gg n_0 + p_0$ the time constant for the density decay is τ_∞, and for $\Delta n \ll n_0 + p_0$ it is τ_0, the relaxation curve must lie between the exponentials $\exp(-t/\tau_\infty)$ and $\exp(-t/\tau_0)$.

Figure 76A gives the relaxation curve plotted using Eq. (25.3) for the injection level $\Delta n_{st}/(n_0 + p_0) = 4$ when the ratio of the time constants is $\tau_0/\tau_\infty = 5$. The same figure includes the two exponentials $\exp(-t/\tau_0)$ and

$\exp(-t/\tau_\infty)$.   Initially the relaxation curve (in the case of a sufficiently high injection level) is close to the former exponential, but with increase of the recombination rate it moves toward the latter exponential. We note that this occurs smoothly without inflection points, since Eq. (25.3) shows that $d^2t/d\,(\Delta n)^2$ does not change sign at any value of τ_0/τ_∞ and $\Delta n/(n_0 + p_0)$.

The experimentally observed photoconductivity decay curves of germanium illuminated with square pulses [23] correspond qualitatively to the solution of Eq. (25.3). They are nonexponential if the injection level is sufficiently high, and lie between the exponentials corresponding to the extreme values of the lifetime (cf. the oscillogram shown in Fig. 76B).

Figure 77 gives the results of a quantitative check of Eq. (25.5). The continuous curve represents the theoretical dependence $\tau_{inst} = f(\Delta n)$ plotted according to Eq. (25.5). The points, which fit this curve quite well, were obtained experimentally for germanium.

B. High Trap Concentration

This case has been considered in § 23B for steady-state conditions. Obviously the excitation cannot greatly affect the electron and hole densities at the M levels (and consequently it cannot affect the rate of capture by these levels which governs the lifetime, as shown § 23B). Consequently the electron and hole lifetimes remain constant during the relaxation process and they are equal to the steady-state values given by Eqs. (23.53) and (23.54). Since, however, $\tau_n \neq \tau_p$, the relaxation curves in the case of excitation with, for example, square pulses are rather complex. If, for example, $\tau_n \ll \tau_p$ then, on the one hand, the hole component of the photoconductivity $\Delta\sigma_p = e\mu_p\,\Delta p = e\mu_p\beta kl\tau_p$ is greater than the electron component $\Delta\sigma_n = e\mu_n\,\Delta n = e\mu_n\beta kl\tau_n$* and on the other hand the relaxation of the electron component is governed by the shorter lifetime τ_n and therefore proceeds faster than the relaxation of the hole component. Consequently, the relaxation curves are kinked (Fig. 78). The very steep portions at the beginning of the rise curve and at the end of the decay curve represent the "fast" electron component of the photoconductivity, while the other parts of the curve represent the slower hole component.

---

* It is assumed that the mobilities μ_n and μ_p are not too different.

Chapter VI

PROCESSES OF NONEQUILIBRIUM CARRIER TRAPPING

§26. RECOMBINATION CENTERS AND TRAPPING CENTERS

When several types of trap are present in a semiconductor it becomes very difficult to describe the recombination processes quantitatively, particularly under non-steady-state conditions (relaxation curves). However, in some special cases we can use semiquantitative treatments based on a plausible classification of the traps according to the role which they play in recombination processes.

In considering above the statistics of recombination through traps, we allowed for two types of transition: "capture" of electrons and holes by traps and "thermal transitions" of captured carriers back to the bands. If a semiconductor has several types of trap then obviously the recombination will be affected most by those traps for which the rate of return thermal transitions is least. The capture by a trap of an electron from the conduction band leads to pair recombination only if this electron is then transferred to the valence band (capture of a hole by the trap). The return thermal transition eliminates this possibility and therefore reduces the rate of recombination.

The traps for which the probability of the return thermal transitions is very high, exchange carriers effectively with only one of the bands. *
Such traps (and the corresponding levels) are known as trapping centers (levels).

In contrast to the trapping centers, the traps from which the thermal transitions are not very likely are known as recombination centers (levels). Such traps can capture electrons and holes (thus giving rise to recombination) without the return thermal transitions.

It is not always possible to divide the traps in this way. However, when such a division is possible we can simplify considerably the qualita-

---

* Under steady-state conditions this corresponds to the case in which the rates of capture by the levels and the return thermal transitions back to the band are equal ("quasi-detailed equilibrium"). Under these conditions there is a single quasi-Fermi level for the band and the capture levels.

tive and quantitative analyses of the recombination phenomena in the presence of several types of trap.

If a trap captures a charge then we can easily determine the ratio k of the probability of the capture of a charge with opposite sign (i.e., the probability of recombination) to the probability of a thermal transition. If, for example, a level M captures an electron, then the probability of the return of this electron to the conduction band is equal to $\gamma_n N_c M$, according to Eq. (22.14), and the probability of recombination with a hole is $\gamma_p p$, and consequently

$$k_n = \frac{\gamma_p p}{\gamma_n N_{cM}} = \frac{\gamma_p p}{\gamma_n N_c e^{-\frac{\Delta\mathscr{E}_M}{kT}}}.$$ (26.1)

Similarly for a trap which has captured a hole, we have

$$k_p = \frac{\gamma_n n}{\gamma_p P_{vM}} = \frac{\gamma_n n}{\gamma_p P_v e^{-\frac{\Delta\mathscr{E} - \Delta\mathscr{E}_M}{kT}}}.$$ (26.2)

It follows that the traps for which the thermal transition probability is higher than the probability of capture of a carrier of opposite sign (i.e., k < 1) are the trapping centers. The traps for which k > 1 are the recombination centers. From Eqs. (26.1) and (26.2) it is clear that the classification of the traps is governed not only by the characteristics of the traps themselves (γ_n, γ_p, $\Delta\mathscr{E}_M$), but also by temperature and density of those carriers which are needed to achieve recombination. Consequently a change of the illumination intensity [and therefore of p or n in Eqs. (26.1) or (26.2)] or of temperature may transform traps from recombination centers into trapping levels or conversely.

A.   Demarcation Levels

At a given illumination intensity and temperature the value of k is governed by the distance of the trap levels from the conduction or valence band edge; it follows from Eqs. (26.1) and (26.2) that the traps lying near these edges have a higher probability of becoming trapping centers, while the traps in the middle of the forbidden band are more likely to act as recombination centers.

Sometimes, when a large variety of traps is present, the concept of "demarcation levels" [2], shown by continuous lines in Fig. 79, is used in discussion. The positions of these levels should be selected so that the traps which coincide with them are characterized by k = 1, i.e., equal probabilities of thermal transitions and of recombination. Under these conditions the traps lying between the upper demarcation level and the conduction band are trapping centers for electrons. The electrons cap-

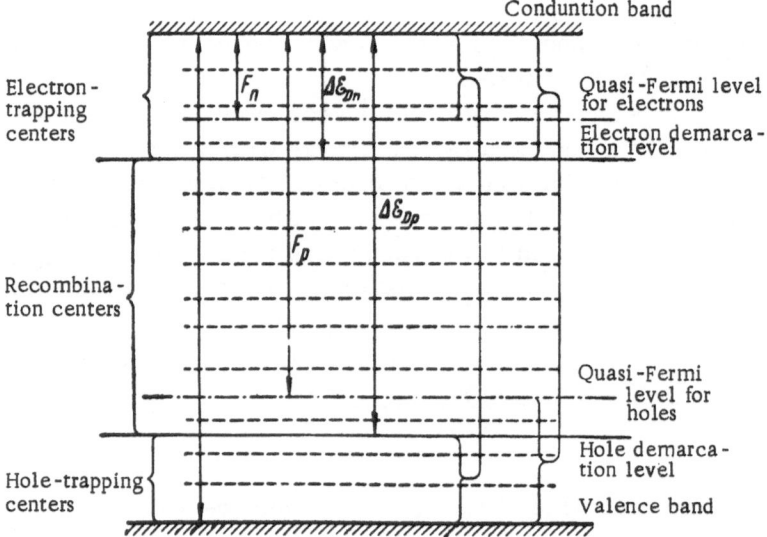

Fig. 79. Positions of the demarcation levels and quasi-Fermi levels in the
forbidden band of a semiconductor.

tured by these centers from the conduction band are practically always
returned by thermal transitions to that band. Trapping centers for holes
lie between the lower demarcation level and the valence band. The traps
between the two demarcation levels act as recombination centers and carriers captured by these centers suffer recombination.

The positions of the electron and hole demarcation levels ($-\Delta\mathscr{E}_{Dn}$
and $-\Delta\mathscr{E}_{Dp}$) may be found from the conditions

$$k_n = \frac{\gamma_p p}{\gamma_n N_c e^{-\frac{\Delta\mathscr{E}_{Dn}}{kT}}} = 1, \qquad (26.3)$$

$$k_p = \frac{\gamma_n n}{\gamma_p P_v e^{\frac{-\Delta\mathscr{E}+\Delta\mathscr{E}_{Dp}}{kT}}} = 1. \qquad (26.4)$$

Replacing p and n by the expressions in (3.1) and (3.2) which include
the Fermi quasi-levels, we obtain

$$-\Delta\mathscr{E}_{Dn} = -\Delta\mathscr{E} - F_p - kT \ln \frac{\gamma_n N_c}{\gamma_p P_v}, \qquad (26.5)$$

$$-\Delta\mathscr{E}_{Dp} = -\Delta\mathscr{E} - F_n + kT \ln \frac{\gamma_p P_v}{\gamma_n N_c}. \qquad (26.6)$$

If the logarithmic terms in Eqs. (26.5) and (26.6) are neglected in the
first approximation, it becomes clear that the hole demarcation level is
separated from the valence band by a distance equal to the separation of

the quasi-Fermi level for electrons from the conduction band. Similarly, the electron demarcation level is separated from the conduction band by an interval equal to the separation of the quasi-Fermi level for holes from the valence band.

This "cross-over" relationship between the demarcation levels and the quasi-Fermi levels is fully expected: the larger the number of holes in the valence band (and consequently the higher the quasi-Fermi level for holes in that band) the more likely the capture of these holes, which competes with thermal transitions of electrons. Consequently, with increase of p only the shallowest traps near the conduction band continue to act as electron trapping centers. Thus, a shift of the quasi-Fermi level for holes toward the valence band should displace the electron demarcation level toward the conduction band.

The demarcation levels divide the forbidden band into three regions: a region of electron trapping levels near the conduction band, a region of hole trapping levels near the valence band, and a central region in which the traps act as recombination centers both for electrons and holes.

Since the positions of the demarcation levels depend on the ratio of the capture cross sections [cf. Eqs. (26.5) and (26.6)], it follows that the demarcation levels have meaning only for those trap levels for which this ratio does not vary too greatly [weak variation is unimportant since the cross-section ratio enters Eqs. (26.5) and (26.6) as a logarithm]. Moreover, the trapping levels filled at equilibrium (for example, the electron trapping levels above the appropriate demarcation level but below the quasi-Fermi level) should obviously act as recombination centers, because they cannot capture electrons, while hole capture leads to recombination (hence, they are hole recombination levels); for details see § 30.

The above discussion shows clearly that the division of levels by demarcation levels is a very approximate qualitative method for analyzing the situation in special cases. The classification of traps as recombination centers or trapping levels does not determine uniquely their role in the process of recombination. If a semiconductor has only shallow traps acting as trapping levels with few or no recombination centers, then obviously the band electrons and holes must recombine through trapping levels. Such recombination is a slow process because a trapped charge returns many times back to the band before it finally manages to recombine with a charge of opposite sign. Nevertheless in this case the recombination occurs at the trapping levels. *

Thus, the nature of recombination processes and the role of various types of trap in these processes are both governed by the conditions in

* Essentially such recombination through trapping levels occurs in a certain range of temperatures (regions 2 and 3 in Fig. 69) in the case of one type of trap [cf. Eq. (23.28)].

the crystal as a whole. However, the division of traps into trapping levels and recombination centers is often very useful.

Let us now consider the role of trapping levels in recombination processes. Trapping levels affect the non-steady-state conditions very strongly by altering the nature of the nonequilibrium conductivity relaxation. However, even under steady-state conditions the presence of trapping levels has a strong effect on the photoconductivity. The problem of the influence of trapping levels on the steady-state characteristics and the nonequilibrium conductivity relaxation is discussed below for the cases of unipolar [Chap. VI(i)] and ambipolar [Chap. VI(ii)] intrinsic photoconductivity. These discussions require certain definite models. For this purpose we use in Chap. VI(i) a model with "high concentration of recombination centers" (cf. §23B), which explains in a natural way the unipolar nature of photoconductivity due to excitation with fundamental region wavelengths, and in Chap. VI(ii) a model with "low concentration of recombination centers" (cf. §23A), which is the most suitable for ambipolar photoconductivity.

(i) Unipolar Photoconductivity

§27. INFLUENCE OF TRAPPING LEVELS ON STEADY-STATE CHARACTERISTICS AND ON THE RELAXATION OF NONEQUILIBRIUM CONDUCTIVITY (LOW DEGREE OF POPULATION OF THE TRAPPING LEVELS – LINEAR CASE)

The problem of the influence of trapping will be considered for the simple but important case of a high excitation level ($n_0 \ll \Delta n \approx n$, $p_0 \ll \Delta p \approx p$).

For this purpose we shall consider the system shown in Fig. 80.

Let us assume that the electrons and holes liberated by illumination recombine at recombination centers S. We shall also assume that the concentration of these centers is very high, so that the illumination has practically no effect on their population, i.e., the concentrations of the filled s and empty (S − s) centers are constant. *

Under these conditions the electron and hole lifetimes are different but constant:

$$\tau_n = \frac{1}{\gamma_{ns}(S-s)} = \text{const,} \qquad (27.1)$$

---

* This case corresponds to a very high trap concentration (§23B).

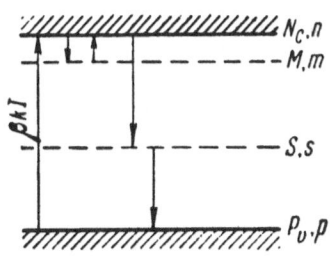

Fig. 80. System of transitions in the presence of one type of recombination center (S) and one type of electron trapping center (M).

$$\tau_p = \frac{1}{\gamma_{ps}\, s} = \text{const.} \qquad (27.2)$$

Let us assume that $\tau_n \gg \tau_p$. This means that holes are captured very rapidly by the S centers and the nonequilibrium conductivity is governed by carriers of one sign (in our case electrons), i.e., we are dealing with the "unipolar" case.

We shall consider below the influence of trapping on the relaxation of the nonequilibrium electron density.

Consider the case where, apart from the S levels which act as recombination centers (we can neglect thermal transitions from these centers, Fig. 80), the semiconductor also has unfilled levels M near the conduction band. In this model (which represents conditions frequently encountered in poorly conducting semiconductors such as CdS, etc.) the M levels act as trapping levels.

These M levels cannot capture holes directly since the holes are localized at the numerous S levels separated spatially from the M levels. Consequently the electrons captured by such levels from the conduction band can only return to the same band (by thermal excitation). Thus the M levels then act as absolute ($k \approx 0$) trapping levels.

We shall first assume that under all conditions the electron population of the trapping levels is small. * Then the change of the electron density in the conduction band per unit time may be written in the form

$$\frac{dn}{dt} = \beta\, kI - \frac{n}{\tau_n} - \gamma_n\, nM + \gamma_n\, m N_{cM}; \qquad (27.3)$$

here τ_n is constant in agreement with Eq. (27.1).

Moreover, bearing in mind that the total number of electrons in the conduction band and at the M levels decreases only by capture of the electrons from the band by the recombination centers, we can write

$$\frac{d\,(n+m)}{dt} = \beta\, kI - \frac{n}{\tau_n}. \qquad (27.4)$$

It follows from Eq. (27.4) that in the steady-state case

$$n_{\text{st}} = \beta k\, I \tau_n, \qquad (27.5)$$

*M $\gg$ m, linear case. A treatment without this restriction is given in Chap. VI(iii).

i.e., the density is the same as in the absence of trapping levels. It follows that in the case considered the trapping levels are not active under steady-state conditions.

The problem of non-steady-state conditions will be discussed for the following two extreme cases:

1. The case where the establishment of equilibrium between the trapping levels and the conduction band requires an effective time θ which is much shorter than the lifetime τ_n: $\tau_n \gg \theta$; the traps corresponding to these levels will be called multiple trapping levels (centers) or α-type levels;

2. The opposite case: $\tau_n \ll \theta$; these traps will be called single trapping levels or β-type levels.

A. Relaxation in the Presence of α-Type Trapping Centers

We shall find the time dependence of the electron density n in the conduction band, if before illumination n = m = 0 and the illumination begins at t = 0. *

The total relaxation period may be divided into two parts:

a. "Short" times $t \ll \tau_n$, for which we can neglect the term n/τ_n in Eqs. (27.3) and (27.4) compared with the terms $\gamma_n nM$ and βkI;

b. "Long" times $t \gg \theta$, during which thermal equilibrium is established between the conduction band and the M levels, so that subsequent relaxation proceeds under equilibrium conditions. †

For the first part of the relaxation period, we have, instead of Eqs. (27.3) and (27.4)

$$\frac{dn}{dt} = \beta\,kI - \gamma_n\,nM + \gamma_n\,mN_{cM}\,, \qquad (27.6)$$

$$\frac{d\,(m+n)}{dt} = \beta\,kI. \qquad (27.7)$$

Using the initial conditions n = m = 0 when t = 0, we find from Eq. (27.7)

$$m + n = \beta k\,It. \qquad (27.8)$$

---

* Here we are analyzing the relaxation after excitation with square light pulses.

† The two parts of the relaxation period overlap in the region $\theta < t < \tau_n$.

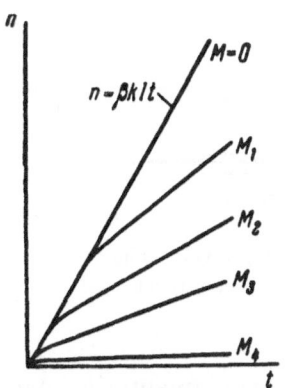

Fig. 81. Initial stage of relaxation in the presence of α-type
trapping levels, $M_1 < M_2 < M_3 < M_4$.

Substituting m from Eq. (27.8) into (27.6)
we obtain

$$\frac{dn}{dt} = k\beta I - \gamma_n n (M + N_{cM}) + \beta k I \gamma_n N_{cM} t. \quad (27.9)$$

The solution of the above equation has the
form [10, 14]

$$n = \beta k I \left[\gamma_n M \theta^2 \left(1 - e^{-\frac{t}{\theta}}\right) + \gamma_n N_{cM} \theta t \right], \quad (27.10)$$

where

$$\theta = \frac{1}{\gamma_n (M + N_{cM})}. \quad (27.10a)$$

The above solution is the sum of an exponential term and a linear term. Figure
81 shows a series of such dependences for various values of M. When
$t \ll \theta$ we find that the expansion of Eq. (27.10) as a series gives, in the
first approximation,

$$n = \beta k I t. \quad (27.11)$$

Consequently the initial parts of the curves given in Fig. 81 are not
affected by recombination or trapping. We can determine the true quantum yield β from these parts of the curves.

A kink is found in the curves at $t \approx \theta$. This is because during the
time θ equilibrium is established between the conduction band and the M
levels, so that subsequent relaxation of n occurs under approximately
equilibrium conditions. From Eq. (27.10) we obtain for $t \gg \theta$ [1, 10,14]*

$$n = \beta k I t \frac{N_{cM}}{N_{cM} + M}. \quad (27.12)$$

This means that the slope of the straight line is now less than the
slope given by Eq. (27.11) and the reduction of the slope is in the ratio
$N_{cM}/(N_{cM} + M)$. This is because only a fraction of the electrons excited to the upper band are in the conduction band (the fraction is given
by the ratio just quoted) and the other electrons are at the M levels. The
lower the value of M the closer the slope to the initial value (i.e., the
value from which we can find the true quantum yield β).

---

* Here we are neglecting a small initial jump of n during a time θ.

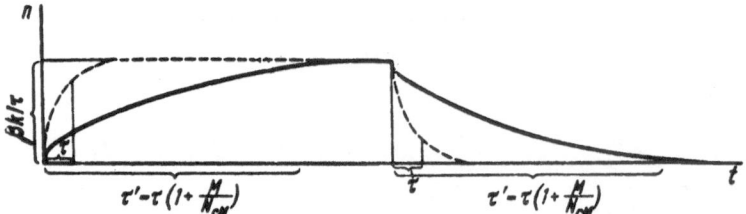

Fig. 82. Relaxation curves in the presence of α-type trapping levels. The dashed curves represent relaxation under the same conditions but without trapping levels.

Sometimes the initial slope cannot be observed experimentally. We can then determine the "phenomenological quantum yield" β', which differs from the true yield by the factor $N_{cM}/(N_{cM} + M)$:

$$\beta' = \frac{\beta}{1 + M/N_{cM}}. \tag{27.13}$$

Let us now consider the solution of Eqs. (27.3) and (27.4) when $t > \theta$. We can assume that the ratio of n to m now corresponds to thermal equilibrium, i.e., in the absence of degeneracy it is given by

$$\frac{n}{m} = \frac{N_c}{M} e^{-\frac{\Delta \mathcal{S}_M}{kT}} = \frac{N_{cM}}{M}, \tag{27.14}$$

and in place of Eq. (27.4) we can write

$$\frac{d}{dt}\left(n + n\frac{M}{N_{cM}}\right) = \beta k I - \frac{n}{\tau_n}, \tag{27.15}$$

or

$$\frac{dn}{dt} = \frac{N_{cM}}{N_{cM} + M}\, \beta\, k I - \frac{n}{\tau_n}\, \frac{N_{cM}}{M + N_{cM}}. \tag{27.16}$$

Assuming that n = 0 when t = 0, we find [10]

$$n = \beta k I \tau_n \left(1 - e^{-\frac{t}{\tau'}}\right), \tag{27.17}$$

where

$$\tau' = \tau_n \frac{M + N_{cM}}{N_{cM}} = \tau_n \left(1 + \frac{M}{N_{cM}}\right) = \tau_n \left(1 + \frac{M}{N_c} e^{\frac{\Delta \mathcal{S}_M}{kT}}\right). \tag{27.18}$$

The expression (27.17), which describes the relaxation of n for $t > \theta$, is in the form of an ascending exponential. However, in the presence of α-type trapping centers the time constant of this exponential τ' may be considerably greater than the lifetime τ_n [cf. Eq. (27.18)]; then the deeper the positions of the α centers, the higher their concentration, and the

Fig. 83. Oscillogram of a photoconductivity relaxation curve for CdS. The initial jump in the rise curve is clearly visible [10,14].

lower the temperature, the greater the difference between the time constant and the lifetime.

We can easily show that both the decay curve after illumination and the rise curve have initial fast regions, which last a time θ, followed by slow regions represented by the expression

$$n = \beta k I \tau_n e^{-\frac{t}{\tau'}}. \qquad (27.19)$$

Figure 82 shows schematically the rise and decay curves for nonequilibrium conductivity in the presence of α-centers.

Figure 83 shows an oscillogram of the photoconductivity relaxation of CdS, which is in agreement with Fig. 81.

B. Relaxation in the Presence of β-Type Trapping Centers

We shall now consider the other extreme case, in which the establishment of equilibrium between the trapping levels and the conduction band requires a time much greater than the lifetime, i.e., [cf. Eq. (27.10a)]

$$\theta = \frac{1}{\gamma_n M + \gamma_n N_{cM}} \gg \tau_n. \qquad (27.20)^*$$

We shall use the same equations (27.3) and (27.4) and the same initial conditions (n = m = 0 when t = 0). Again, we can divide the total relaxation period into two parts.

Since τ_n is small we can assume that, within the short time τ_n following the illumination, an almost steady-state value of the density n is established in the conduction band, and then the second (slow) process takes place: electrons are captured by the β-type trapping centers.

The influence of the trapping process on the density n is greatest at the beginning of the process (when m is still small and the return thermal transitions are practically absent), and then the influence decreases with time, tending to zero when the rate of thermal transitions becomes comparable with the capture rate.

---

* Obviously the condition (27.20) requires that simultaneously

$$\left. \begin{array}{l} \dfrac{1}{\gamma_n M} \gg \tau, \\[2mm] \dfrac{1}{\gamma_n N_{cM}} \gg \tau. \end{array} \right\} \qquad (27.21)$$

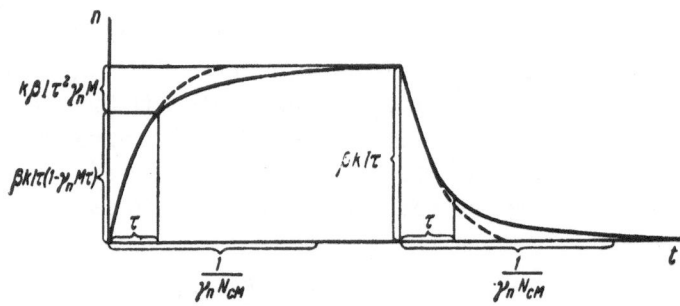

Fig. 84. Relaxation curves in the presence of β -type trapping levels. The dashed curves represent relaxation under the same conditions but without trapping levels.

In the initial relaxation stage we shall neglect the term $\gamma_n n M_{cM}$ in Eq. (27.3); we then find

$$\frac{dn}{dt} = \beta kI - n\left(\frac{1}{\tau_n} - \gamma_n M\right),\tag{27.22}$$

and hence we obtain for n

$$n = \beta kI\tau''\left(1 - e^{-\frac{t}{\tau''}}\right),\tag{27.23}$$

where

$$\tau'' = \frac{\tau_n}{1 + \gamma_n M\tau_n} \simeq \tau_n.\tag{27.24}$$

After a time $\tau'' \approx \tau_n$ the density n becomes

$$n = \beta kI\tau'' \simeq \beta kI\tau_n,\tag{27.25}$$

i.e., it almost reaches the steady-state value.

Next (in the second stage of relaxation) the process of trapping begins under conditions of almost constant carrier density $n \approx n_{st} = \beta kI\tau_n$; under these conditions the electrons at β centers obey the expression

$$\frac{dm}{dt} = \gamma_n M n_{st} - \gamma_n m N_{cM},\tag{27.26}$$

and hence

$$m = \frac{M}{N_{cM}} n_{st} \left(1 - e^{-t\gamma_n N_{cM}}\right).\tag{27.27}$$

Substituting this expression for m into Eq. (27.4) we find, after some transformations, that in the second stage

$$n = \beta kI\tau\left(1 - \gamma_n M\tau e^{-t\gamma_n N_{cM}}\right).\tag{27.28}$$

Calculations show that the decay curves are symmetrical with the rise curves.

Figure 84 shows the rise and decay curves for the electron density in the presence of β-type trapping levels.

Comparing these curves with those for α centers (Fig. 82), we notice their apparent similarity: in both cases the initial fast regions are followed by slow ones.

However, in the case of α centers the initial fast relaxation is governed by the time θ which is necessary for the establishment of equilibrium with the trapping levels, while in the case of β centers this initial relaxation occurs in a time equal to the lifetime τ.

C.   Criterion for Distinguishing α and β Trapping

In the unipolar case we can distinguish α and β trapping experimentally using, for example, the value of the ratio of the amplitude of the slow component Δn_s to the total steady-state nonequilibrium density Δn_{st}.

For α trapping

$$\Delta n_{st} = \beta k I \tau_n,$$

$$\Delta n_s = \Delta n_{st} - \Delta n_f.$$

Here Δn_f is the fast (low-inertia) component of the nonequilibrium density ("the initial jump"). From Eq. (27.10) we find

$$\Delta n_f = \beta k I \theta^2 \gamma_n M = \beta k I \theta \frac{\theta}{1/\gamma_n M}, \qquad (27.29)$$

and hence

$$\Delta n_s = \beta k I \left(\tau_n - \theta \frac{\theta}{1/\gamma_n M} \right) \approx \beta k I \tau_n. \qquad (27.30)$$

The second term in the parentheses is much smaller than the first because $\tau_n \gg \theta$ for α-trapping conditions.

Moreover,

$$\theta = \frac{1}{\gamma_n M + \gamma_n N_{cM}} \leqslant \frac{1}{\gamma_n M}.$$

Thus, in the α-trapping case $\Delta n_s / \Delta n_{st} \approx 1$, i.e., the initial fast jump is only a small fraction of the total change in the density (cf. Fig. 82).

For β trapping

$$\Delta n_s = \beta k I \tau_n \frac{\tau_n}{1/\gamma_n M}, \qquad (27.31)$$

$$\Delta n_{st} = \beta k / \tau_n,$$ (27.32)

$$\frac{\Delta n_s}{\Delta n_{st}} = \frac{\tau_n}{1/\gamma_n M}.$$ (27.33)

Here we have

$$\tau_n \ll \theta = \frac{1}{\gamma_n N_{cM} + \gamma_n M}.$$ (27.34)

Consequently, in the case of β trapping the condition $\tau_n \ll 1/\gamma_n M$ applies even more strongly. Thus, we find that $\Delta n_s / \Delta n_{st} \ll 1$, i.e., in the β-trapping case the slow component of the relaxation represents only a small fraction of the total density change* (cf. Fig. 84).

§28. INFLUENCE OF α TRAPPING ON THE PHENOMENOLOGICAL YIELD AND LIFETIME IN THE GENERAL CASE OF ARBITRARY POPULATION OF THE TRAPPING LEVELS

An approximate analysis shows that if some of the M levels are occupied by electrons then this is apparently equivalent to a reduction of the concentration of the trapping levels M capable of capturing electrons from the conduction band.

We have considered above the case of low population of the levels M ($m \ll M$) and we found for α trapping the following relationships between the true values of τ and β and the measured values τ' and β' (found from the second slope)

$$\beta' = \beta \frac{1}{1 + M/N_{cM}},$$ (28.1)

$$\tau' = \tau \left(1 + \frac{M}{N_{cM}} \right).$$ (28.2)

Similar relationships, which will be very important in later treatment, can easily be obtained for the general case of any population of the M levels.

Assume that we are dealing with α trapping, and that consequently at any instant (with the exception of the short initial time interval θ) the M levels and the conduction band are in quasi-equilibrium, i.e., the rate of transitions from the band to the levels is equal to the rate of return transitions back to the band:

---

*Other characteristics of β trapping are its constancy with time and the lack of influence of additional illumination on the instantaneous lifetime in the decay "tail" of the relaxation curve (Fig. 84).

$$\gamma n (M - m) \cong \gamma m N_{cM}. \tag{28.3}$$

Hence

$$m \cong \frac{nM}{n + N_{cM}}. \tag{28.4}$$

For the rise curve we can write

$$\frac{d}{dt}(n + m) = \beta k I - \frac{n}{\tau} \tag{28.5}$$

or, replacing m by its value from Eq. (28.4)

$$\frac{d}{dt}\left(n + \frac{nM}{n + N_{cM}}\right) = \beta k I - \frac{n}{\tau}. \tag{28.6}$$

After transforming the left-hand side we obtain:

$$\left[1 + \frac{MN_{cM}}{(n + N_{cM})^2}\right]\frac{dn}{dt} = \beta k I - \frac{n}{\tau}. \tag{28.7}$$

The expression (28.7) may be rewritten in the form

$$\frac{dn}{dt} = \beta' k I - \frac{n}{\tau'}, \tag{28.8}$$

where

$$\beta' = \frac{\beta}{1 + \dfrac{MN_{cM}}{(n + N_{cM})^2}} = \frac{\beta}{1 + \alpha}, \tag{28.9}$$

$$\tau' = \tau\left[1 + \frac{MN_{cM}}{(n + N_{cM})^2}\right] = \tau(1 + \alpha). \tag{28.10}$$

From Eq. (28.4) we have

$$\frac{MN_{cM}}{N_{cM} + n} = M - m, \tag{28.11}$$

and therefore the quantity α, which will be called the trapping coefficient, can be written in the form

$$\alpha = \frac{MN_{cM}}{(n + N_{cM})^2} = \frac{M - m}{N_{cM} + n}. \tag{28.12}$$

Trapping is important when $\alpha > 1$, i.e., when M is large and N_{cM} is small. An increase of the population of the levels, i.e., an increase of m [and consequently an increase of n, because n and m are related by the expression (28.4)], reduces the coefficient α and the hence role of trapping.

When $m \ll M$ (and consequently also when $n \ll N_{cM}$) we have $\alpha = M/N_{cM}$ and Eqs. (28.9) and (28.10) transform into Eqs. (28.1) and (28.2).

The expresions (28.8)-(28.12) were obtained for the case $n_0 = m_0 = 0$ (an "insulator"). It can be shown that in general when the equilibrium electron population (or the population produced by constant additional illumination) of the conduction band and M levels is not equal to zero, we obtain similar relationships.

The transport equation analogous to (28.8) is in its simplest form given by

$$\frac{d}{dt}\Delta n = \beta' k I - \frac{\Delta n}{\tau'}, \tag{28.13}$$

where $\beta' = \beta/(1 + \alpha)$, $\tau' = \tau(1 + \alpha)$, and the quantity α is as before by Eq. (28.12).

Equations (28.7), (28.8), and (28.13) are nonlinear [we shall return to their solution in Chap. VI(iii)] because the quantities β' and τ' themselves depend on n or Δn and consequently they are not constant during the whole relaxation process. Therefore the quantities β' and τ' represent the "instantaneous yield" and the "instantaneous lifetime."

The general expression (28.12) for the trapping coefficient will be used frequently in our later treatment. The principal assumption in this calculation was the establishment of dynamic equilibrium (quasi-equilibrium) between the trapping levels and the conduction band [cf. Eq. (28.3)]. Under steady-state conditions this assumption is valid for any trapping levels (including β levels). Consequently the coefficient α under steady-state conditions is an important characteristic of trapping in general (§30), in addition to representing the entire relaxation in the case of α processes.

§29. INFLUENCE OF STEADY ILLUMINATION ON THE RELAXATION OF PHOTOCONDUCTIVITY IN THE PRESENCE OF α TRAPPING

It can be seen from Eq. (28.12) that an increase of the carrier density n reduces α and consequently reduces the role of trapping. The density n is most conveniently altered by means of steady [constant-intensity] illumination. This method is widely used in order to reduce or eliminate completely the influence of trapping on the photoconductivity.

Using strong steady illumination on top of which weak excitation pulses are superimposed, we can make β' and τ' constant during the relaxation processes, the values of these quantities being determined by the intensity of the steady illumination.

The relaxation curves then become exponential, apart from the initial regions, and we can easily determine τ' and β' from them.

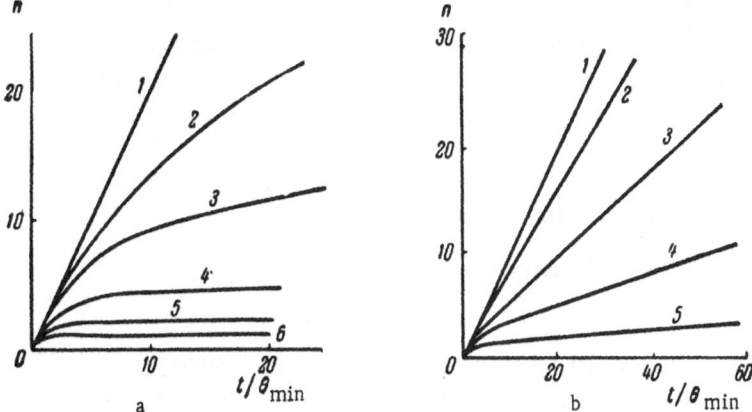

Fig. 85. Calculated initial stages of the relaxation curves [14]: a) $M \gg N_{cM}$;
$(M - m_0)$ was: 1) 0; 2) $0.5N_{cM}$; 3) N_{cM}; 4) $2.5N_{cM}$; 5) $5N_{cM}$; 6) $10N_{cM}$.   b)
$M > N_{cM}$; $(M - m_0)$ was: 1) 0; 2) N_{cM}; 3) $2.5N_{cM}$; 4) $5N_{cM}$; 5) $10N_{cM}$.

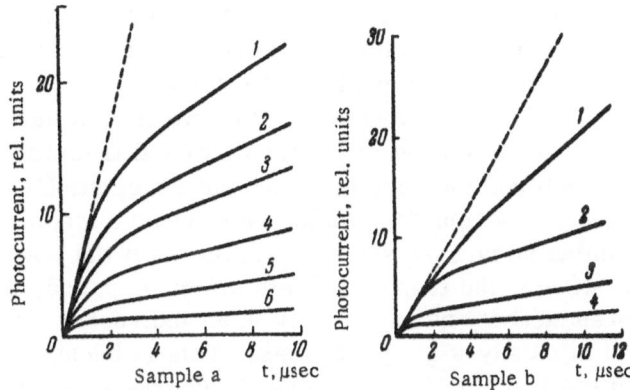

Fig. 86. Initial stages (the first 10^{-5} sec) of the photoconductivity
relaxation for two typical samples of CdS for various intensities
of steady illumination [14]. The steady-illumination intensity is
represented by the electron density n_{ill} produced by this illumina-
tion. Sample a: n_{ill} (cm^{-3}): 1) 2.59×10^{11}; 2) 8.2×10^{10}; 3)
5.97×10^{10}; 4) 2.49×10^{10}; 5) 8.7×10^{9}; 6) 5×10^{9}. Sample b:
n_{ill} (cm^{-3}): 1) 2.5×10^{12}; 2) 7.8×10^{11}; 3) 3.25×10^{11}; 4) 1.3×10^{11}.

Figure 85 shows clearly how the initial stages of the relaxation curves
approximate more closely to the straight line 1, representing the absence
of trapping, as the steady-illumination intensity is increased, keeping M
constant (this reduces $M - m_0$).

Steady illumination is very frequently found to have this type of influence in experimental work [19]. The curves in Fig. 86 show that steady illumination does not affect greatly the slope of the initial "first" stage of the rise, but the slope of the "second" stage, which determines β' $= \beta(1 + \alpha)$, increases on increase of the steady-illumination intensity (because α decreases), and approaches a limit which is the slope in the first stage giving the true yield β. The experimental curves in Fig. 86 are similar to the calculated ones in Fig. 85.

(ii) Ambipolar Photoconductivity

§30. INFLUENCE OF TRAPPING LEVELS ON STEADY-STATE PHOTOCONDUCTIVITY AND STEADY-STATE LIFETIMES OF ELECTRONS AND HOLES

When the concentration of recombination centers S is high (§27) the trapping levels are not active under steady-state conditions [cf., for example, Eq. (27.5)]: they affect only the relaxation processes. At first sight it would seem that in other cases too the trapping levels should not affect the steady-state electron and hole lifetimes and consequently they should not have any effect on the photoconductivity.

Since the trapping levels give rise to only two transitions (carrier capture from the conduction band and return thermal transitions back to the band, cf. Fig. 80) and they are balanced out under steady-state conditions: $\gamma m N_c M = \gamma (M - m)n$,* it would seem that the trapping process is effectively eliminated and the system behaves as if the trapping levels were absent. However, the trapping levels affect the steady-state lifetimes and photoconductivity quite strongly in an indirect way without being directly involved in steady-state processes. This is because the accumulation of charge at the trapping levels alters the charge (i.e., the electron and hole densities) at the recombination levels by reason of the neutrality condition, and this affects the steady-state values of τ_n, τ_p, and $\Delta \sigma$.†

We shall first determine qualitatively how τ_n and τ_p change in the presence of the trapping levels and for this purpose we shall consider the simple case of high excitation level and low concentration of recombination centers (Fig. 87).

---

* Consequently in the steady-state transport equations we can cancel both terms which represent transitions involving the M levels.

† Such an effect was absent in the system shown in Fig. 80 because we assumed there that the illumination cannot greatly alter the electron (hole) population of the S level, since the densities of the occupied and free levels are very great.

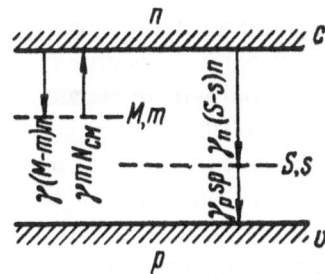

Fig. 87.   System of transitions.

Under steady-state conditions the rate of electron capture $\gamma_n p_S n$ by the recombination centers S is equal to the rate of hole capture $\gamma_p n_S p$ from the valence band:

$$\gamma_n p_S n = \gamma_p n_S p. \qquad (30.1)$$

Here n_S and p_S are, respectively, the electron and hole densities at the S centers; n and p are the densities of electrons and holes in the bands; γ_n and γ_p are the capture coefficients.

It follows that the electron lifetime in the conduction band is given by*

$$\tau_n = \frac{1}{\gamma_n p_S}, \qquad (30.2)$$

and the hole lifetime in the valence band is

$$\tau_p = \frac{1}{\gamma_p n_S}. \qquad (30.3)$$

Moreover, n_S and p_S are related by

$$n_S + p_S = S, \qquad (30.4)$$

where S is the concentration of the recombination centers. The last expression indicates that any reduction of the electron density at the S levels causes an increase of the hole density at these levels or, as shown by Eqs. (30.2) and (30.3), a reduction of n_S causes a reduction of τ_n and an increase of τ_p. (A reduction of p_S obviously produces the opposite result.) However, under steady-state conditions the electron and hole densities at the S centers are closely related to the electron and hole densities in the bands. Thus, if the density of electrons in the conduction band is reduced by some means, then the flow of these electrons $\gamma_n p_S n$ to the S levels is reduced and consequently the density of holes p_S increases and the density of electrons n_S decreases. Therefore we have a reduction of $\tau_n = 1/\gamma_n p_S$ and an increase of $\tau_p = 1/\gamma_p n_S$.

It is easy now to visualize how the trapping levels affect τ_n and τ_p by considering the following hypothetical experiment. Assume that in an illuminated semiconductor, which has only recombination centers and is under steady-state conditions, we suddenly introduce a certain number

---

*We can neglect thermal transitions from the S levels to the band at high excitation levels.

of trapping levels. * This will disturb the steady state and the following transient process will occur: the trapping levels will begin to capture electrons, the density of electrons in the conduction band will decrease, their flow to the recombination centers will become smaller and consequently τ_n will decrease and τ_p will increase. A similar analysis may also be carried out for the capture of holes from the valence band by the trapping levels.

The following general conclusions may be drawn about the nature of the influence of the trapping levels on the lifetime of the nonequilibrium electrons and holes in the steady-state cases: the capture of carriers by the trapping levels reduces the lifetime of carriers that are captured, but it increases the lifetime of the carriers which are not being trapped.

Let us now consider the problem in a quantitative manner [16].

When the excitation level is high, the equilibrium carrier densities n_0, p_0, and (we will assume) m_0 are negligibly small compared with the corresponding nonequilibrium densities Δn, Δp, and Δm, and therefore $\Delta n \approx n$, $\Delta p \approx p$, and $\Delta m \approx m$. Then the system of differential equations which describe the electron transitions in a system with one type of recombination level S and one type of electron trapping level M in the forbidden band, may be written as follows:

$$
\left.
\begin{aligned}
\frac{dn}{dt} &= \beta kI - \gamma n \,(M - m) + \gamma m N_{cM} - \gamma_n n \,(S - s), &\text{(a)} \\[6pt]
\frac{dm}{dt} &= \gamma n \,(M - m) - \gamma m N_{cM}, &\text{(b)} \\[6pt]
\frac{ds}{dt} &= \gamma_n n \,(S - s) - \gamma_p p s, &\text{(c)} \\[6pt]
\frac{dp}{dt} &= \beta kI - \gamma_p p s. &\text{(d)}
\end{aligned}
\right\} \qquad (30.5)
$$

We shall assume that the concentration of recombination centers S is low. In the absence of trapping levels (M = 0) the neutrality condition gives

$$n = p, \quad \text{i.e.,} \quad \tau_n = \tau_p = \tau_0. \qquad (30.6)$$

We shall now consider the effect of introducing the trapping levels M on the lifetimes τ_n and τ_p. The neutrality condition becomes

$$n + m = p,$$

* We are assuming that acceptor levels (free of electrons) are formed. Otherwise we would have the trivial case of impurity compensation.

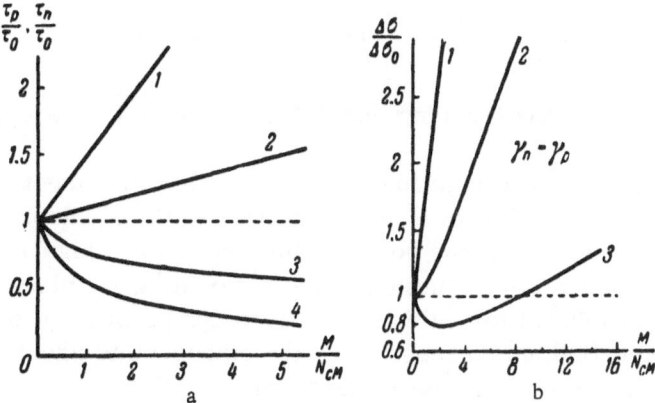

Fig. 88. a): 1) τ_p when $\gamma_n = \gamma_p$; 2) τ_p when $\gamma'_p = 0.1$; 3) τ_n when $\gamma_n = \gamma_p$; 4) τ_n when $\gamma'_p = 0.1$; b): 1) $\mu'_p = 0.9$; 2) $\mu'_p = \mu'_n = 0.5$; 3) $\mu'_p = 0.1$; the values of γ' and μ' are given by Eq. (30.15).

and consequently either n < p or

$$\tau_n < \tau_p \qquad (30.7)$$

[cf. Eqs. (30.7) and (30.3)].

Under steady-state conditions we find from Eq. (30.5b)

$$\frac{m}{n} = \frac{M}{N_{cM} + n}. \qquad (30.8)$$

Restricting ourselves to the case of low population of the trapping levels (M ≫ m), we have

$$\frac{m}{n} = \frac{M}{N_{cM}}. \qquad (30.9)$$

Substituting into Eq. (30.7) the value of m from Eq. (30.9), we find

$$\frac{n}{p} = \frac{\tau_n}{\tau_p} = \frac{N_{cM}}{M + N_{cM}}. \qquad (30.10)$$

The values of τ_n and τ_p in the presence of the M levels may be written as follows:

$$\tau_n = \tau_0 \frac{1 + \dfrac{\gamma_p}{\gamma_n + \gamma_p} \dfrac{M}{N_{cM}}}{1 + \dfrac{M}{N_{cM}}} = \tau_0 \frac{1 + \dfrac{\gamma_p}{\gamma_n + \gamma_p} \alpha}{1 + \alpha}, \qquad (30.11)$$

$$\tau_p = \tau_0 \left(1 + \frac{\gamma_p}{\gamma_n + \gamma_p}\frac{M}{N_{cM}}\right) = \tau_0 \left(1 + \frac{\gamma_p}{\gamma_n + \gamma_p}\alpha\right), \qquad (30.12)$$

where $\alpha = M/N_{cM}$ is the trapping coefficient given by the general expression (28.12).

Considering the above relationships we see that the direct result of introducing the trapping levels M is a reduction of the electron lifetime ($\tau_n < \tau_0$) and an increase of the hole lifetime ($\tau_p > \tau_0$). Figure 88a shows the values of τ_n and τ_p as a function of the ratio M/N_{cM}, plotted from Eqs. (30.11) and (30.12) for various relationships between γ_n and γ_p.

We shall now find an expression for the steady-state photoconductivity. Its value $\Delta\sigma$ can be written in the form

$$\Delta\sigma = e\beta kl \, (\mu_n \tau_n + \mu_p \tau_p). \qquad (30.13)$$

Substituting into Eq. (30.13) the values of τ_n and τ_p from Eqs. (30.12) and (30.11), we find

$$\Delta\sigma = \Delta\sigma_0 \left(1 + \gamma'_p \frac{M}{N_{cM}}\right)\left(\frac{\mu'_n}{1 + \dfrac{M}{N_{cM}}} + \mu'_p\right) \qquad (30.14)\,*$$

(here $\Delta\sigma_0$ is the steady-state photoconductivity in the absence of the trapping levels),

$$\gamma'_p = \frac{\gamma_p}{\gamma_n + \gamma_p}; \qquad \mu'_n = \frac{\mu_n}{\mu_n + \mu_p}; \qquad \mu'_p = \frac{\mu_p}{\mu_n + \mu_p}. \qquad (30.15)$$

Figure 88 (a and b) shows that the presence of the trapping levels may very strongly affect the steady-state values of τ_n, τ_p, and $\Delta\sigma$.

It should be pointed out that, in contrast to recombination centers, which affect the nonequilibrium processes when present even in very low concentrations,† the trapping levels influence the steady-state conditions only when their concentration M, or more exactly the charge accumulated at the trapping levels $m = n \dfrac{M/N_{cM}}{1 + n/N_{cM}}$ [cf. Eq. (30.8)], is so large that it

---

* At low temperatures the expressions (30.11), (30.12), and (30.14) are no longer valid ($\tau_p \to \infty$). To obtain expressions valid for these temperatures we should allow for the "nonlinearity" of the trapping level population in the derivation of the formulas for τ_n and τ_p.

† We mean concentrations such that the rate of recombination through traps is greater than the rate of direct recombination between the bands. In semiconductors with a wide forbidden band this corresponds to negligibly small trap concentrations that need not be allowed for in the neutrality condition.

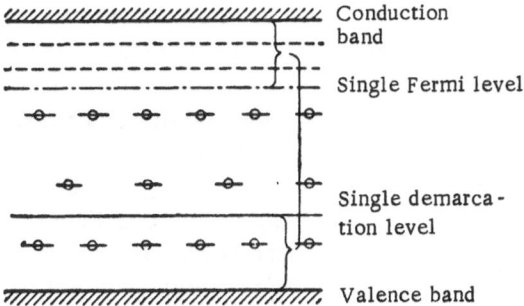

Fig. 89. Positions of the demarcation and Fermi levels
in the case of weak excitation.

it must be allowed for in the neutrality condition (30.7). (In the special case of low population of the M levels this corresponds to the condition $M/N_{cM} \gtrsim 1$.)

We shall now consider the influence of the trapping levels at a low excitation level and at a low concentration of the recombination levels. However, we must first consider more carefully the problem of dividing the traps into trapping levels and recombination levels in the case of weak excitation. For simplicity we shall consider the case of very weak excitation when $\Delta n = \Delta p \begin{cases} \ll n_0 \\ \ll p_0 \end{cases}$ and all the densities are governed by the equilibrium conditions and a single Fermi level F. Consequently if we neglect the difference between the capture cross sections for electrons and holes (γ_n, γ_p) and between the effective densities of states in the bands (N_c, P_v), then it follows from Eqs. (26.5) and (26.6) that $\Delta\mathcal{E}_{D_n} = \Delta\mathcal{E}_{D_p}$, i.e., the forbidden band of the semiconductor has only one demarcation level (Fig. 89)

$$-\Delta\mathcal{E}_D = -\Delta\mathcal{E} - F.$$

All the levels above this demarcation level act simultaneously as trapping levels for the conduction-band electrons and as recombination levels for the valence-band holes. (The opposite is true of the levels below $\Delta\mathcal{E}_D$.) This means that an electron captured by such a level has a higher probability of being thermally ejected to the conduction band (trapping process) than of dropping to the valence band, while a captured hole is more likely to drop into the conduction band (recombination process) than to jump thermally to the valence band.

Since each trap acts simultaneously as a trapping level (for carriers of one sign) and a recombination level (for carriers of the other sign), its actual role is governed by the carriers which are more likely to be captured by the trap under given conditions (i.e., for a given position of the Fermi level).

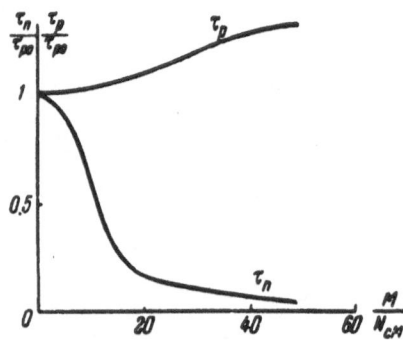

Fig. 90.   Fermi level above the trapping levels: $n_0 \gg p_0$; $N_d = 10N_{cM} = 100N_{cS}$.

Let us consider the traps between the conduction band and the Fermi level (Fig. 89). These traps are almost empty and therefore they readily capture electrons, for which they act as trapping levels, because they lie above the demarcation level. These traps capture practically no holes (for which they act as recombination centers). Consequently such traps act as trapping levels. On the other hand, the traps lying between the Fermi and demarcation levels act as recombination levels because they are filled with electrons and consequently they capture holes, for which they act as recombination levels, since they are above the demarcation levels. The traps between the valence band and the demarcation level, which are filled with electrons, capture holes and act as trapping levels for holes, because they are below the demarcation level.

In an intrinsic semiconductor in which the Fermi level and consequently the demarcation level lie in the middle of the forbidden band, practically all the traps act as trapping levels. In a strongly doped semiconductor the traps (with the exception of those very close to the band) act as recombination centers.

These considerations show clearly the arbitrary and purely relative nature of the division of traps into the two categories. In this division the governing factors are not so much the trap parameters and not even their position in the forbidden band, as the properties of the substance (the Fermi level position, etc.). Sometimes the difference between the electron and hole capture cross sections of the traps may be important (this difference may be very large).

In some cases it is useful to introduce the concept of "absolute trapping levels," which is a convenient abstraction in considering many processes connected with the phenomenon of trapping. For such levels the capture cross section for carriers of one sign is assumed to be zero, and consequently these carriers are assumed to undergo thermal exchange with one band only.

Calculation of the steady-state values of τ_n and τ_p in the presence of absolute trapping levels, in the case of weak excitation and low trap concentration S in the system shown in Fig. 87, leads to the following expressions:

$$\tau_n = \tau_0 \frac{p_0 + n_0}{p_0 + n_0 \left[1 + \dfrac{M N_{cM}}{(N_{cM} + n_0)^2} \right]} = \tau_0 \frac{p_0 + n_0}{p_0 + n_0 (1 + a)}, \qquad (30.16)$$

$$\tau_p = \tau_0 \frac{(p_0 + n_0)\left[1 + \dfrac{MN_{cM}}{(N_{cM} + n_0)^2}\right]}{p_0 + n_0\left[1 + \dfrac{MN_{cM}}{(N_{cM} + n_0)^2}\right]} = \tau_0 \frac{(p_0 + n_0)(1 + \alpha)}{p_0 + n_0(1 + \alpha)}; \qquad (30.17)$$

here τ_0 denotes the electron-hole pair lifetime, given by the expression (23.12). We must assume that in Eqs. (30.16) and (30.17) the quantities n_0, p_0, and τ_0 are functions of the trapping level density M because the introduction of the trapping levels changes, in general, the Fermi level position. To illustrate this point, Fig. 90 gives the dependences of τ_n and τ_p on M for a particular case. Here, τ_p rises and τ_n falls with increase of M, as in the case of strong excitation.

As a rule, the introduction of trapping levels reduces the lifetime of the trapped carriers and increases the lifetime of the carriers of opposite sign; we should stress, however, that there are some cases in which the strong influence of the trapping levels on the Fermi level position invalidates this rule.

§ 31. INFLUENCE OF TRAPPING LEVELS ON THE RELAXATION OF NONEQUILIBRIUM AMBIPOLAR CONDUCTIVITY

For simplicity we shall restrict ourselves to the case of α trapping at a high excitation level. The system of transitions (Fig. 80) and the main restrictions remain the same as in the unipolar case. However, in the unipolar case we assumed that the recombination center concentration is high (S $\gg \Delta n$, Δp), while for the ambipolar case we shall assume this concentration is low (S $\ll \Delta n$, Δp). This difference is not arbitrary but reflects the real conditions under which the unipolar and ambipolar types of conduction appear.

Ambipolar photoconductivity is unlikely in the case of high trap concentration because the electron and hole lifetimes (and consequently Δn and Δp) have similar values only if the trap levels lie in a narrow range of energies $\sim kT$ near the equilibrium Fermi level. When the S levels lie far from the Fermi level the population of the former is strongly "asymmetric" and consequently the lifetimes and densities are very different for electrons and holes in the band (the unipolar case).

When the trap concentration S is low the neutrality condition may be written in the form $\Delta n = \Delta p$, i.e., we are dealing with ambipolar photoconductivity. (So far we have not considered the role of the trapping levels.)

We are interested in the time dependence of Δn and Δp assuming that $\Delta n \approx n = 0$, $\Delta p \approx p = 0$, $\Delta m \approx m = 0$ and that illumination begins at t = 0.

Up to $t \approx \theta$ the capture of electrons by the trapping levels is not important and the electron and hole densities in the bands increase linearly with time:

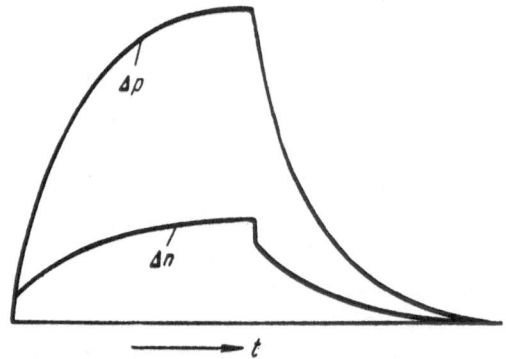

Fig. 91. Relaxation curves of the trapped (Δn) and un-trapped (Δp) carriers after illumination with square light pulses.

$$\Delta n = \Delta p = \beta k I t \qquad (31.1)$$

(Fig. 91).

When $t > \theta$ the relaxation proceeds under conditions of equilibrium between the conduction band and the trapping levels M, i.e., the electrons transferred from the valence band are shared by the conduction band and the M levels in the ratio

$$\frac{\Delta n}{\Delta m} = \frac{N_{cM}}{M}. \qquad (31.2)$$

Thus, the electrons (the trapped carriers) behave as in the unipolar case (§ 27).

The relaxation curve for electrons has a kink at $t \approx \theta$ (Fig. 91). The relaxation is completed after a time

$$\tau' = \tau_n \left(1 + \frac{M}{N_{cM}} \right), \qquad (31.3)$$

and the relaxation law is given by Eq. (27.17), i.e.,

$$\Delta n = \beta k I \tau_n (1 - e^{-t/\tau'}). \qquad (31.4)$$

The quantity τ_n now represents the steady-state electron lifetime in the presence of the trapping levels.

According to Eq. (30.11)

$$\tau_n = \tau_0 \frac{1 + \dfrac{\gamma_p}{\gamma_p + \gamma_n} \alpha}{1 + \alpha}, \qquad (31.5)$$

where τ_0 is the lifetime when M = 0.

Substituting Eq. (31.5) into Eqs. (31.3) and (31.4), we obtain

$$\tau' = \tau_0 \left(1 + \frac{\gamma_n}{\gamma_n + \gamma_p} \alpha \right), \tag{31.6}$$

$$\Delta n = \beta k I \tau_0 \frac{1 + \dfrac{\gamma_n}{\gamma_n + \gamma_p} \alpha}{1 + \alpha} \left(1 - e^{-\frac{t}{\tau'}} \right). \tag{31.7}$$

Thus, the relaxation of the trapped carriers lags by a factor of $\left(1 + \frac{\gamma_n}{\gamma_n + \gamma_p} \alpha \right)$ and the steady-state density decreases by a factor of $\left(\frac{1 + \frac{\gamma_n}{\gamma_n + \gamma_p} \alpha}{1 + \alpha} \right)$ (cf. Fig. 88a, curves 3 and 4).

The behavior of the carriers which are not being trapped (in our case holes) is considerably different from that in the unipolar case where their density was assumed to be zero.

The neutrality condition

$$\Delta p = \Delta n + \Delta m \tag{31.8}$$

and Eq. (31.2) give

$$\Delta p = \left(1 + \frac{M}{N_{cM}} \right) \Delta n. \tag{31.9}$$

Thus, Δp is greater than Δn and when $M/N_{cM} \gg 1$ the photoconductivity is purely p type. *

Using Eqs. (31.4) and (31.3), we can rewrite Eq. (31.9) in the form

$$\Delta p = \beta k I \tau' \left(1 - e^{-\frac{t}{\tau'}} \right). \tag{31.10}$$

When $t \ll t'$, $\Delta p = \beta k I t$, i.e., Eq. (31.10) becomes identical with Eq. (31.1). Thus, in contrast to the electrons, the hole density rises exponentially without any kinks (cf. Fig. 91), its time constant being $\tau' = \tau_0 \left(1 + \frac{\gamma_n}{\gamma_n + \gamma_p} \alpha \right)$ until it reaches the steady-state value [cf. Eq. (30.12)]

$$\Delta p_{st} = \beta k I \tau' = \beta k I \tau_0 \left(1 + \frac{\gamma_n}{\gamma_n + \gamma_p} \alpha \right). \tag{31.11}$$

---

* Thus, the model which is ambipolar when there are no trapping levels may change to "unipolar" photoconductivity when the trapping level density is high.

Thus, the presence of the trapping levels in the ambipolar case gives rise
to a considerable increase in the density of untrapped carriers (in the

ratio $\frac{N_{cM} + M}{N_{cM}}$) and of the photoconductivity. This is because the elec-

tron trapping greatly increases the hole lifetime (§ 30).

When $M/N_{cM} \gg 1$ the photoconductivity, being mainly due to holes,
relaxes in such a way that the initial jump (due to electron trapping) may
be even more noticeable than in the unipolar case.

It is interesting to note that the smaller this jump (which indicates
the presence of the trapping levels) the "stronger" the trapping effect and
its influence on the measured photoconductivity. Therefore, experiments in which the jump is not observed at all are precisely those cases
where the trapping effect is particularly important. Under these conditions the detection of trapping requires methods other than the observation
of the jump. Among such methods the foremost is the study of the influence of additional steady illumination on the nature of the photoconductivity relaxation after pulse excitation.

(iii) "Nonlinear" Relaxation Processes in the Case of High Degree of Population of the Trapping Levels

We have considered above the influence of the capture of carriers by
the trapping levels on the photoconductivity relaxation in the relatively
simple "linear" case when the population of these levels changes little
during relaxation.

In the present subchapter we shall consider, for the unipolar photoconductivity, the kinetics for any excitation level when the population of
the trapping levels may vary considerably. We shall show that an analysis of the relaxation curves when the trapping-level population changes
considerably gives us important information on the role of these levels
at various stages of the relaxation process.

It should be noted that the phenomena related to the "nonlinear population" of the local levels, in the processes of recombination and trapping
of the nonequilibrium carriers, play an important role in the photocon-
duction mechanism of many semiconductors, and may be used as the basis
for interpreting various complex effects (for example, superlinearity, infrared quenching of the photoconductivity, etc.).

§ 32. INFLUENCE OF HIGH DEGREE OF POPULATION OF THE TRAPPING LEVELS ON THE INITIAL STAGES OF PHOTOCONDUCTIVITY RISE

As before, we shall consider the band structure of a photoconductor (Fig. 92), which has, in its forbidden band, recombination centers S which are responsible for the much longer electron lifetime τ_n in the conduction band compared with the hole lifetime τ_p in the valence band; thus photoconductivity may be regarded as purely electronic. Apart from these centers the forbidden band also contains electron trapping levels (concentration M). We shall assume that these are the multiple trapping levels, i.e., the time θ for the establishment of equilibrium between these levels and the conduction band is considerably shorter than the electron lifetime τ_n. It is assumed that the equilibrium carrier density in the band is negligibly small. We have shown in § 27 that in the linear case, when the electron density m at the trapping levels is low compared with M, the rise of the electron density n in the conduction band during the initial stages of the illumination (when $t \ll \tau$) may be described by Eq. (27.10).

The rise curve consists of an initial exponential region with a time constant θ, representing the establishment of equilibrium between the conduction band and the M levels, and having an initial slope βkI, corresponding to the true quantum yield; this is followed by a rectilinear region with slope $\beta kI \dfrac{N_{cM}}{M + N_{cM}}$, corresponding to the "phenomenological" quantum yield which represents the rise of the density n in the case of approximate equilibrium (Fig. 93, curve 1).

We shall consider how the nature of the rise changes when the M levels are filled to a considerable degree during the relaxation (rise) but continue to act as multiple trapping levels.

An exact solution of this problem is very complex. The principal characteristics of the rise can, however, be found by solving the problem by the following approximation.

When $t \ll \tau_n$, the recombination of electrons may be neglected, and the kinetics of the rise of n on commencing illumination may be described by the following system of differential equations:

$$\frac{d(n+m)}{dt} = \beta kI, \qquad (32.1)$$

$$\frac{dm}{dt} = \gamma n(M - m) - \gamma N_{cM}m. \qquad (32.2)$$

If we consider the time $t \gg \theta$ when quasi-equilibrium between the conduction band and the M levels is established, then (as shown in § 28)

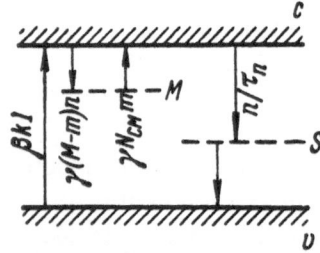

Fig. 92. System of transitions.

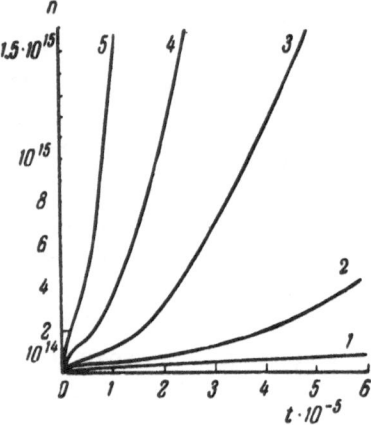

Fig. 93. Initial stages of the theoretical dependence n(t) for various illumination intensities I (in %): 1) 2; 2) 10; 3) 25; 4) 50; 5) 100.

we may assume that the current of electrons captured by the M levels at any instant is approximately balanced by the current of electrons from these levels to the conduction band (due to thermal excitation), i.e.,

$$\gamma n (M - m) \cong \gamma N_{cM} m, \qquad (32.3)$$

and hence

$$m \cong \frac{Mn}{N_{cM} + n}. \qquad (32.4)$$

Substituting m from Eq. (32.4) into Eq. (32.1), we obtain, instead of the system (32.1) and (32.2), a single differential equation

$$\left[1 + \frac{MN_{cM}}{(N_{cM} + n)^2} \right] dn = \beta k I dt, \qquad (32.5)$$

which, when integrated for the initial condition n = 0 when t = 0, gives the following dependence of the free electron density on time:

$$n = -\frac{1}{2}(M + N_{cM} - \beta k I t)$$

$$+ \sqrt{\frac{1}{4}(M + N_{cM} - \beta k I t)^2 + \beta k I t N_{cM}}. \qquad (32.6)$$

At short times, when

$$t \ll \frac{M + N_{cM}}{\beta k I}, \qquad (32.7)$$

the above formula reduces to

$$n = \beta k I \frac{N_{cM}}{M + N_{cM}} t, \qquad (32.8)$$

which, to within a term $\beta k I \theta^2 \gamma_n M (1 - e^{-t/\theta})$, which under these conditions is independent of time (when t ≫ θ) and equal to the height of the first jump in the rise curve, is identical with Eq. (27.10) for the rise of n in the case of linear filling of the trapping levels.

Hence we may conclude that in the case of a high degree of population the following complete expression for the photoelectron density is valid [17]:

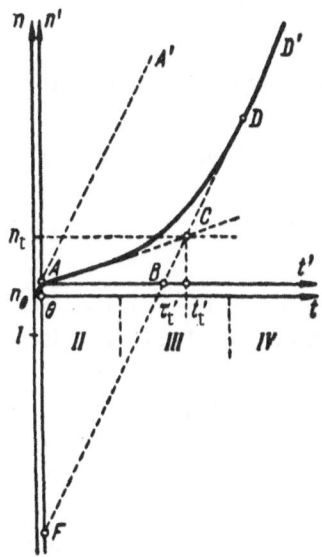

Fig. 94. Initial stage of the dependence n(t) for "nonlinear" filling of the trapping levels. The trapping level parameters may be determined from the coordinates of the points A, B, C, F, and the slopes of the straight lines AA', AC, and FD [17].

$$n = \beta k I \theta^2 \gamma_n M \left(1 - e^{-\frac{t}{\theta}}\right)$$
$$+ \frac{1}{2} \beta k I \left(\tau_t - t\right) \left[\pm \sqrt{1 + \frac{4 N_{cM} t}{\beta k I \left(\tau_t - t\right)^2}} - 1\right], \quad (32.9)$$

where

$$\tau_t = \frac{M + N_{cM}}{\beta k I} \qquad (32.10)$$

is a quantity having the meaning of the effective time for the filling of the trapping levels.

Figure 94 shows schematically the relaxation curve for the photoelectron density in the initial stage of the rise ($t \ll \tau_n$) when the filling of the trapping levels is nonlinear. In the regions denoted by I and II, i.e., when $t \ll \tau_t$, we are dealing with relaxation in the linear case of trapping-level filling, given by Eq. (27.10). In region III (where t is close to τ_t) the trapping levels become highly populated and they are finally filled in region IV, the relaxation proceeding with a slope equal to the slope of the initial rise in region I, i.e., it is governed only by the rate of generation.

It follows from Eq. (32.10) that τ_t decreases on increase of the illumination intensity I. Figure 93, which gives the rise curves for various values of I, shows that the kink of the relaxation curve characteristic of linear population may be observed quite clearly at illumination intensities such that

$$\tau_t = \frac{M + N_{cM}}{\beta k I} \gg \theta \qquad (32.11)$$

(curve 3 in Fig. 93).

At higher illumination intensities the trapping levels are filled in a time τ_t which is close to θ and the relaxation curve is practically a straight line with slope $\beta k I$. The rise curve of n (Fig. 94) has several characteristic points from which the trapping level parameters may be determined.

We can find the quantity $\beta k I$ from the initial slope of the rise curve in region I (the straight line AA') and from the slope of the straight line DD' in region IV. The slope of the straight line AC in region II is $\beta k l \dfrac{N_{cM}}{M + N_{cM}}$.   The abscissa of the point of intersection of the lines AA'

and AC (the point A) gives the value of the quantity θ, while the ordinate
of this point is β kIθ.

When $t \gg \tau_t$ the dependence n(t) in region IV is given approximately
by the straight line

$$n = \beta k I t - (M + N_{cM}).\tag{32.12}$$

Consequently, if we shift the origin of coordinates to the point A, we
can then easily determine τ_t as the value of t' at the point of intersection
of the axis At' with the continuation of the line DD' into region IV (this
point is denoted by B). The ordinate of the intersection of this line
with the An' axis gives $-(M + N_{cM})$, denoted by F. The intersection of
the straight lines AC and D'F (the point denoted by C) has the abscissa
$t'_t = \tau_t [1 + (N_{cM}/M)]$ and the ordinate $n_t = (M + N_{cM})(N_{cM}/M)$. It is interesting that the quantity n_t is independent of the illumination intensity and
lies on a straight line parallel to the time axis in the family of relaxation
curves for various illumination intensities. It should be noted that the
abscissa of the point C may be determined from the intersection of the
line $n = n_t$ with the straight line in region II.

These relationships can be used to determine the value of M, N_{cM},
and γ in various ways.

§33. INFLUENCE OF TRAPPING LEVELS ON THE GENERAL NATURE OF THE PHOTOCONDUCTIVITY RELAXATION CURVES

The presence of carrier trapping affects not only the initial stages
but also the whole process of rise until the steady state is established.

We shall consider below the shape of the relaxation curve for unipolar photoconductivity in the case of linear recombination (the electron
lifetime being τ_n = const) and multiple nonlinear trapping. Assuming,
as before, that the equilibrium electron density is negligibly small compared with the nonequilibrium density, and allowing for the linear electron recombination, we can write down the following differential equations for the photoconductivity rise and decay respectively:

$$\frac{d}{dt}(n + m) = \beta k I - \frac{n}{\tau_n},\tag{33.1}$$

$$\frac{d}{dt}(n + m) = -\frac{n}{\tau_n}.\tag{33.2}$$

Using Eq. (32.4), the above equations may be rewritten as follows [17]

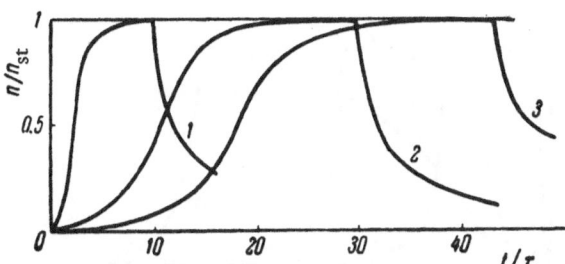

Fig. 95. Theoretical curves with S-shaped rise portions (due to the nonlinear filling of the trapping levels) for various intensities of illumination. $M = 10^{11}$ cm^{-3}; N$_{cM}$ = 10^9 cm^{-3}; 1) βkIτ_n = 10^{11} cm^{-3}; 2) 10^{10} cm^{-3}; 3) 5×10^9 cm^{-3}.

$$\left[\frac{1}{\beta kI - n/\tau_n} + \frac{MN_{cM}}{(N_{cM} + n)^2 (\beta kI - n/\tau_n)} \right] dn = dt, \tag{33.3}$$

$$\frac{dn}{dt} \left[1 + \frac{MN_{cM}}{(n + N_{cM})^2} \right] = -\frac{n}{\tau_n}. \tag{33.4}$$

Integrating Eq. (33.3) for the initial condition n = 0 when t = 0, and Eq. (33.2) for the initial condition n = βkIτ_n = n_{st} when t = 0, we obtain for the rise curve

$$\frac{t}{\tau_n} = \ln (1 - n')$$

$$+ \frac{MN_{cM}}{(N_{cM} + \beta kI\tau_n)^2} \ln \frac{\frac{N_{cM}}{\beta kI\tau_n} + n'}{(1 - n') N_{cM}} + \frac{Mn'}{(N_{cM} + \beta kI\tau_n)\left(\frac{N_{cM}}{\beta kI\tau_n} + n' \right)}, \tag{33.5}$$

and for the decay curve

$$\frac{t}{\tau_n} = - \ln n'$$

$$+ \frac{M}{N_{cM}} \ln \frac{n'\left(\frac{N_{cM}}{\beta kI\tau_n} + 1 \right)}{\frac{N_{cM}}{\beta kI\tau_n} + n'} + \frac{M(1 - n')}{\left(\frac{N_{cM}}{\beta kI\tau_n} + n' \right)(N_{cM} + \beta kI\tau_n)}, \tag{33.6}$$

where $n' = n/\beta kI\tau_n$.

Equations (33.5) and (33.6) give the shapes of the rise and decay curves for the photoconductivity in the nonlinear multiple trapping case, accurate to within the initial jump (Fig. 94) which is frequently not observed experimentally. We note that the condition for multiple trapping is satisfied at any illumination intensity if the inequality $\theta \ll \tau_n$ is initially quite rigorously satisfied and the temperature is not too low.

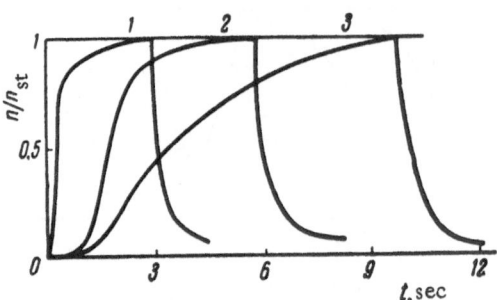

Fig. 96. Photocurrent relaxation curves of CdS for various intensities of illumination I (in %): 1) 100; 2) 5; 3) 1 [17].

Figure 95 gives the dependences of n/n_{st} on time for various illumination intensities. It shows that the rise curves have a pronounced S-shaped form,* and the rise curves are strongly nonexponential (the instantaneous lifetimes for the latter curves increase with time).

We shall now consider the condition for the rise curve to have a pronounced S-shaped form. We have shown above that the deviation from linearity of the slope which corresponds to the "phenomenological" yield (the deviation being an increase of the slope), due to the nonlinear nature of the trapping level filling, occurs in a time τ_t given by Eq. (32.10). It is easily seen that for the S-shaped form to appear the inequality $\tau_t < \tau_n$ must be satisfied. If the value of τ_t is close to or greater than τ_n, then carrier recombination will play an important role before the nonlinear rise begins, and this will lead to a reduction of the slope and disappearance of the S shape. On the other hand, the condition of Eq. (32.11) must also be satisfied. Consequently, the S-shaped form appears when

$$\theta \ll \tau_t \ll \tau_n. \qquad (33.7)$$

We shall consider briefly the characteristic features of the photoconductivity decay curve. It follows from Eq. (33.4) that the instantaneous relaxation time τ_{inst} may be written in the form

$$\tau_{inst} = -\frac{n}{\dfrac{dn}{dt}} = \tau_n \left[1 + \frac{M-m}{N_{cM}+n} \right] = \tau_n \left[1 + \frac{M}{N_{cM}} \frac{1}{(1+n/N_{cM})^2} \right]. \qquad (33.8)$$

The above formula shows that when the filling of the trapping levels is rapid the second term in brackets may be much smaller than unity and

---

* We note that an S-shaped form of the rise curve is obtained when the initial stage of relaxation does not appear (cf. Fig. 94). If the initial jump is allowed for, the form of the relaxation curve is found to be more complex and represents essentially a "two-step" process.

τ_{inst} may approach the value of the lifetime. As time elapses and the trapping levels are emptied, multiple trapping becomes increasingly important (the second term increases when n decreases) and this slows down the decay. After a considerable time from the end of illumination τ_{inst} approaches a constant value, equal to τ'

$$\tau_{inst} \rightarrow \tau' = \tau_n \left(1 + \frac{M}{N_{cM}}\right).$$

By investigation of τ_{inst} it is frequently possible to determine τ_n and the trapping level parameters.

In conclusion we note that the S-shaped photoconductivity rise curve (together with the dependence of the "yield" on the excitation level) is sometimes used as an evidence of the two-step nature of the electron excitation in CdS. However, the above treatment shows that these features may also be explained by the influence of the trapping levels when they become highly populated during relaxation. The oscillograms for CdS given in Fig. 96, with their pronounced S-shaped rise portions, are similar to the theoretical curves of Fig. 95.

§ 34. CONCLUSIONS

The above discussion of the influence of the trapping levels on the relaxation curves does not apply in all cases.

We have simply illustrated for important special cases (but not the only possible ones) the nature of the influence of the trapping processes on the nonequilibrium kinetics, and we have attempted to demonstrate the method of approach to such processes. The nonequilibrium conductivity relaxation curves observed experimentally for various semiconducting materials are frequently similar to the curves shown in Figs. 82, 84, and 95. We may assume that many effects may be explained using the treatment of trapping processes given above.

A thorough study of the nonequilibrium processes should not be limited to the steady-state regime; the most complete information on the recombination processes may be obtained by studying the nonequilibrium conductivity kinetics, in which the effects of trapping appear very clearly.

Chapter VII

RECOMBINATION THROUGH MULTIPLY-CHARGED CENTERS

So far we have considered "simple" centers which have only one energy level in the forbidden band. In many cases the energy spectrum of impurities in a semiconductor is more complex and has several levels in the forbidden band. In particular, many impurities which act as effective traps in germanium and silicon have complex energy spectra.

Below we shall consider the problem of recombination through complex traps. First of all we shall consider the general features of the energy spectra of complex centers.

§ 35. ENERGY SPECTRA OF COMPLEX CENTERS

Simple donors and acceptors can have only two charged states (Fig. 97).

A donor center is neutral when its level is occupied by an electron, and is singly charged positive when this level is free. Similarly a simple acceptor center can be neutral or singly charged negative. Moreover, a simple donor or acceptor, irrespective of whether it is charged or neutral, has only one energy level, as shown in Fig. 97.

Turning now to complex impurities we shall consider the case of gold in germanium [1].

Impurity centers of gold in germanium can have five different charged states: neutral; singly charged positive; singly, doubly, and triply charged negative. The energy structure of these five states is shown in Fig. 98. The four energy levels of the gold impurity are shown dashed. The thick lines indicate the "active" levels (free or occupied with an electron) which

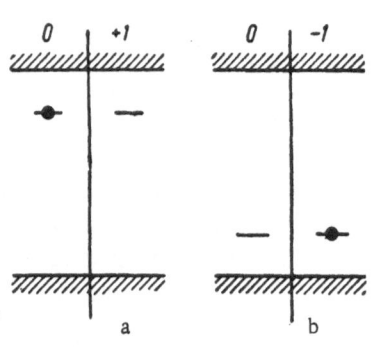

Fig. 97. Possible charged states of simple centers; a) Donor; b) acceptor. The numbers at the top of the figure denote the charge state.

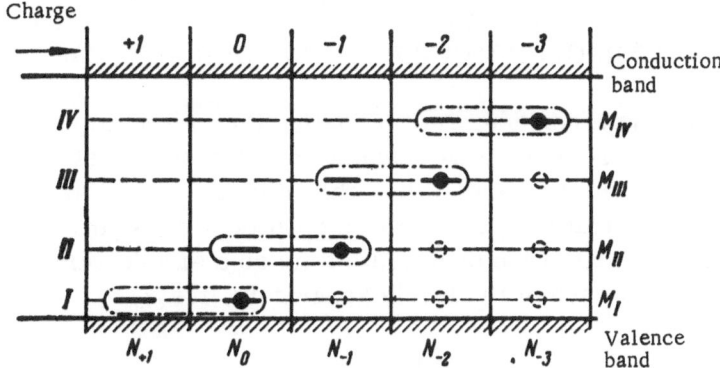

Fig. 98. System of levels of an impurity center with five possible charged states (for example, gold in germanium). "Active" levels are shown by thick lines.

are detected experimentally in the given charged state. The fact that not all the four levels, but only one or two of them, are detected experimentally in each charged state is related to the strong interaction between the impurity center charges, and may be accounted for as follows. Figure 98 shows that the neutral state of the impurity center corresponds to the lowest level I occupied by an electron, and to the free level II. If such a center captures one more electron at one of the higher levels, we must assume that the energy of the level I can no longer be detected experimentally because of the strong interaction between the two electrons and their indistinguishability.

If, assuming that the electrons are distinguishable, we transfer the lower electron (at the level I) to the conduction band, then the electron from the level II drops to I and transfers the energy liberated (strong interaction) to the electron from the level I. Consequently, the ionization requires an energy corresponding to the separation of the level II from the conduction band.

It follows then that the lowest filled level is not an active level. The ionization energy of any of the electrons in a singly charged negative center is given by the level II. Apart from this level there is also another active free level III (third from the valence band) which represents the energy of attachment of another (third) electron. Such attachment produces a doubly charged negative center, the electron energies of which are represented by the level III, and the energy for the attachment of the next electron is represented by the level IV. Thus the existence of a given active level is determined by whether the level below it is free or is occupied by an electron. For this reason the energy spectrum of an impurity center with strongly interacting charges is not an assembly of independent levels. Obviously the statistics applicable to such impurities has its own distinctive features [2].

§ 36. LIFETIME IN THE CASE OF RECOMBINATION THROUGH MULTIPLY-CHARGED CENTERS

We shall now consider recombination through multiply-charged centers or, more precisely, the calculation of the steady-state lifetime [5, 9].

We shall restrict ourselves to the case of low excitation level* and low trap concentration. Under these conditions it is possible (§ 24) to find the reciprocal of the overall lifetime as a simple sum of the reciprocal lifetimes of each of the levels separately, i.e.,

$$\frac{1}{\tau} = \sum_l \frac{1}{\tau_l}. \tag{36.1}$$

In contrast to the case of several different impurities (§ 24), when the density of levels of different types is constant, these densities are interrelated in the case of multiply-charged centers and they may vary with variation of temperature, Fermi level, etc. [2].

Since each of the τ_i values in Eq. (36.1) may be found using the Shockley-Read formula for simple centers:

$$\tau_l = \tau_{p0}^l \frac{n_0 + N_{cM}^l}{n_0 + p_0} + \tau_{n0}^l \frac{p_0 + P_{vM}^l}{n_0 + p_0}, \tag{36.2}$$

where

$$\tau_{p0}^l = \frac{1}{\gamma_{p_l} M_l}, \quad \tau_{n0}^l = \frac{1}{\gamma_{n_l} M_l},$$

the problem can be reduced essentially to the determination of the quantities M_i, i.e., the densities of the "active" energy levels of the given type which, as pointed out above, depend on temperature, Fermi level, etc. Having determined M_i we can use the formula (36.2) to find τ_i and then calculate τ by means of Eq. (36.1).

We shall demonstrate the method of determining M_i in the case of a gold impurity in germanium. The total gold concentration N may be expressed as the sum of the concentrations of centers in various charge states, i.e.,

$$N = N_{+1} + N_0 + N_{-1} + N_{-2} + N_{-3} \tag{36.3}$$

(the subscripts of N denote the charge of the center, cf. Fig. 98).

It is clear from Fig. 98 that the required densities of the active centers M_i are related to the concentrations of the centers in various charge states:

---

*This means that we are considering the recombination when the electron and hole densities in the bands and at the levels are determined by the conditions at equilibrium.

$$M_{\mathrm{I}} = N_{+1} + N_0, \tag{36.4}$$

$$M_{\mathrm{II}} = N_0 + N_{-1}, \tag{36.5}$$

$$M_{\mathrm{III}} = N_{-1} + N_{-2}, \tag{36.6}$$

$$M_{\mathrm{IV}} = N_{-2} + N_{-3}. \tag{36.7}$$

Thus, the problem reduces to finding the concentrations

$$N_{+1}, \; N_0, \; N_{-1}, \; N_{-2}, \; N_{-3}.$$

This may be done on the basis of the following considerations. The centers with charge -3 have the upper level IV filled by an electron. Consequently, the concentration of such centers may be written in the form of the product of the density of the IV levels and the probability of their being filled by an electron:

$$N_{-3} = M_{\mathrm{IV}} f_{\mathrm{IV}} = (N_{-3} + N_{-2}) f_{\mathrm{IV}}. \tag{36.8}$$

Similarly:

$$N_{-2} = M_{\mathrm{III}} f_{\mathrm{III}} = (N_{-2} + N_{-1}) f_{\mathrm{III}}, \tag{36.9}$$

$$N_{-1} = M_{\mathrm{II}} f_{\mathrm{II}} = (N_{-1} + N_0) f_{\mathrm{II}}, \tag{36.10}$$

$$N_0 = M_{\mathrm{I}} f_{\mathrm{I}} = (N_0 + N_{+1}) f_{\mathrm{I}}. \tag{36.11}$$

Finally, the concentration of the centers with charge $+1$ is equal to the density of holes at the levels I, i.e.,

$$N_{+1} = M_{\mathrm{I}} (1 - f_{\mathrm{I}}) = (N_0 + N_{+1})(1 - f_{\mathrm{I}}). \tag{36.12}$$

Solving the system of equations (36.8)-(36.12), we find the concentrations of the centers in various charge states:

$$N_{-3} = N \Big[1 + \frac{1 - f_{\mathrm{IV}}}{f_{\mathrm{IV}}} + \frac{(1 - f_{\mathrm{IV}})(1 - f_{\mathrm{III}})}{f_{\mathrm{IV}} f_{\mathrm{III}}} \\ + \frac{(1 - f_{\mathrm{IV}})(1 - f_{\mathrm{III}})(1 - f_{\mathrm{II}})}{f_{\mathrm{IV}} f_{\mathrm{III}} f_{\mathrm{II}}} + \frac{(1 - f_{\mathrm{IV}})(1 - f_{\mathrm{III}})(1 - f_{\mathrm{II}})(1 - f_{\mathrm{I}})}{f_{\mathrm{IV}} f_{\mathrm{III}} f_{\mathrm{II}} f_{\mathrm{I}}} \Big]^{-1}, \tag{36.13}$$

$$N_{-2} = N \Big[\frac{f_{\mathrm{IV}}}{1 - f_{\mathrm{IV}}} + 1 + \frac{1 - f_{\mathrm{III}}}{f_{\mathrm{III}}} + \frac{(1 - f_{\mathrm{III}})(1 - f_{\mathrm{II}})}{f_{\mathrm{III}} f_{\mathrm{II}}} \\ + \frac{(1 - f_{\mathrm{III}})(1 - f_{\mathrm{II}})(1 - f_{\mathrm{I}})}{f_{\mathrm{III}} f_{\mathrm{II}} f_{\mathrm{I}}} \Big]^{-1}, \tag{36.14}$$

$$N_{-1} = N \left[\frac{f_{III}f_{IV}}{(1-f_{IV})(1-f_{III})} + \frac{f_{III}}{1-f_{III}} + 1 \right.$$
$$\left. + \frac{1-f_{II}}{f_{II}} + \frac{(1-f_{II})(1-f_{I})}{f_{II}f_{I}} \right]^{-1}, \qquad (36.15)$$

$$N_0 = N \left[\frac{f_{II}f_{III}f_{IV}}{(1-f_{IV})(1-f_{III})(1-f_{II})} + \frac{f_{III}f_{II}}{(1-f_{III})(1-f_{II})} \right.$$
$$\left. + \frac{f_{II}}{1-f_{II}} + 1 + \frac{1-f_{I}}{f_{I}} \right]^{-1}, \qquad (36.16)$$

$$N_{+1} = N \left[\frac{f_{I}f_{II}f_{III}f_{IV}}{(1-f_{IV})(1-f_{III})(1-f_{II})(1-f_{I})} \right.$$
$$\left. + \frac{f_{I}f_{II}f_{III}}{(1-f_{III})(1-f_{II})(1-f_{I})} + \frac{f_{I}f_{II}}{(1-f_{II})(1-f_{I})} + \frac{f_{I}}{(1-f_{I})} + 1 \right]^{-1}. \quad (36.17)$$

It follows from Eqs. (36.8)-(36.12) that having divided Eqs. (36.13)-(36.17) by f_{IV}, f_{III}, f_{II}, f_{I}, respectively, we find the required expressions for M_{IV}, M_{III}, M_{II}, M_{I}.

The expressions obtained for the concentrations of the centers with a given charge state and for the densities of the levels seem at first sight very complex, but they simplify considerably in actual cases when the levels I, II, III, and IV are separated from one another by large energy gaps (much greater than kT).

To illustrate this point we shall consider, for example, the case when the Fermi level position is between the second and third levels but well away from both. Then, obviously,

$$\left. \begin{array}{l} f_I \approx f_{II} \approx 1, \\ f_{III} \approx f_{IV} \approx 0 \end{array} \right\} \qquad (36.18)$$

and from Eqs. (36.13)-(36.17) we obtain

$$N_{-3} \approx 0, \quad N_{-2} \approx 0, \quad N_{-1} \approx N, \quad N_0 \approx 0, \quad N_{+1} \approx 0,$$

i.e., all the centers are in the charge state −1.

Then the level density is

$$M_I \approx 0, \quad M_{II} \approx N, \quad M_{III} \approx N, \quad M_{IV} \approx 0.$$

Thus, in the case considered the active levels are the levels II, completely filled with electrons, and the empty levels III. The same conclusion can be reached by direct analysis of the system in Fig. 98 when the Fermi level lies between the levels II and III.

If the Fermi level coincides, for example, with the levels III, then obviously half these levels are occupied by electrons and the other half is empty. Consequently, the charge state for half of these centers is −2 and for the other half it is −1. The same conclusion follows from Eqs. (36.13)-(36.17).

Figure 99a shows the dependence of the effective charge of the centers on the Fermi level position on the assumption that the separations of

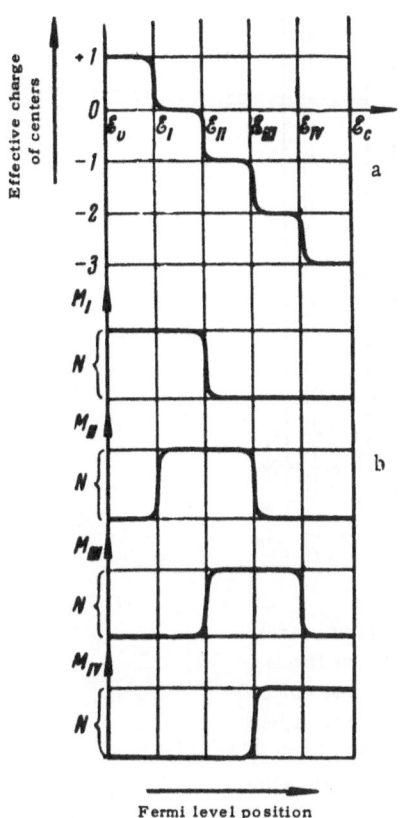

Fig. 99. a) Dependence of the effective charge of centers on the Fermi level position; b) dependences of the densities of the levels I (M_I), II (M_{II}), III (M_{III}), IV (M_{IV}) on the Fermi level position.

the levels are large compared with kT. The dependence is in the form of a curve with sharp steps: when the Fermi level is between two neighboring levels, practically all the centers are in the same charge state, corresponding to the lower levels being filled and the higher levels being empty; when the Fermi level crosses any level of a center, its charge state changes discontinuously; when the Fermi level actually coincides with a level of a center then we have equal numbers of centers in two neighboring charge states.

We have stressed above that to determine the lifetime [using Eqs. (36.1) and (36.2)] it is necessary to know the densities of active levels of various types.

Figure 99b indicates that when the position of the Fermi level is close to the allowed bands (i.e., when this level is in the energy ranges $\mathscr{E}_v - \mathscr{E}_I$ or $\mathscr{E}_{IV} - \mathscr{E}_c$) levels of only one type are active and the lifetime can be calculated using the simple Shockley-Read formula (36.2).

If the Fermi level is far from the edges of the forbidden band (i.e., between $\mathscr{E}_I$ and $\mathscr{E}_{IV}$), then depending on its exact position either one or another pair of levels is active. Then the value of τ should be calculated using the expressions (36.2) and (36.1), and the sum in Eq. (36.1) reduces in practice to a binomial*

$$\frac{1}{\tau} = \frac{1}{\tau_i} + \frac{1}{\tau_{i+1}}. \qquad (36.19)$$

Figure 100 gives a typical dependence of τ on the Fermi level position for the case of four levels considered here. In plotting the curves

* It is understood that this is valid only for well-separated levels ($\mathscr{E}_{i+1} - \mathscr{E}_i \gg kT$). Otherwise the number of levels which simultaneously take part in recombination may be greater than two.

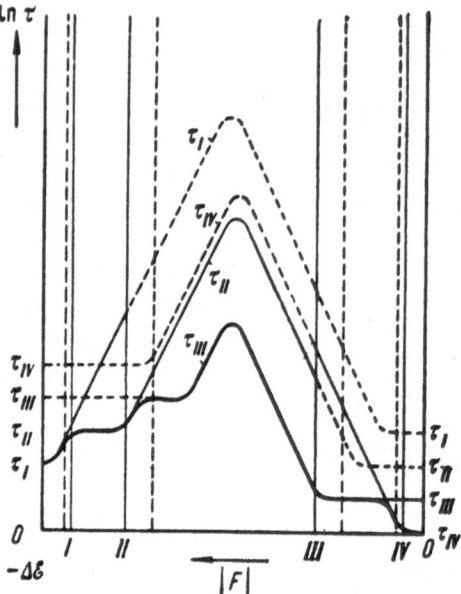

Fig. 100.   Dependence of τ on the Fermi level
position for an impurity center with four levels
(gold in germanium); the thick curve was calcu-
lated. τ_I, τ_{II}, τ_{III}, and τ_{IV} represent the dependences for each of the levels on the assumption that
the number of levels of different type is the same.
The dashed lines define regions where the density
of a given level is zero (cf. Fig. 99b) and consequently $\tau_i = \infty$.

of Fig. 100 we assumed that $\Delta\mathcal{E}$ = 0.7 eV, kT = 0.025 eV (room temperature). The levels in Fig. 100 are shown in accordance with the available
data for gold in germanium. When the charge changes by unity the electron (hole) capture cross section changes by a factor of $\sqrt{10}$.

Figure 100 gives four curves illustrating the variation of the lifetime
with shift of the Fermi level for the four levels taken separately, in accordance with the Shockley-Read formula and assuming that the numbers
of levels of different type are equal. *

Two points should be borne in mind when plotting the dependence of
the effective lifetime on the Fermi level position from these four curves:

* Curves of this type have a certain characteristic feature: at energies close to the band edges
the values of τ for different levels vary monotonically with the position of the level (in agreement with the assumption that the cross section depends only on the charge of the center). However, when the Fermi level is far from the band edges, i.e., when thermal transitions are important as well as capture, the value of τ is small for deep levels (III and II) and large for
shallow levels (IV and I).

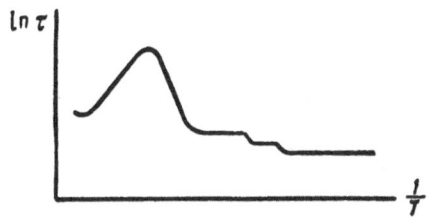

Fig. 101. Calculated temperature dependence of the lifetime for recombination through multicharged centers.

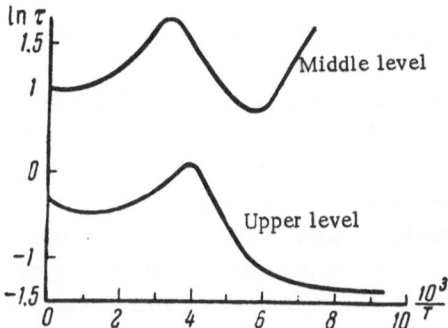

Fig. 102. Temperature dependence of partial lifetimes for the upper and middle levels of copper (obtained by calculation). It was assumed that when the charge changes by unity the cross section changes by a factor of 10 [8].

1. The density of levels of different type is not constant but varies as shown in Fig. 99b;
2. The terms with the lowest τ_i play an important role in the calculation of τ using Eq. (36.1).

When allowance is made for these two points the dependence of τ on the Fermi level position is found to be that given by the thick curve in Fig. 100. As expected the relatively simple curve for each separate level (Fig. 69) is now replaced by a complex dependence, consisting of several rectilinear regions representing different levels of a multicharged center.

Similarly the dependence of τ on temperature becomes more complex [18]. Figure 101 gives an example of a calculated temperature dependence for a center with three acceptor levels. At low temperatures the curve has several plateaus.

It follows that at different temperatures recombination is governed by different levels. One should remember, however, that if the capture cross sections of various levels differ very strongly and the energy gaps between them are sufficiently great, we can frequently have a situation in which only one of the levels of the center dominates the recombination over a wide range of temperatures. Figure 102 shows two temperature dependences of τ for n-type germanium containing copper, corresponding to the two upper levels of copper (in n-type germanium the third level from the top is completely filled and inactive), obtained after making some reasonable assumptions.

It is clear that, in the example given in Fig. 102, over a wide range of temperatures the whole recombination is governed only by the upper level of copper for which the dependence of τ lies considerably lower.

Chapter VIII

DIRECT (INTERBAND) RECOMBINATION

In the preceding chapters we have discussed in detail the statistics of recombination processes involving impurity centers.

Apart from recombination through traps, we can also have "direct recombination," i.e., the recombination of free holes and electrons. In contrast to impurity recombination, the rate of which may vary considerably for different samples of the same material, direct interband recombination is obviously the same for different samples when the conditions (electron and hole densities in the bands, and temperature) are the same. This is one of the characteristic features of interband recombination.

We shall consider the two most important mechanisms of direct recombination, which have been confirmed experimentally:

1) radiative recombination,
2) impact recombination. *

§ 37. RADIATIVE RECOMBINATION
OF FREE ELECTRONS AND HOLES

We shall consider the direct recombination of electrons from the conduction band and holes from the valence band when each recombination act is accompanied by the emission of a photon which carries away the liberated energy.

To calculate the rate of such recombination we can compare the processes of radiative recombination of nonequilibrium carriers with the similar processes for equilibrium carriers. The equilibrium processes may be considered on the following basis [2]. Under the conditions of thermodynamic equilibrium (when the principle of detailed balancing is valid) the radiant energy emitted in any range of wavelengths should be equal to the amount of energy absorbed in the same wavelength range; we can there-

---

* Both these mechanisms are also possible when an electron or a hole is captured by an impurity center. Here we shall consider them only in connection with interband recombination.

fore replace the calculation of the recombination rate for the equilibrium carriers by a calculation of the intensity of absorption. We must know, however, what fraction of the absorbed quanta is used to form electron-hole pairs. The principle of detailed balancing indicates that only this fraction corresponds to the radiation following recombination of electrons and holes, i.e., the process opposite to pair formation.

In semiconductors, electron-hole pairs are formed following the absorption of light quanta in the fundamental absorption region, and there are sufficient grounds for assuming that the quantum yield for this process is unity (for germanium this has been proved experimentally in the wavelength range 1-2 μ).

Consequently if we know the intensity of photon absorption (the number of photons absorbed per unit time in unit volume) in the "fundamental absorption" region under conditions of thermodynamic equilibrium, we can find the intensity of radiative recombination of the equilibrium carriers and consequently the recombination coefficient γ_r for such carriers.

If we assume that this coefficient γ_r can also be used to represent the radiative recombination of the nonequilibrium carriers (this will be proved below), then the problem can be regarded as solved. Thus, in contrast to recombination through traps, where theoretical calculations of the recombination rate require a knowledge of the microscopic mechanism of electron or hole capture, radiative recombination can be calculated in full from general thermodynamic considerations using experimental absorption data.

We shall now consider some quantitative relationships. Let us assume that at thermodynamic equilibrium the number of fundamental-band photons absorbed per unit time in unit volume and the number of photons emitted as the result of recombination are equal and both denoted by R. Obviously the rate of the bimolecular radiative recombination should be proportional to the densities of holes and electrons. Consequently at equilibrium R is given by

$$R = \gamma_r \, n_0 p_0 = \gamma_r \, n_i^2. \qquad (37.1)$$

where, as usual, n_0 and p_0 are the equilibrium densities of electrons and holes respectively, n_i is the density of carriers in an intrinsic semiconductor, γ_r is the radiative recombination coefficient.

From Eq. (37.1) we find

$$\gamma_r = \frac{R}{n_i^2}. \qquad (37.2)$$

If the electron and hole densities deviate from the equilibrium values but their energy distributions in the bands are still the same as at equilibrium, then the equilibrium and nonequilibrium carriers are indistinguishable and there are no reasons for assuming that the process of radiative recombination of the nonequilibrium electrons and holes may be different from the recombination of an equilibrium pair. Consequently the radiative recombination coefficient for the nonequilibrium carriers is the same as in the case of equilibrium conditions, i.e., it is equal to γ_r.

Thus, if we denote the recombination rate in the absence of equilibrium by R_c, then

$$R_c = \gamma_r np = \frac{np}{n_i^2} R. \tag{37.3}$$

We shall now find the lifetimes of the nonequilibrium electrons and holes taking part in radiative recombination. By definition

$$\tau_n = \frac{\Delta n}{\Delta R_c} = \frac{\Delta n}{R_c - R}, \tag{37.4}$$

$$\tau_p = \frac{\Delta p}{\Delta R_c} = \frac{\Delta p}{R_c - R}. \tag{37.5}$$

Using Eq. (37.3), we find

$$\Delta R_c = R_c - R = R \frac{np - n_i^2}{n_i^2}. \tag{37.6}$$

Substituting into Eq. (37.6) the expressions

$$\left. \begin{array}{l} n = n_0 + \Delta n, \\ p = p_0 + \Delta p, \\ n_i^2 = n_0 p_0, \end{array} \right\} \tag{37.7}$$

we obtain

$$\Delta R_c = R \frac{n_0 \Delta p + p_0 \Delta n + \Delta n \Delta p}{n_0 p_0}. \tag{37.8}$$

Substituting Eq. (37.8) into Eqs. (37.4) and (37.5), we obtain expressions for τ_n and τ_p in their general form.

Further analysis of the expressions obtained can be carried out for actual individual cases. We shall consider first the case of a low impurity concentration when the neutrality condition may be written in the form $\Delta n = \Delta p$. Then

$$\Delta R_c = R \frac{(n_0 + p_0) \Delta n + (\Delta n)^2}{n_i^2}, \tag{37.9}$$

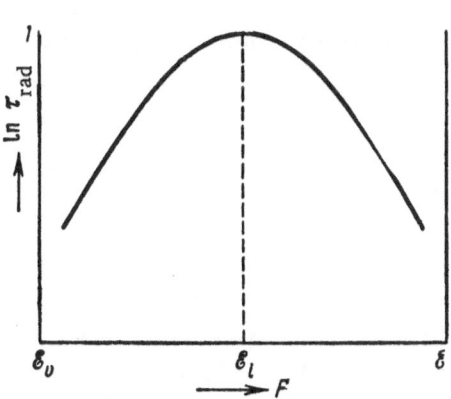

Fig. 103. Dependence of the radiative recombination lifetime on the Fermi level
position.

$$\tau_n = \tau_p = \frac{1}{R} \frac{n_i^2}{(n_0 + p_0) + \Delta n} \cdot \quad (37.10)$$

When the excitation level is low $(n_0 + p_0) \gg \Delta n$, we find from Eq. (37.10) that the recombination velocity is linearly related to Δn (linear recombination), and $\tau_n = \tau_p$ are given by the equilibrium parameters:

$$\Delta R_c = R \frac{n_0 + p_0}{n_i^2} \Delta n, \quad (37.11)$$

$$\tau_n = \tau_p = \frac{1}{R} \frac{n_i^2}{n_0 + p_0} = \frac{1}{\gamma_r (n_0 + p_0)} . \quad (37.12)$$

For a heavily doped material, when either $p_0 \gg n_0$ or $n_0 \gg p_0$, we find from Eq. (37.12):
for an n-type semiconductor

$$\tau^n = \frac{p_0}{R} , \quad (37.13)$$

for a p-type semiconductor

$$\tau^p = \frac{n_0}{R} . \quad (37.14)$$

Introducing the quantity τ_i representing the effective lifetime in an intrinsic material [cf. Eq. (37.12)]

$$\tau_i = \frac{n_i}{2R} , \quad (37.15)$$

we obtain:
for an n-type semiconductor

$$\tau^n = 2 \frac{p_0}{n_i} \tau_i = 2 \frac{n_i}{n_0} \tau_i = \frac{p_0}{R} = \frac{1}{\gamma_r n_0} , \quad (37.16)$$

for a p-type semiconductor

$$\tau^p = 2 \frac{n_0}{n_i} \tau_i = 2 \frac{n_i}{p_0} \tau_i = \frac{n_0}{R} = \frac{1}{\gamma_r p_0} . \quad (37.17)$$

Figure 103 gives the dependence of the radiative lifetime, for a low excitation level, on the Fermi level position. The lifetime has the highest value (τ_i) for an intrinsic semiconductor.

When the excitation level is high $\Delta n \gg (n_0 + p_0)$, the recombination velocity is proportional to the square of the nonequilibrium density [quadratic recombination, cf. Eq. (37.9)]:

$$\Delta R_c = \frac{R}{n_i^2} (\Delta n)^2, \qquad (37.18)$$

$$\tau_n = \tau_p = \frac{n_i^2}{R} \frac{1}{\Delta n}. \qquad (37.19)$$

Thus, in contrast to recombination through traps, where even at a high excitation level τ is independent of Δn (in the case of low trap concentration), radiative recombination is a typical bimolecular (quadratic process).

It is evident that the relaxation curves of the nonequilibrium conductivity excited by pulses should be exponential at a low intensity of pulse excitation, becoming hyperbolic tangents (for the rise) and hyperbolas (for the decay) when the excitation level is high (cf. Fig. 4 in Chap. I). It is clear from Eqs. (37.4), (37.5), (37.10), (37.12), and (37.19) that all the principal parameters of the radiative recombination are governed by the value of R, i.e., by the intensity of absorption of thermal radiation in the range of wavelengths represented by the fundamental absorption band.

The value of R can be written as

$$R = \int P_{(\nu)} \rho_{(\nu)} \, d\nu, \qquad (37.20)$$

where $\rho_{(\nu)} d\nu$ is the photon density in the crystal over the frequency range $d\nu$, and $P_{(\nu)}$ is the probability of the absorption of a photon of frequency ν, per unit time.

Shockley and Van Roosbroeck [2] showed that Eq. (37.20) can be represented in the form

$$R = 1.785 \cdot 10^{22} \left(\frac{T}{300}\right)^4 \int_0^\infty \frac{n^3 \kappa u^3}{e^u - 1} \, du \quad cm^{-3} \cdot sec^{-1}, \qquad (37.21)$$

where $u = h\nu/kT$, n is the refractive index, and κ is the absorption coefficient.

The integrand in Eq. (37.21) consists of the product of a universal function $u^3/(e^u - 1)$ and a quantity $n^3 \kappa$, which varies from material to material. Thus, we can calculate R for a given material if we know the dependences of its optical constants κ and n on the wavelength or the frequency.

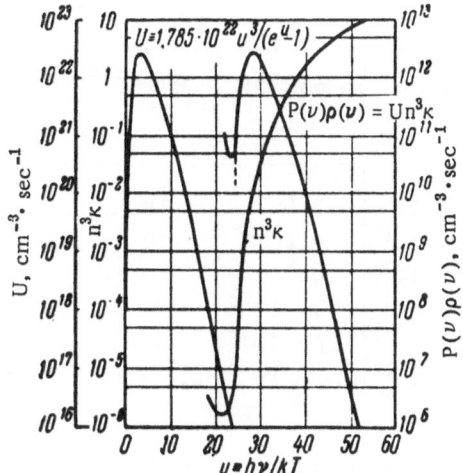

Fig. 104. Dependence of $n^3\kappa$ and of the radiative recombination velocity per unit frequency interval $P(\nu)\rho(\nu)$ on $u = h\nu/kT$ for germanium at 300°K [2].

Figure 104 gives, for germanium, the calculated frequency dependences of: $n^3\kappa$ (from the known experimental data), $u^3/(e^u-1)$, and the whole integrand in Eq. (37.21) (the latter dependence obviously represents the spectral distribution of the radiative recombination intensity).

Figure 104 shows that the maximum of the radiative recombination intensity lies in the fundamental absorption band near its long-wavelength edge. * Further into the band the probability of recombination decreases, in spite of an increase of the absorption, because the photon density in the thermal radiation spectrum decreases rapidly.

Graphical integration of the curve gives the value

$$R = 1.57 \cdot 10^{13} \ \text{cm}^{-3} \text{sec}^{-1}.$$

Hence, using Eq. (37.15), we can determine "radiative" lifetime τ_{rad} for pure germanium at 300°K. This lifetime was found to be 0.75 sec. †

This very high value of τ_{rad} (the usual measured lifetimes of the nonequilibrium carriers in germanium amount to 10^{-3}-10^{-4} sec) shows that direct radiative recombination is not the main mechanism in germanium. ‡

The same can be said about germanium with the usual amounts of dopants. Thus, when $\rho = 10 \ \Omega \cdot \text{cm}$, $\tau_{rad} = 0.22$ sec, i.e., it is still very high.

However, in heavily doped samples the radiative recombination mechanism may be important. Thus, for example, when $n_0 = 10^{19} \ \text{cm}^{-3}$ we find from Eq. (37.16) that **

* In the region beyond the edge the quantities $n^3\kappa$ and the whole integrand increase somewhat because of the absorption by free carriers. Since, according to the principle of detailed balancing, radiative recombination is connected only with the absorption which produces the interband transitions, this increase should not be included and the curve should be as shown dashed in Fig. 104. Because of this the lower integration limit in Eq. (37.21) should not be zero but $h\nu$, which corresponds to the fundamental absorption edge.
† More accurate data on the absorption yield $\tau_i = 0.3$ sec.
‡ The principal mechanism is recombination through traps.
** No correction is made for the "degeneracy" which occurs at these impurity concentrations in germanium.

Table 3

$\Delta \mathcal{E}_0$, eV	Substance	n_i, cm^{-3}	R, cm$^{-3} \cdot$ sec^{-1}	$\tau_i = n_i/2R$, sec	τobs, sec (max. value)
1.1	Si	$1.4 \cdot 10^{10}$	$2 \cdot 10^9$	3.5	10^{-3}
0.7	Ge	$2.4 \cdot 10^{13}$	$3.7 \cdot 10^{13}$	0.3	10^{-3}
0.37	PbS	$3 \cdot 10^{15}$	$1.4 \cdot 10^{20}$	10^{-5}	$9 \cdot 10^{-6}$
0.22	PbSe	$4 \cdot 10^{17}$	$3.3 \cdot 10^{22}$	$3 \cdot 10^{-6}$	−
0.27	PbTe	$6 \cdot 10^{16}$	$1.8 \cdot 10^{22}$	$1.7 \cdot 10^{-6}$	−
0.18	InSb	$2.2 \cdot 10^{16}$	$2.6 \cdot 10^{22}$	$0.4 \cdot 10^{-6}$	$0.1 \cdot 10^{-6}$

$$\tau_{rad} = 2\tau_i \frac{n_i}{n_0} = 2 \cdot 0.75 \frac{2.4 \cdot 10^{13}}{10^{19}} = 3.6 \cdot 10^{-6} \text{ sec.}$$

Obviously under these conditions radiative recombination in germanium may compete with recombination through traps if the trap concentration does not increase with the degree of doping.

The value of τ_i has also been calculated for other materials for which the optical constants near the absorption edge are known relatively well.*

The results of calculation os τ_i at room temperature are listed in Table 3, which also includes the highest observed values of the lifetimes [5, 10].

Table 3 shows that the radiative lifetime decreases, i.e., the role of radiative recombination becomes more important, with reduction of the forbidden bandwidth.   In substances such as InSb, where the calculated τ_i is close to τ measured experimentally, radiative recombination may be the main recombination mechanism at the sample purities presently achievable.

Figure 105 shows the temperature dependence of τ_i for germanium calculated assuming that the quantity $n^3\kappa = f(\nu)$ varies only because the forbidden bandwidth varies with temperature and consequently the absorption edge is shifted.   The value of τ_i rises rapidly on cooling because, in Eq. (37.15), the reduction of n_i on cooling is slower than the reduction of R.   However, this does not mean that under real conditions the radiative recombination becomes weak at low temperatures.

At low temperatures, even when the impurity concentration is low, it is very difficult to obtain the conditions for intrinsic conduction, and τ should be calculated from Eqs. (37.16) or (37.17), i.e., for an n-type semiconductor $\tau^n = 1/\gamma_r n_0$, and for a p-type semiconductor $\tau^p = 1/\gamma_r p_0$.

---

* We note that in the case of semiconductors with a steeper absorption edge (for example, $A_{III}B_V$ compounds) but with the same carrier densities the radiative recombination lifetime is considerably shorter than in germanium.

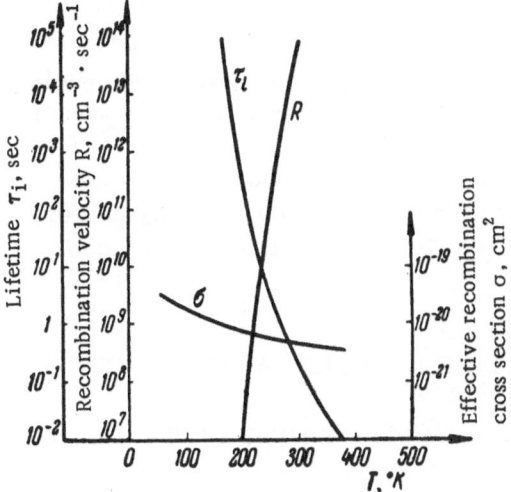

Fig. 105. Temperature dependence of the radiative recombination velocity R the lifetime τ_i (for intrinsic germanium) and the effective recombination cross section σ [2, 18].

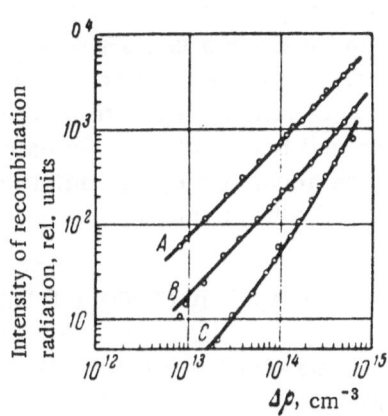

Fig. 106. Dependence of the radiative recombination velocity on the carrier density: ○ represents the experimental points; the continuous curves are theoretical.

If n_0 or p_0 does not vary with temperature (for example, when the ionization of impurities is complete), the dependence of τ^n or τ^p on temperature is governed by the temperature dependence of γ_r. The coefficient γ_r may be represented in the form

$$\gamma_r = v\sigma, \qquad (37.22)$$

where v is the thermal velocity of carriers, and σ is the radiative recombination cross section. It follows from Eq. (37.1) that

$$\sigma = \frac{R}{n_0 p_0 v} = \frac{R}{n_i^2 v}. \qquad (37.23)$$

Figure 105 gives the temperature dependence of σ for germanium; it is seen that σ (and consequently γ_r) increases on cooling, i.e., the radiative lifetime τ^n or τ^p decreases, and consequently the intensity of radiative recombination rises somewhat on cooling.

Experimental studies of radiative recombination in many semiconductors have, in general, confirmed the main theoretical predictions.

Figure 106 gives, on double logarithmic scale, the dependence of the recombination radiation intensity ΔR_c on the excitation level Δp in the case of germanium [9]. This dependence, given by Eq. (37.9), should be in the form of straight lines with slope equal to unity at low excitation levels ($\Delta p \ll n_0 + p_0$) and with slope equal to 2 at high excitation levels

($\Delta p \gg n_0 + p_0$). Figure 106 shows that the experimental points fit the theoretical curves well and that in the case of high-conductivity samples these curves degenerate into straight lines (A and B) with the unit slope, while the curve C for a sample with higher resistivity demonstrates the transition from unit slope to the slope represented by 2.

In conclusion we note that not only the recombination of free electrons and holes but also the capture of free carriers by impurity centers may be accompanied by radiation. Such "radiative capture" has been detected experimentally.

§ 38. INVESTIGATION OF RADIATIVE RECOMBINATION SPECTRA

The radiative recombination spectrum of germanium at 300°K is shown in Fig. 107 [1, 10].

It can be seen from the theoretical curves of Fig. 104 that the radiation maximum should be close to $h\nu/kT = 27.5$, i.e., at $\lambda = 1.76 \mu$. The experimental curve of Fig. 107 has a distinct maximum at this wavelength. However, it also has a maximum at shorter wavelengths near 1.52μ. The latter maximum lies in the region where the absorption in germanium rises steeply (near 0.81 eV) due to direct phototransitions* (cf. Fig. 57). Thus the long- and the short-wavelength maxima represent respectively indirect transitions (with the participation of phonons) and direct radiative transitions.

The radiative recombination spectra of two other semiconductors are given in Fig. 108 [3]. In both cases the radiation maximum lies close to the fundamental absorption edge. Figure 109 gives the radiative recombination spectrum of gallium arsenide [17, 19]. The long-wavelength maxima represent radiative transitions to impurity levels. The short-wavelength maximum corresponds to direct recombination radiation (or possibly to radiative transitions involving shallow impurity levels).†

A.   Induced Radiative Recombination

So far we have always assumed that the photons emitted after recombination leave the substance without being absorbed and consequently have no influence on the nonequilibrium processes. Allowance for the absorption of these photons may in some cases be very important. We shall briefly consider this problem. The emitted photons may, on ab-

---

* This maximum is absent in the calculated curve of Fig. 104 plotted using earlier data on the absorption spectrum of germanium.

† There are other ways of interpreting the spectrum in Fig. 109, for example, by means of impurity recombination.

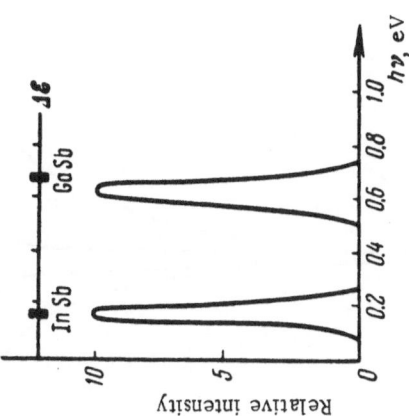

Fig. 108. Recombination radiation spectra of InSb and GaSb at 300°K.

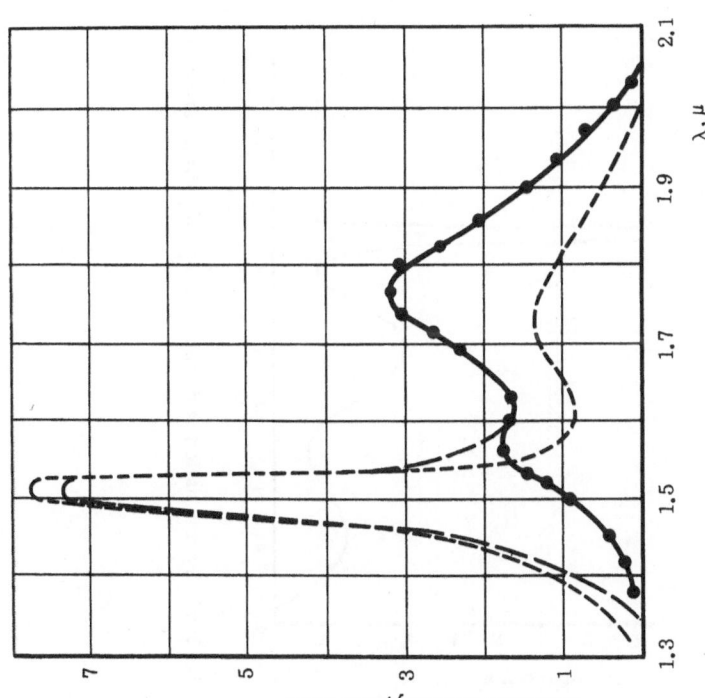

Fig. 107. Direct recombination radiation spectrum for a thin germanium plate: —— experimental curve; — — — experimental curve corrected for self-absorption; – - – - – curve plotted from the absorption data and Shockley-Van Roosbroeck theory.

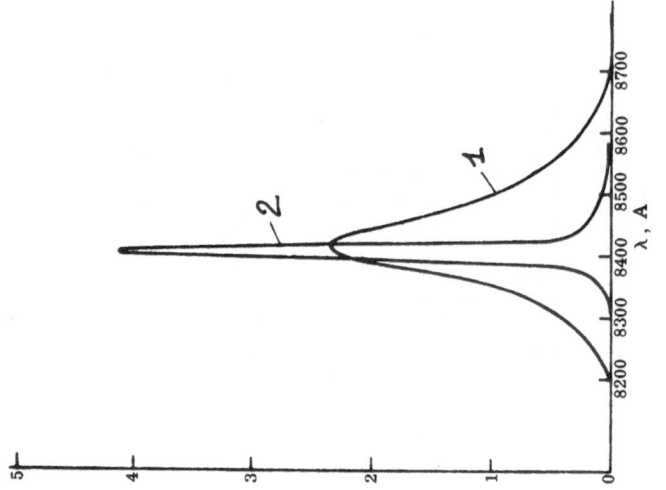

Fig. 110. Spectrum of the radiation of a p-n junction in gallium arsenide [22]: 1) Under normal conditions (spontaneous radiation); 2) at high injection levels (induced radiation) under generation conditions.

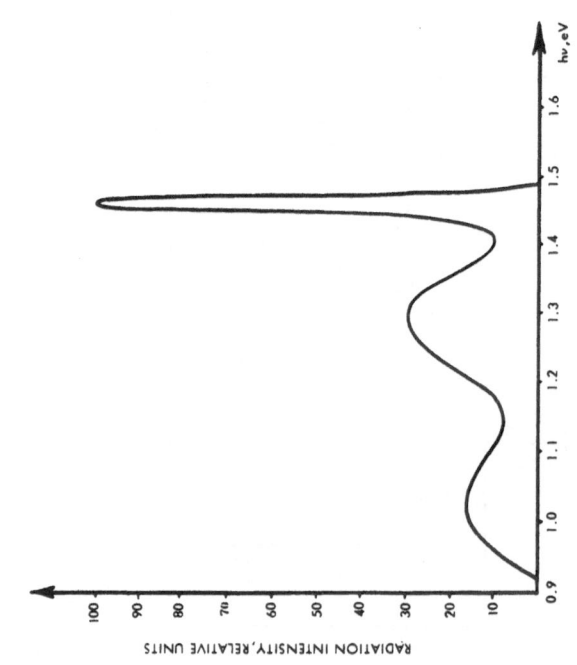

Fig. 109. Recombination radiation spectrum of gallium arsenide.

sorption, transfer electrons from the valence to the conduction band. This should appear as an apparent slowing down of recombination and reduction of the intensity (and shift of the spectrum) of the radiation emerging from the sample.

However, apart from this "normal" process there is a definite probability of the reverse process: a photon may, on absorption, transfer an electron from the conduction to the valence band. Such "induced recombination" gives rise not to absorption but to the liberation of energy (negative absorption): an interaction act of this type gives rise to two photons instead of one. This second process accelerates the recombination and raises the intensity of the radiation emerging from the sample. The resultant effect is obviously governed by the difference of the rates of the two processes. It is easily seen that the sign of this difference depends on the population of the levels (bands) with electrons and holes.

We shall consider electron transitions between two energy levels. The intensity of the "normal" process (positive absorption) is obviously proportional to the electron density at the lower level and the number of vacancies (holes) at the upper level; the intensity of the reverse process (negative absorption) is governed by the density of holes at the lower level and of electrons at the upper level.

Einstein used very general considerations to show that the coefficients of proportionality for both these processes are identical. Consequently if the population of the lower level is greater than that of the upper level the positive absorption will predominate. Otherwise the absorption is negative ("population inversion").

Under normal conditions (at equilibrium and at small departures from it) the electron population of the lower levels is always higher than that of the upper levels and therefore we have "normal" positive absorption.

However, if the equilibrium is disturbed in some way to produce population inversion, then negative absorption should be observed, with intensification of the radiative recombination due to induced transitions. In the case of population inversion (frequently called "negative temperature state") the "absorption" of a photon produces a further photon and these two photons give rise to four photons, etc. Thus, we can produce avalanche increase of the radiation intensity. This radiation is highly uniform in its properties because the photon produced by an induced recombination transition is identical with the photon which causes this transition. This allows us to use systems with inverse population of the levels as generators of monochromatic, coherent, very narrow radiation beams ("lasers").

Various types of crystal and gas lasers have been developed in which transitions between discrete atomic levels are used.

To produce population inversion in such devices, "optical pumping" is used, i.e., the laser system is subjected to intense illumination with photons whose energy is greater than that of the emitted photons.

There is considerable interest in the development of semiconductor lasers using interband radiative transitions. In this case population inversion may be achieved by the injection of electrons and holes at a p-n junction. In devices of this type the electrical energy is converted directly into high-quality radiant energy [20, 21, 17, 22].

In conclusion we note that the spectrum of the interband radiative recombination becomes much narrower in the case of population inversion. This is because the "mean free path of photons" between the acts of induced recombination is shorter for the photons at the maximum of the emission band than for photons at the edges of this band. Consequently avalanche multiplication of photons at the band maximum occurs more rapidly than multiplication of the photons at the edges.

A particularly strong reduction of the line width and an increase in the degree of directivity of the radiation should occur when two opposite faces of a radiating volume are plane-parallel mirrors (one of them semi-transparent). Then that part of the radiation which is propagated at right angles to the mirrors is constantly being returned to the radiating volume.

Consequently, the concentration of photons moving along the selected direction is much greater than the concentration of photons (generated, for example, by spontaneous recombination) moving in other directions. Hence, the former group of photons domaintes the initiation of induced-transition cascades, and practically all the recombination radiation generates only the "useful" photons moving in the required direction and emerging from the semitransparent mirror.

The increase of the time spent by the photons in the radiating volume, achieved by the use of mirrors, reduces very considerably the radiation line width (Fig. 110).

§39. INFLUENCE OF TRAPPING ON RADIATIVE RECOMBINATION

Let us assume that radiative recombination between the bands is the predominant mechanism and that it governs the nonequilibrium carrier density.

The radiative lifetime under steady-state conditions and the time constant governing the relaxation processes may again be strongly affected by the presence of trapping levels.

We shall prove this for a low excitation level using the transport equation (cf. the transition system in Fig. 111)

$$\frac{d}{dt}(\Delta n + \Delta m) = \beta k I - \Delta R_c. \tag{39.1}$$

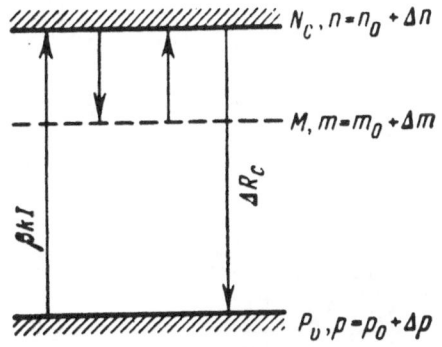

Fig. 111. Transition diagram.

From Eq. (37.8) we find

$$\Delta R_c = \frac{R}{n_i^2} (n_0 \Delta p + p_0 \Delta n), \quad (39.2)$$

or, using the neutrality condition $\Delta p = \Delta n + \Delta m$,

$$\Delta R_c = \frac{R}{n_i^2} [\Delta n (n_0 + p_0) + \Delta m \, n_0]. \quad (39.3)$$

Using Eq. (39.3), Eq. (39.1) can be written as follows

$$\frac{d}{dt} (\Delta n + \Delta m) = \beta k I - \frac{R}{n_i^2} [\Delta n (n_0 + p_0) + \Delta m \, n_0]. \quad (39.4)$$

We shall consider first the steady-state conditions when Δn and Δm are related by [cf. Eq. (28.12)]

$$\alpha = \frac{\Delta m_{st}}{\Delta n_{st}} = \frac{M N_{cM}}{(N_{cM} + n_0)^2}. \quad (39.5)$$

Then we find from Eq. (39.4) that

$$\Delta n_{st} = \frac{\beta k I}{\gamma_r} \frac{1}{[(1 + \alpha) n_0 + p_0]}, \quad (39.6)$$

$$\Delta p_{st} = \Delta n_{st} + \Delta m_{st} = \Delta n_{st} (1 + \alpha) = \frac{\beta k I}{\gamma_r} \frac{1 + \alpha}{(1 + \alpha) n_0 + p_0}. \quad (39.7)$$

[In Eqs. (39.6) and (39.7) R/n_i^2 is replaced by γ_r.]

Consequently the steady-state lifetimes of electrons and holes (which are no longer equal in the presence of trapping levels) are

$$\tau_n^{st} = \frac{\Delta n_{st}}{\beta k I} = \frac{1}{\gamma_r [(1 + \alpha) n_0 + p_0]}, \quad (39.8)$$

$$\tau_p^{st} = \frac{\Delta p_{st}}{\beta k I} = \frac{1 + \alpha}{\gamma_r [(1 + \alpha) n_0 + p_0]}. \quad (39.9)$$

Comparing the above expressions with Eq. (37.12) we see that trapping reduces the lifetime (and the steady-state density) of the trapped carriers and increases the lifetime of the untrapped carriers.

Figure 112 gives the dependences of τ_n^{st} and τ_p^{st} (expressed in units of τ_0, the lifetime in the absence of trapping) on α, i.e., on the trapping-level density, expressed in units of $N_{cM}/(N_{cM} + n_0)^2$ [cf. Eq. (39.5)].

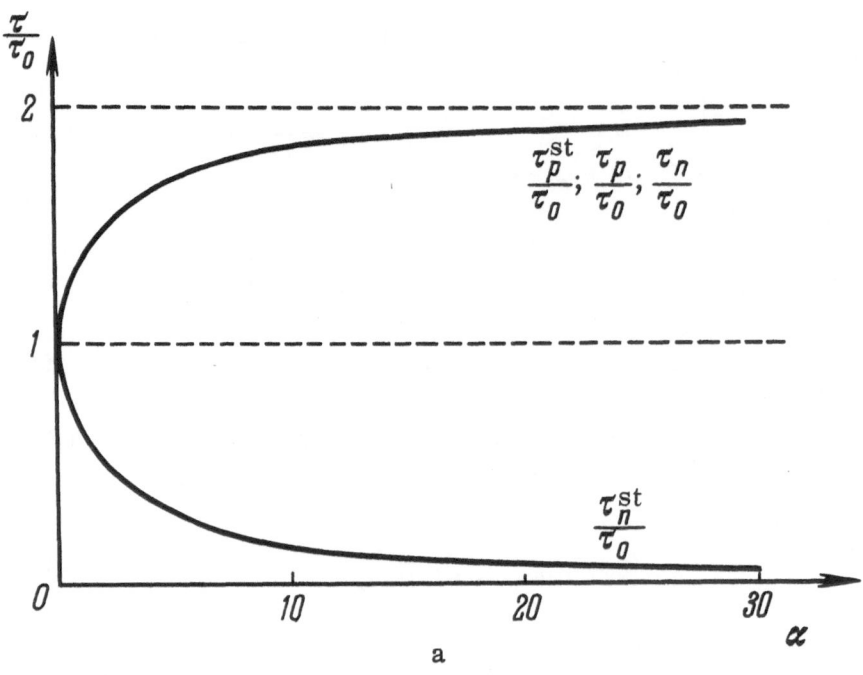

a

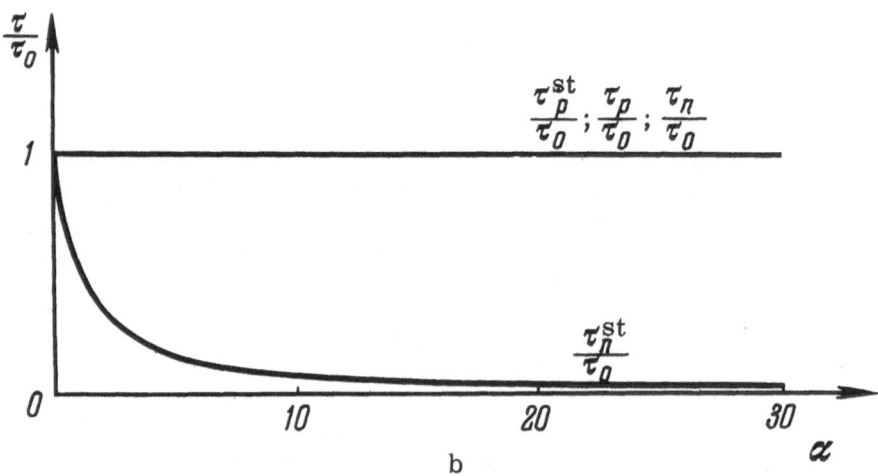

b

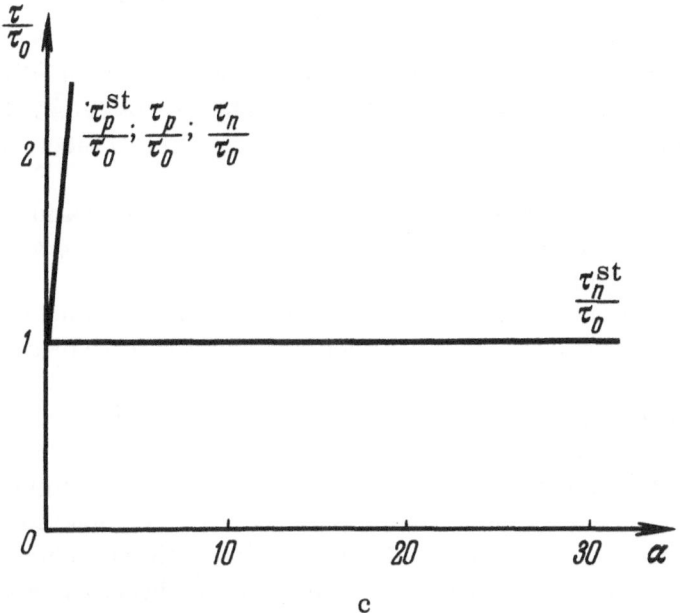

Fig. 112. Dependence of the lifetime on the trapping level density: a) Intrinsic semiconductor, $n_0 = p_0 = n_i$, $\tau_0 = 1/2\gamma_r n_i$; b) n-type semiconductor, $n_0 \gg p_0$, $\tau_0 = 1/\gamma_r n_0$, trapping of majority carriers; c) p-type semiconductor, $p_0 \gg n_0$, $\tau_0 = 1/\gamma_r p_0$, trapping of minority carriers.

We shall consider next the transient conditions in the case of α-trapping, i.e., we shall assume that the quasi-equilibrium is retained between the conduction band and the M levels during relaxation and that the relationship between Δm and Δn is at any moment given approximately by Eq. (39.5).

Then Eq. (39.4) may be rewritten in the form

$$\frac{d}{dt}\Delta n = \frac{\beta kI}{1+\alpha} - \gamma_r \Delta n \left[\frac{n_0(1+\alpha)+p_0}{1+\alpha}\right]. \tag{39.10}$$

It follows directly from Eq. (39.10) that trapping reduces the measured photoconductivity yield [which becomes $\beta kI/(1 + \alpha)$ instead of βkI] and increases the relaxation time:

$$\tau_n = \frac{1+\alpha}{\gamma_r[n_0(1+\alpha)+p_0]} = \tau_n^{st}(1+\alpha). \tag{39.11}$$

Using the condition $\Delta p = \Delta n(1 + \alpha)$, we easily find that

$$\tau_p = \frac{1+\alpha}{\gamma_n[n_0(1+\alpha)+p_0]} = \tau_p^{st}. \tag{39.12}$$

Thus the trapped and untrapped carriers have the same relaxation time constant (which is identical with the steady-state lifetime of the untrapped carriers), and this is always greater than τ_0 (Fig. 112).

§40. IMPACT RECOMBINATION

The energy evolved on recombination of a band electron with a band hole may be transferred to a third carrier. This is possible if three particles collide (interact) simultaneously, for example two electrons and a hole or two holes and an electron.

The result of such an interaction may be the recombination of one electron and one hole and the transfer of the third carrier to a higher energy level in the band.

Figures 113a and 113b show two possible variants of impact recombination, with energy transfer to an electron and a hole respectively.

It is easily seen that the impact recombination process is the converse of impact ionization, in which an electron-hole pair is generated at the expense of the kinetic energy of an electron (Fig. 113c) or a hole (Fig. 113d). It would seem that this should give us a means of finding the impact recombination coefficient, in the same way as was done for radiative recombination, i.e., by equating (in accordance with the principle of detailed balancing) the impact recombination and ionization rates at equilibrium. However, the probability of impact ionization can be found much less accurately than the probability of photoionization (which may be calculated relatively accurately from the experimental data on absorption by using the Planck distribution).

This, however, does not prevent us from finding the main features of the process which can be derived from the transport equations for impact recombination.

Let us write down the transition rate in the case of impact recombination.

Obviously the probability of a triple collision of two electrons and a hole (Fig. 113a) is proportional to $n^2 p$, and for two holes and an electron (Fig. 113b) it is proportional to $p^2 n$.

The rate of change of the electron and hole densities due to impact recombination can be written in the form [14]

$$\frac{dn}{dt} = \frac{dp}{dt} = -\eta_n n^2 p - \eta_p p^2 n, \tag{40.1}$$

where η_n and η_p are the impact recombination coefficients for energy transfer to an electron and a hole respectively. These coefficients have different dimensions from the recombination coefficients in all the other cases considered so far, where they were denoted by γ.

Substituting in Eq. (40.1)

$$\begin{aligned} n &= n_0 + \Delta n, \\ p &= p_0 + \Delta p, \end{aligned} \tag{40.2}$$

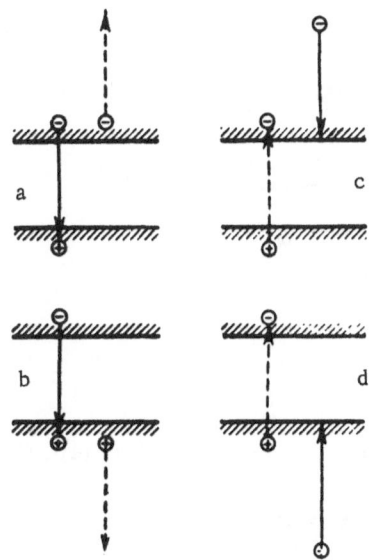

Fig. 113. Two possible variants of energy transfer after impact recombination: a) Transfer to an electron; b) transfer to a hole; c and d) the corresponding converse processes of impact ionization.

we obtain, on arranging the terms in order of increasing powers of Δn (or Δp),

$$-\frac{d}{dt}\Delta n = -\frac{d}{dt}\Delta p$$

$$= \Delta n\left(\eta_p p_0^2 + 2\eta_n n_0 p_0\right)$$

$$+ \Delta p\left(\eta_n n_0^2 + 2\eta_p n_0 p_0\right)$$

$$+ 2\Delta n\,\Delta p\left(\eta_n n_0 + \eta_p p_0\right)$$

$$+ (\Delta n)^2\,\eta_n p_0 + (\Delta p)^2\,\eta_p n_0$$

$$+ (\Delta n)^2\,\Delta p\,\eta_n + (\Delta p)^2\,\Delta n\,\eta_p. \quad (40.3)$$

We shall now find the lifetime for the case of impact recombination and determine the nature of the relaxation curves for the limiting cases of high and low excitation levels.

A. High Excitation Level

$$\Delta n\,(\Delta p) \gg n_0 + p_0$$

Dropping from Eq. (40.3) all terms except the last two which contain the highest powers of Δn (or Δp), we obtain

$$-\frac{d}{dt}\Delta n = -\frac{d}{dt}\Delta p = \eta_n\,(\Delta n)^2\,\Delta p + \eta_p\,(\Delta p)^2\,\Delta n. \quad (40.4)$$

When the excitation level is high and the trap concentration not too great (the latter condition is necessary to make sure that the impact ionization is considerable) we can assume that $\Delta p \approx \Delta n$. Then

$$-\frac{d}{dt}\Delta n = (\Delta n)^3\,(\eta_n + \eta_p). \quad (40.5)$$

The instantaneous lifetime is then

$$\tau_{\text{inst}} = \frac{\Delta n}{-\dfrac{d}{dt}\Delta n} = \frac{1}{(\eta_n + \eta_p)\,(\Delta n)^2}, \quad (40.6)$$

which indicates that the lifetime depends very strongly (quadratic dependence) on the carrier density.

We shall now find the relaxation law for square modulation of the excitation intensity. The rise curve is found from the equation

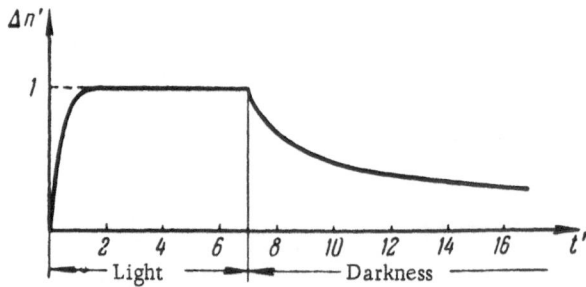

Fig. 114. Photoconductivity relaxation in the impact recombination case.

$$\frac{d}{dt}\,\Delta n = \beta kI - \Delta n^3\,(\eta_n + \eta_p),\qquad (40.7)$$

the solution of which, for the initial condition $\Delta n = 0$ when $t = 0$, is

$$t' = \frac{1}{3}\,\ln\frac{\sqrt{1 + \Delta n' + (\Delta n')^2}}{1 - \Delta n'} + \frac{1}{\sqrt{3}}\,\tan^{-1}\frac{\sqrt{3}\,\Delta n'}{2 + \Delta n'},\qquad (40.8)$$

where

$$t' = \frac{t}{(\beta kI)^{-2/3}\,(\eta_n + \eta_p)^{-1/3}}\,,\ \text{and}\ \ \Delta n' = \left(\frac{\eta_n + \eta_p}{\beta kI}\right)^{1/3}\Delta n.$$

The steady-state value of the density is

$$\Delta n_{\text{st}} = \left(\frac{\beta kI}{\eta_n + \eta_p}\right)^{1/3} \sim I^{1/3}.\qquad (40.9)$$

The decay curve can be found from Eq. (40.5). Assuming that at $t = 0$

$$\Delta n = \Delta n_{\text{st}} = \left(\frac{\beta kI}{\eta_n + \eta_p}\right)^{1/3},$$

we find that

$$\Delta n' = \frac{1}{\sqrt{1 + 2t'}}.\qquad (40.10)$$

Figure 114 shows the relaxation curves for the case considered. Here we have "cubic recombination," for which the asymmetry of the rise and decay is even more marked than in the "quadratic recombination" case (cf. Fig. 4 in Chap. I).

B. Low Excitation Level

$$\Delta n\,(\Delta p) \ll n_0 + p_0$$

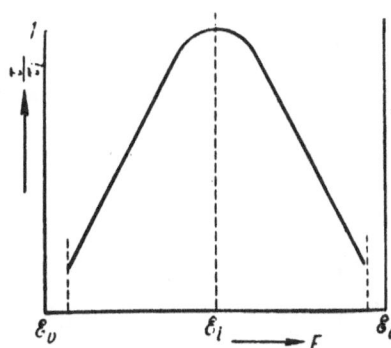

Fig. 115. Dependence of the lifetime on the Fermi level position in the case of impact recombination.

Retaining only the first two terms in Eq. (40.3), we find that the electron lifetime is given by

$$\frac{1}{\tau_n} = -\frac{1}{\Delta n}\frac{d}{dt}\Delta n =$$

$$\eta_p p_0^2 + 2\eta_n n_0 p_0$$

$$\frac{\Delta p}{\Delta n}\left(\eta_n n_0^2 + 2\eta_p n_0 p_0\right). \qquad (40.11)$$

The hole lifetime is obviously

$$\tau_p = \frac{\Delta p}{\Delta n}\tau_n. \qquad (40.12)$$

If

$$\Delta n = \Delta p. \quad \text{then} \quad \tau_n = \tau_p.$$

For the case of a strongly n-type semiconductor ($n_0 \gg p_0$) we find from Eq. (40.11) that

$$\tau_n = \tau_p = \frac{1}{\eta_n n_0^2}. \qquad (40.13)$$

For a strongly p-type semiconductor ($p_0 \gg n_0$)

$$\tau_n = \tau_p = \frac{1}{\eta_p p_0^2}. \qquad (40.14)$$

Thus the recombination depends strongly on the density of majority carriers, and collisions between the latter increase the rate of this recombination and reduce γ.

When the excitation level is low the value of τ is independent of Δn (and the intensity of light) and, consequently, the lux-ampere characteristic should be linear. (At higher excitation levels this characteristic becomes [cf. Eq. (40.9)] a dependence of the type $\Delta n \sim \sqrt[3]{I}$.)

Figure 115 gives on a logarithmic scale the dependence of the lifetime on the Fermi level position on the assumption that η_n and η_p are not very different. The maximum represents intrinsic conduction where τ for $p_0 = n_0 = n_i$ is

$$\tau_i = \frac{1}{3(\eta_n + \eta_p)}\frac{1}{n_i^2}. \qquad (40.15)$$

Substituting for n_i its value in terms of temperature,

$$n_i = \sqrt{N_c P_v}\, e^{-\frac{\Delta \mathscr{E}}{2kT}},$$

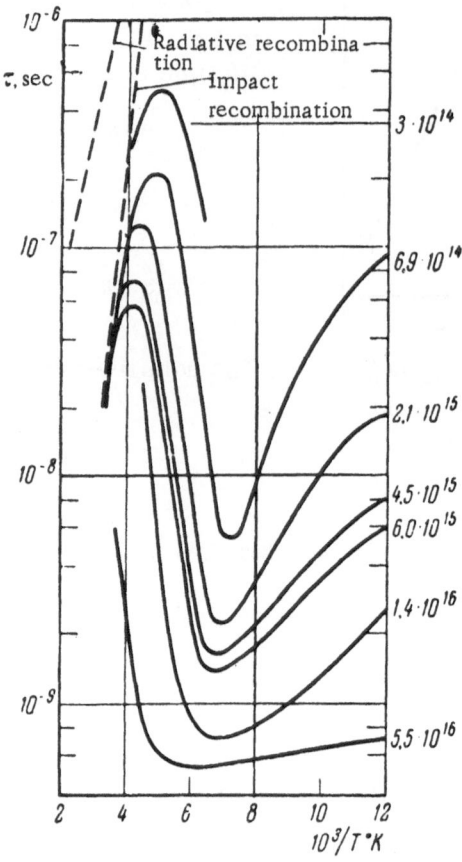

Fig. 116. Temperature dependence of the lifetime
in p-type InSb containing different amounts of acceptors, indicated by numbers on the right [12].

we obtain the temperature dependence of τ_i*

$$\tau_i = \frac{1}{3\,(\eta_n + \eta_p)\,N_c\,P_v}\,e^{\frac{\Delta\mathscr{E}}{kT}}\ .\tag{40.16}$$

* Allowance for the theoretically predicted temperature dependence of η_n and η_p gives rise
to a more complex expression [8] in which

$$\tau_i \sim \exp\!\left(\frac{1 + 2\dfrac{m_e}{m_h}}{1 + \dfrac{m_e}{m_h}}\,\frac{\Delta\mathscr{E}}{kT}\right).\tag{40.17}$$

However, if $m_e/m_h \ll 1$ (for example in InSb), the temperature dependence of τ_i is in practice governed by the term $\exp\left(\frac{\Delta\mathscr{E}}{kT}\right)$.

There have been few experimental studies of impact recombination. We shall consider the results obtained for InSb at high temperatures. Figure 116 gives a number of experimental curves for the temperature dependence of the lifetime in InSb. The general nature of the dependence can apparently be accounted for by Shockley–Read recombination through traps (cf. Fig. 69c). *

However, at high temperatures (to the left of the maximum) this explanation conflicts with the fact that all the experimental curves merge into a single curve. If recombination is indeed due to traps then, because the trap concentration is different in different samples, the curves should not merge. This contradiction can be explained by assuming that to the left of the maximum we are dealing with one of the "intrinsic recombination" mechanisms which do not depend on the presence of traps.

The dashed lines in Fig. 116 represent two theoretical temperature dependences of τ, with slopes corresponding to two known intrinsic recombination mechanisms: radiative recombination and impact recombination. It is evident that the impact recombination curve matches the experimental data well.

Impact recombination is strongly enhanced when the carrier density increases and, therefore, one would expect that it would be important in semiconductors with a narrow energy gap at high temperatures. In the example considered above these two conditions are satisfied by InSb.

---

* Increase of τ on cooling is due to trapping.

Chapter IX

IMPURITY PHOTOCONDUCTIVITY

§41. CHARACTERISTIC FEATURES
OF IMPURITY PHOTOCONDUCTIVITY

As with thermal excitation, light may produce transitions between impurity levels and the allowed bands, as well as interband transitions. The absorption of light and the photoconductivity due to such transitions are called "impurity" effects. *

Since the ionization of levels in the forbidden band requires photons of lower energy than for the interband transitions, the long-wavelength edge of the impurity absorption and photoconductivity is shifted toward long wavelengths compared with the fundamental edge (Fig. 117).

On the other hand, since the impurity concentrations are usually many orders of magnitude lower than the concentrations of the host lattice atoms, the impurity absorption and consequently the impurity photoexcitation are considerably weaker than the fundamental absorption.

Consequently in that region of the spectrum where both impurity and fundamental absorptions occur, the former is unimportant and can be neglected. This region of the

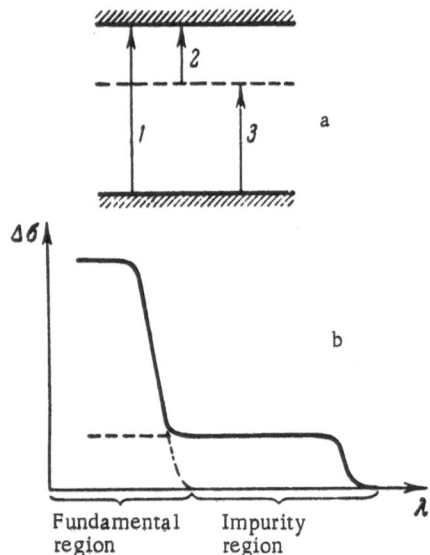

Fig. 117. a) Transitions in the case of fundamental (1) and impurity (2, 3) excitation; b) schematic representation of the photoconductivity spectrum.

* Essentially, impurity absorption and photo-conductivity are processes related to the photoionization of any levels in the forbidden band, not only levels belonging to impurity centers but also levels of structural defects.

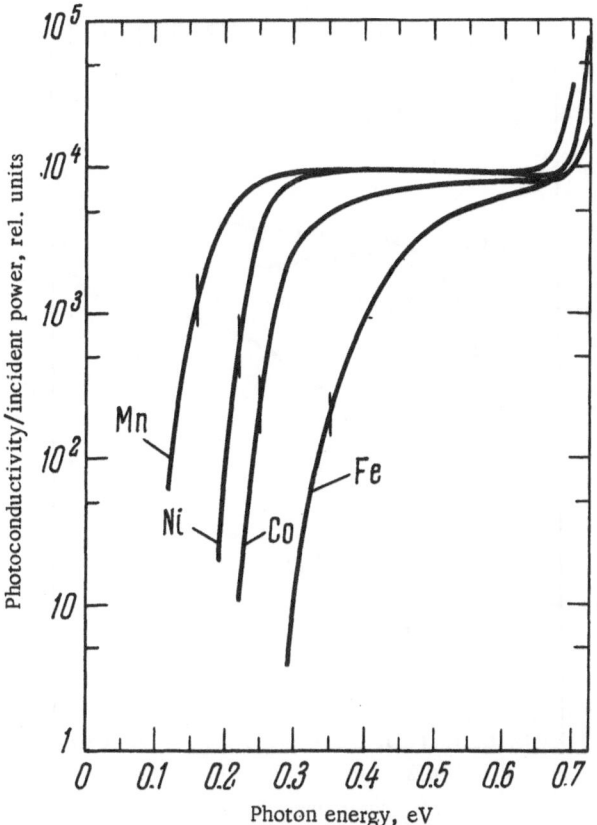

Fig. 118. Photoconductivity spectra of p-type germanium crystals, doped with Mn, Fe, Co, or Ni, at liquid nitrogen temperature [12]. The vertical lines indicate the thermal ionization energy. The scales are different for the various curves.

spectrum is known as the fundamental absorption and photoeffect region (Fig. 117b). The adjoining region where the fundamental absorption is absent and only the impurity effects are active* is known as the impurity absorption and photoeffect region.

Figures 118-121 [12, 19] give the experimental curves of the photoconductivity spectra for germanium containing various impurities. The experimental data indicate that the minimum energy $h\nu_{min}$ (determined from the long-wavelength edge) for impurity phototransitions is close to the thermal ionization energy. Figure 122 shows the positions of the levels of various impurities in germanium.

Two special features of impurity photoconductivity, which distinguish it from the fundamental effect are worth detailed study.

* More precisely the impurity absorption plus the absorption by free carriers.

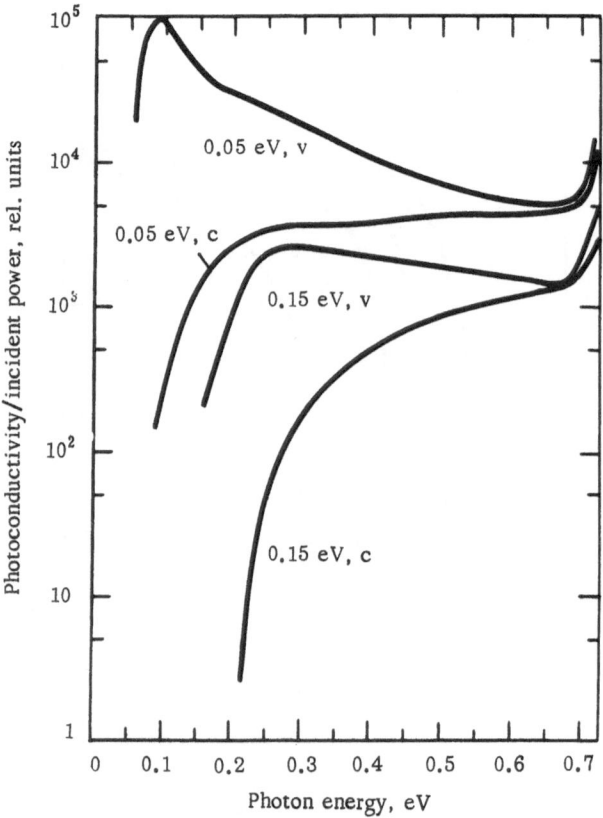

Fig. 119. Photoconductivity spectra of Au-doped germanium
[3]. The annotation of the curves indicates the impurity levels
near which the Fermi level is located (the letters "c" and "v"
indicate the band from which the energy is counted). The
spectrum appearing on excitation of carriers from the levels
at 0.15 eV and 0.20 eV was recorded at 77°K, while the other
spectra were recorded at 20°K. The scales are different for
the various curves.

In contrast to the fundamental photoconductivity, in which the excitation always produces initially equal numbers of free electrons and holes,
the impurity excitation generates either electrons only or holes only. This
does not mean that the fundamental photoconductivity is always ambipolar
and the impurity photoconductivity always unipolar. The large difference between the electron and hole lifetimes may produce conditions in
which even in the case of fundamental excitation the photoconductivity is
practically unipolar.

On the other hand the impurity excitation may sometimes generate
free electrons and holes. For example in the system shown in Fig. 117a
the light may transfer electrons from the impurity levels to the conduction

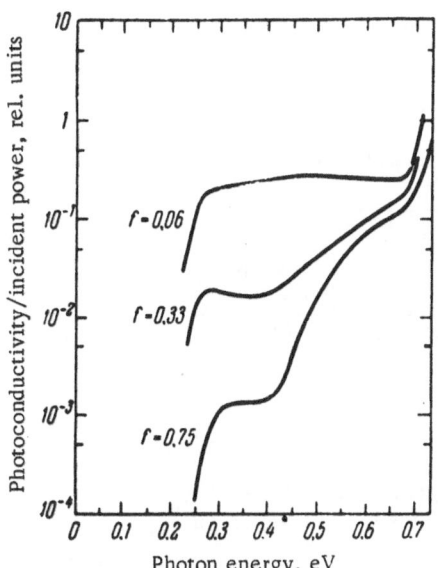

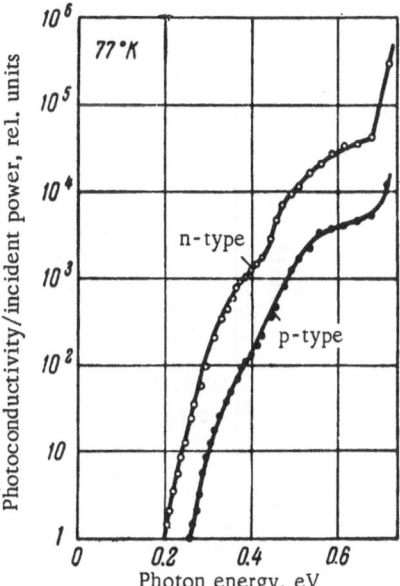

Fig. 120. Photoconductivity spectra of Ni-doped p-type germanium samples at 77°K [12]. The Fermi level for all the samples was close to the level $E_v + 0.23$ eV. The samples differ from one another only in the value of the population function f.

Fig. 121. Photoconductivity spectrum of germanium, irradiated with 3 MeV electrons, and arising from the excitation of various levels of radiation defects. The scales of the two curves are different.

band and the resultant reduction of the electron density at the impurity levels may disturb the thermal equilibrium between these levels and the valence band. The new equilibrium corresponds to a higher hole density in the valence band. Thus, in principle, the impurity excitation may also give rise to ambipolar photoconductivity.* However, in most cases realized experimentally the disturbance of the thermal equilibrium referred to above has no effect and the impurity photoconductivity is essentially unipolar (cf. § 43 for the criterion of unipolarity).

The second important feature of the impurity photoconductivity is related to the dependence of the absorption coefficient in the impurity region on the intensity of the exciting light. In general the probability of photon absorption and the related electron transition from levels of any given type to levels of another type are both proportional to the electron density at the levels of the first type and the hole density at the levels of the second type.

If these levels are in the valence and conduction bands, then, as a rule, even illumination with light of the highest possible intensity is un-

* Ambipolar conditions may also be produced by double optical transitions of electrons (for example, from the valence band to an impurity level and from this level to the conduction band).

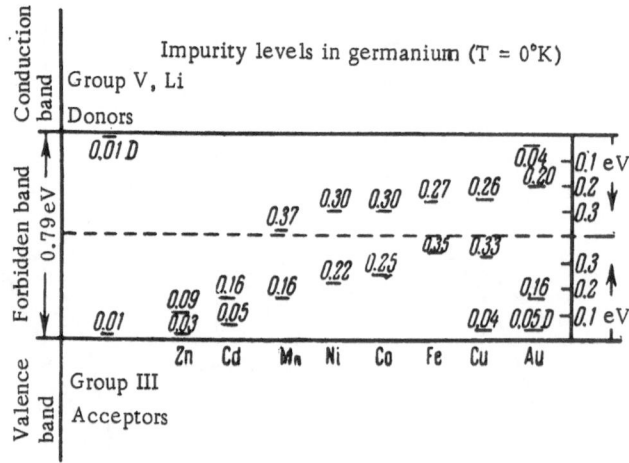

Fig. 122. Positions of deep levels in germanium [12]. All the levels are of acceptor type, except for the lowest level of Au.

able to alter greatly the effective electron density in the valence band and the density of holes in the conduction band. Consequently the absorption coefficient in the fundamental excitation case is independent of the illumination intensity.

The situation is different for the impurity excitation. If, for example, illumination transfers electrons from impurity centers to the conduction band, then sometimes even at moderate illumination intensities the impurity centers are emptied to a considerable extent and consequently the absorption coefficient is altered. Since the rate of generation is βkI and k is not constant (it decreases with increase of I) then the obvious assumption, used in discussing fundamental photoconductivity, that the generation is proportional to the intensity of illumination, is in general inapplicable to the impurity photoconductivity. This means that some other dependences [for example, $\Delta\sigma = f(I)$] as well as the impurity photoconductivity relaxation, exhibit important special properties.

Below (in §42) we shall consider quantitatively the impurity photoconductivity process for one type of level. The relationships obtained may be used to analyze the more complex cases.

§42. IMPURITY PHOTOCONDUCTIVITY RELATED TO ONE TYPE OF LEVEL [13-15]

We shall consider an n-type semiconductor in which, as well as shallow donor levels N_d, there is one type of acceptor level M (Fig. 123). Due to the compensation effect electrons from the donors drop to the acceptors below them and at absolute zero they fill them completely (when $N_d \geqq M$, Fig. 123a) or partly (when $N_d < M$, Fig. 123b).

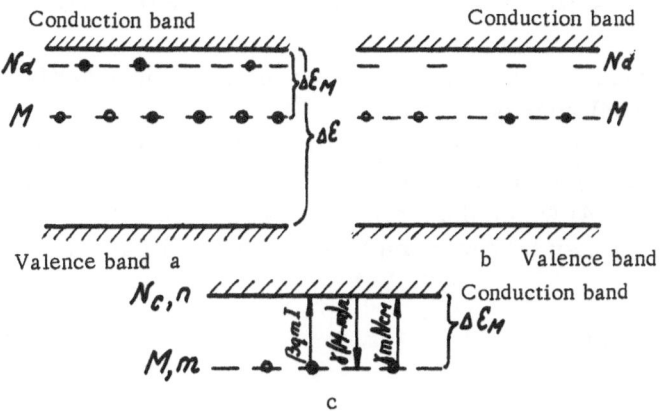

Fig. 123. Equilibrium populations of levels.

When this semiconductor is illuminated with light such that $\Delta \mathscr{E}_M \lesssim h\nu < \Delta \mathscr{E}$ impurity photoconductivity appears due to transitions of electrons from the M levels to the conduction band. We are interested in the range of temperatures at which the shallow levels N_d are completely ionized. Then they need not be allowed for directly in the transport equations for electron transitions* and consequently instead of using the systems of Fig. 123a and 123b, we can use the system shown in Fig. 123c (here the N_d levels and the valence band are not shown).

We shall write down the transport equations for the system in Fig. 123c. The thermal transition rate is as usual given by $\gamma m N_{cM}$. The phototransition rate is $\beta kI = \beta qmI$, where β is the quantum yield and q = k/m is the cross section for the capture of a photon by an electron at an impurity center.

Then we have

$$\frac{dn}{dt} = \gamma m N_{cM} + \beta qmI - \gamma (M - m) n, \qquad (42.1)$$

$$\frac{d}{dt}(m + n) = 0 \quad \text{or} \quad m + n = \text{const.} \qquad (42.2)$$

As usual, we shall express n and m in the form

$$\left. \begin{array}{l} n = n_0 + \Delta n, \\ m = m_0 + \Delta m \end{array} \right\} \qquad (42.3)$$

and we shall assume that $-\Delta n = \Delta m$ and m + n = m_0 + n_0; after simple transformations we obtain:

_____

* Although the role of these transitions in the initial conditions is very important.

$$\frac{d}{dt}\Delta n = \beta q I (m_0 - \Delta n) - \gamma \Delta n [n_0 + \Delta n + N_{cM} + (M - m_0)]. \qquad (42.4)$$

Thus, in general, the transport processes of photoconductivity are described by a nonlinear equation.

The most important difference from the fundamental photoconductivity case is the variation of the generation rate even if I = const. The first term in Eq. (42.4) which determines the generation rate, includes Δn. Equation (42.4) may be rewritten by collecting all the terms containing Δn:

$$\frac{d}{dt}\Delta n = \beta q I m_0 - \gamma \Delta n \left[\frac{\beta q I}{\gamma} + n_0 + \Delta n + N_{cM} + (M - m_0)\right]. \qquad (42.5)$$

The above equation can be written as follows:

$$\frac{d}{dt}\Delta n = \beta q m_0 I - \frac{\Delta n}{\tau^i_{inst}}, \qquad (42.6)$$

where

$$\tau^i_{inst} = \frac{1}{\gamma\left[\frac{\beta q I}{\gamma} + n_0 + \Delta n + N_{cM} + (M - m_0)\right]}, \qquad (42.7)$$

and has the meaning of the instantaneous relaxation time during illumination.

The quantity τ^i_{inst} is not the instantaneous lifetime in the usual sense because it is governed not only by the recombination processes * but also by the nonequilibrium excitation [the term $\beta q I/\gamma$ in the denominator of Eq. (42.7)].

It is easily seen that after illumination has ceased the expression for the instantaneous relaxation time differs only by the absence of the term $\beta q I/\gamma$ in the numerator (because I = 0), i.e.,

$$\tau^d_{inst} = \frac{1}{\gamma[n_0 + \Delta n + N_{cM} + (M - m_0)]}. \qquad (42.8)†$$

* This applies to the capture and thermal transitions.
† In the expressions (42.7) and (42.8) the values of the equilibrium carrier densities m_0 and n_0 are related by $m_0 = \dfrac{M}{\dfrac{N_{cM}}{n_0} + 1}$, and consequently, for example, Eq. (42.7) can also be represented in the form:

$$\tau^i_{inst} = \frac{1}{\gamma\left[(n_0 + N_{cM})\left(1 + \dfrac{MN_{cM}}{(N_{cM} + n_0)^2}\right) + \Delta n\right] + \beta q I} = \frac{1}{\gamma[(1 + \alpha)(n + N_{cM}) - \Delta n] + \beta q I},$$

$$(42.9)$$

where α is the trapping coefficient (cf. § 28).

Consequently we always have $\tau^i_{inst} \leq \tau^d_{inst}$, i.e., the impurity photo-conductivity kinetics have a characteristic asymmetry: the photoconductivity rise is faster than the decay. *

Comparison of Eqs. (42.7) and (42.8) shows that

$$\frac{1}{\tau^i_{inst}} - \frac{1}{\tau^d_{inst}} = \beta qI, \qquad (42.10)$$

i.e., the difference of the reciprocals of the instantaneous values of τ for the rise and decay, taken at the same density Δn, is always equal to βqI. It is usually easy to obtain conditions in which the competing absorption processes are unimportant and $\beta \cong 1$. Then Eq. (42.10) provides us with the possibility of determining an important parameter of the excitation process (the photon capture cross section) from the difference between the rise and decay kinetics of the impurity photoconductivity. We shall consider this problem in greater detail below.

Solutions of Eq. (42.5) for the case of excitation with square pulses give the density changes for the rise Δn_i at the beginning of illumination and the decay Δn_d at the end of illumination $(I=0)$:†

$$\Delta n_i = A \tanh(\gamma At + B) - C, \qquad (42.11)$$

where

$$A = \sqrt{C^2 + m_0 \frac{qI}{\gamma}}^{\ddagger}$$

$$B = \frac{1}{2} \ln\left(1 + \frac{2C}{\Delta n_{st}}\right),$$

$$C = \frac{1}{2}\left(N_{cM} + M - m_0 + n_0 + \frac{qI}{\gamma}\right),$$

$$\Delta n_d = \frac{\Delta n_{st} \exp\left(-\frac{t}{\tau_0}\right)}{1 + \gamma \Delta n_{st} \tau_0 \left[1 - \exp\left(-\frac{t}{\tau_0}\right)\right]}, \qquad (42.12)$$

and $\tau_0 = \dfrac{1}{\gamma(N_{cM} + M - m_0 + n_0)}$.

The steady-state density of the nonequilibrium carriers is

* The asymmetry in the nonlinear case is also found in fundamental photoconductivity (cf., for example, §5B). However, in the case of impurity photoconductivity the asymmetry is governed by the excitation kinetics and not by the recombination.

† In our case the value of the impurity photoconductivity $\Delta\sigma$ is directly related to the non-equilibrium density of carriers of one sign $\Delta\sigma = e\mu_n \Delta n$, where e and μ_n are, respectively, the charge and mobility of an electron. Therefore from now on we shall use the value of the nonequilibrium carrier density.

‡ Here and later we shall assume that $\beta = 1$.

$$\Delta n_{st} = \frac{N_{cM} + M - m_0 + n_0 + \frac{qI}{\gamma}}{2}$$

$$\times \left[\sqrt{1 + \frac{4m_0 qI}{\gamma \left(N_{cM} + M - m_0 + n_0 + \frac{qI}{\gamma} \right)^2}} - 1 \right].$$

(42.13)

A. Lux-Ampere Characteristics

We shall consider in greater detail the dependence of the steady-state carrier density on the illumination intensity. Assuming that in the case of low and high illumination intensities

$$\frac{4m_0 qI}{\gamma \left(N_{cM} + M - m_0 + n_0 + \frac{qI}{\gamma} \right)^2} \ll 1,$$

we find that at low intensities

$$\Delta n_{st} = \frac{m_0 qI}{\gamma (N_{cM} + M - m_0 + n_0)},$$

(42.14)

and at high intensities

$$\Delta n_{st} \cong m_0.$$

(42.15)

Thus the lux-ampere characteristic of the impurity photoconductivity at low intensities is mainly linear, with a tendency to saturation at higher intensities. The latter can be understood if one bears in mind that at sufficiently high illumination intensities all the electrons can be transferred from the impurity centers to the conduction band by the action of light. Such saturation of the lux-ampere characteristic due to depletion of the photoeffect centers is a characteristic feature of impurity photoconductivity and does not appear in fundamental photoconductivity at the usual intensities. * Figure 124 gives, on double logarithmic scale, the calculated lux-ampere characteristics of the impurity photoconductivity at two temperatures for a particular case.

B. Relaxation Curves

The formulas (42.11)-(42.13) show that the relaxation curves of the impurity photoconductivity are in general complex and this makes it dif-

---

* In principle complete saturation cannot occur because of the appearance of individual recombination transitions (§ 38).

However, in the case of excitation from local levels the role of the induced transitions is not great if $M \ll N_c$.

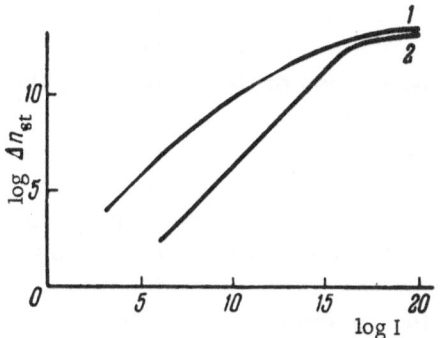

Fig. 124. Dependence of the steady-state carrier density on the illumination inten-sity [15]. 1) T = 77°K; 2) T = 190°K. In plotting the curves in Figs. 124-126 we as-sumed that $\Delta\mathscr{E}_M = 0.2$ eV; $M = 10^{12}$ cm^{-3}; $q/\gamma = 5 \times 10^{-4}$ sec · cm^{-1}; $M - m_0 = n_0$.

ficult to interpret them directly or to use them to determine the local-level parameters. Therefore we shall consider certain special cases in which the characteristic features of the impurity photoconductivity kinetics appear most simply. Equa-tion (42.5) becomes linear and its solutions are exponentials if

$$N_{cM} + M - m_0 + n_0 + \frac{qI}{\gamma} \gg \Delta n,$$

$$(42.16)$$

which represents the condition τ_{inst} = const [cf. Eqs. (42.7) and (42.8)].

The condition (42.16) is satisfied in two special cases:

1) low excitation level (i.e., low illumination intensity I and a cor-respondingly low Δn);

2) weak population of local centers: $M \gg m_0 \gg \Delta n$. Then the con-dition of Eq. (42.16) is satisfied for any intensity because no matter how high the illumination intensity the value of Δn cannot exceed m_0 [cf. Eq. (42.15)].

In the latter case the solutions of Eq. (42.6) have the form

$$\Delta n_i = \Delta n_{st} \left[1 - \exp\left(-\frac{t}{\tau_c}\right) \right],$$

$$(42.17)$$

$$\Delta n_d = \Delta n_{st} \exp\left(-\frac{t}{\tau_T}\right),$$

$$(42.18)$$

where

$$\tau_i = \frac{1}{\gamma(N_{cM} + M + n_0) + qI}$$

$$(42.19)$$

is the relaxation time constant for the photoconductivity rise;

$$\tau_d = \frac{1}{\gamma(M + N_{cM} + n_0)}$$

$$(42.20)$$

is the relaxation time constant for the photoconductivity decay, and

$$\Delta n_{st} = m_0 qI \tau_i$$

$$(42.21)$$

is the steady-state density.

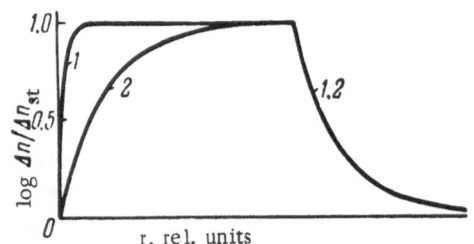

Fig. 125. Calculated photoconductivity relaxation curves for excitation with square light
pulses [15]. 1) High intensity I_1; 2) low intensity I_2; $I_1/I_2 = 150$; $T = 77°K$.

Equations (42.17)-(42.20) indicate that in the case of weak population of the local levels the impurity photoconductivity relaxation is given by a simple exponential law. Then the time constant for the rise τ_i is always smaller than the time constant for the decay τ_d, which makes the relaxation curves asymmetrical, particularly at high illumination intensities. In the case of low intensities, when the condition $\gamma (N_{cM} + M + n_0) \gg qI$ is satisfied, the rise and decay time constants differ only slightly. These features of the impurity photoconductivity kinetics are shown clearly in Fig. 125.

At a low excitation level and any degree of level population the solutions of Eq. (42.6) have the form

$$\Delta n_i = \Delta n_{st}\left[1 - \exp\left(-\frac{t}{\tau_i}\right)\right], \tag{42.22}$$

$$\Delta n_d = \Delta n_{st}\exp\left(-\frac{t}{\tau_d}\right), \tag{42.23}$$

where

$$\tau_i = \frac{1}{\gamma (N_{cM} + M - m_0 + n_0) + qI}, \tag{42.24}$$

$$\tau_d = \frac{1}{\gamma (N_{cM} + M - m_0 + n_0)}, \tag{42.25}$$

$$\Delta n_{st} = m_0 qI\tau_i. \tag{42.26}$$

Obviously the above discussion on the asymmetry of the relaxation curves and on the dependence of the constants τ_d and τ_i on the intensity of illumination is, in principle, also applicable in the case of low intensity.

The relationships (42.19) and (42.20), or (42.24) and (42.25), allow us to determine the impurity center parameters, in particular the value of the cross section for the capture of a photon by an electron at an impurity center, q, and the recombination constant, γ. Having determined the decay time constant and knowing the values of n_0, $M - m_0$, N_{cM}, we find γ from Eq. (42.20) or (42.25). On the other hand, the difference of the reciprocals of the rise and decay constants is related to the illumination intensity by the simple expression

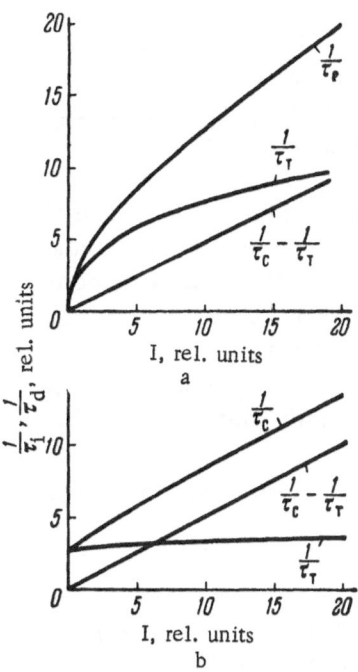

Fig. 126. Dependence of the relaxation time constants on the illumination intensity ($\Delta n = \Delta n_{st}$) [17]. a) $T = 77°K$, nonlinear case; b) $T = 190°K$, linear case.

$$\frac{1}{\tau_i} - \frac{1}{\tau_d} = qI. \qquad (42.27)$$

Consequently, by recording the dependence of this difference on the intensity of illumination, we obtain a straight line the slope of which gives the value of q. As mentioned above [cf. Eq. (42.9)], the difference between the reciprocals of τ_{inst} for the rise and decay, taken at the same density Δn, is also equal to qI in the general (nonlinear) case. This is illustrated in Fig. 126a, which shows the calculated dependences of the reciprocals of the instantaneous rise and decay times on the illumination intensity, as well as the dependence of the difference of these quantities on the illumination intensity, the latter being a straight line the slope of which gives q.

In practice, the cross section for the capture of a photon by an electron at an impurity center can be determined in the most general case from the relaxation curves. Using the relationships

$$\tau_{inst}^i = \frac{\Delta n}{m_0 qI - \dfrac{d\,\Delta n}{dt}}, \qquad (42.28)$$

$$\tau_{inst}^d = -\frac{\Delta n}{\dfrac{d\,\Delta n}{dt}} \qquad (42.29)$$

for equal Δn, we find that

$$\frac{1}{\tau_{inst}^i} - \frac{1}{\tau_{inst}^d} = \frac{1}{\Delta n}\left[m_0 qI - \left(\frac{d\,\Delta n}{dt}\right)_i + \left(\frac{d\,\Delta n}{dt}\right)_d \right].$$

Since $m_0 qI = \left(\dfrac{d\,\Delta n}{dt}\right)_i^{inst}$ represents the initial slope of the rise curve, we finally obtain

$$q = \frac{1}{I\,\Delta n'}\left[\left(\frac{d\,\Delta n}{dt}\right)_i^{init} - \left(\frac{d\,\Delta n}{dt}\right)_i^{\Delta n'} + \left(\frac{d\,\Delta n}{dt}\right)_d^{\Delta n'} \right], \qquad (42.30)$$

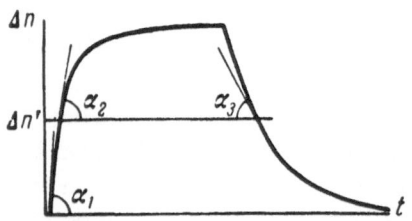

Fig. 127. Determination of the cross section for photon capture; α_1, α_2, α_3 are the angles whose tangents are used to determine q.

where $\Delta n'$ is the nonequilibrium carrier density at the points on the relaxation curve where measurements are taken. Thus the problem of determining q reduces to finding three slopes (Fig. 127): the initial slope of the rise curve and two additional slopes at a selected density $\Delta n'$ on the rise and decay curves respectively.

It is convenient to use $\Delta n' = \Delta n_{st}$ as the carrier density at which the value of q is determined. Then $\left(\dfrac{d \Delta n}{dt} \right)_i^{\Delta n' = \Delta n_{st}} = 0$ and the expression (42.30) simplifies to:

$$q = \frac{1}{I \Delta n_{st}} \left[\left(\frac{d \Delta n}{dt} \right)_i^{init} + \left(\frac{d \Delta n}{dt} \right)_d^{\Delta n_{st}} \right]. \qquad (42.31)$$

Thus to determine q it is sufficient to find the initial slopes of the rise and decay curves.

The method of finding the cross section for the capture of a photon by an electron at an impurity center using the relaxation curves of the impurity photoconductivity allows us to determine q at very low impurity level densities, of the order of 10^{11}-10^{12} cm^{-3}, while the usual optical absorption method requires densities which are several orders of magnitude higher.

In conclusion, we note that the determination of the impurity center concentration M can be carried out at follows: from the Fermi distribution we have

$$M = m_0 \left(\frac{N_{cM}}{n_0} + 1 \right), \qquad (42.32)$$

where the quantity m_0 is found from the relationship

$$m_0 = \frac{\left(\dfrac{d \Delta n}{dt} \right)_i^{init}}{qI}. \qquad (42.33)$$

C. Influence of Trapping

To determine the impurity-center parameters from the impurity photoconductivity kinetics one sometimes uses the dependence of these parameters (for example, q) on the additional illumination or on temperature [20].

These dependences are due to the influence of the trapping processes on the impurity photoconductivity. If, as well as the levels M which are being studied, there are additional levels K which are closer to an allowed band, then the latter levels will always act as the absolute trapping levels. They may exchange carriers only with the conduction band, since carrier exchange with the spatially separated M levels is impossible.

The transport equation is in this case

$$\frac{d}{dt}(n+k) = qmI - \gamma_M n(M-m) + \gamma_M m N_{cM}, \tag{42.34}$$

where

$$n = n_0 + \Delta n, \quad k = k_0 + \Delta k, \quad m = m_0 - \Delta n - \Delta k. \tag{42.35}$$

If the K levels are of the α type, i.e.,

$$\gamma_K k N_{cK} \cong \gamma_K n(K-k), \tag{42.36}$$

then, after transformations, we obtain from Eqs. (42.34)-(42.36)

$$\frac{d}{dt}\Delta n = \frac{1}{1+f} qm_0 I$$

$$- \Delta n \left\{ \frac{1+g}{1+f} qI + \gamma_M \left[\frac{1}{1+f}(M-m_0) + \frac{1+g}{1+f}(n_0 + \Delta n + N_{cM}) \right] \right\}, \tag{42.37}$$

where

$$g = \frac{K - k_0}{N_{cK} + n_0 + \Delta n}, \tag{42.38}$$

$$f = \frac{(K - k_0)(N_{cK} + n_0)}{(N_{cK} + n_0 + \Delta n)^2} = g \frac{N_{cK} + n_0}{N_{cK} + n_0 + \Delta n}. \tag{42.39}$$

Analysis of Eq. (42.37) shows that the initial stage of the impurity photoconductivity rise is S shaped, like the intrinsic photoconductivity in the presence of α-type trapping levels (Figs. 93 and 95). Only when the excitation level is very high or where the additional constant illumination is very intense do the trapping levels cease to affect the relaxation curves of the impurity photoconductivity. In the opposite case of a low excitation level, when $\Delta n \ll N_{cK} + n_0$, Eq. (42.37) becomes

$$\frac{d\,\Delta n}{dt} = \frac{qm_0 I}{1 + \dfrac{K - k_0}{N_{cK} + n_0}} - \Delta n \left[qI + \gamma_M \left(\frac{M - m_0}{1 + \dfrac{K - k_0}{N_{cK} + n_0}} + n_0 + \Delta n + N_{cM} \right) \right], \tag{42.40}$$

i.e., the value of q found from the initial slope of the photo-current rise curve without allowing for the trapping levels is lower by a factor of $\left(1 + \dfrac{K - k_0}{N_{cK} + n_0}\right) = (1 + \alpha_K)$ while the value of q found from the

difference $(1/\tau_i) - (1/\tau_d)$ represents the true value.

§43. CRITERION FOR UNIPOLAR IMPURITY PHOTOCONDUCTIVITY

Impurity excitation may give rise to ambipolar nonequilibrium conductivity for two reasons:

a) due to thermal transitions from a band to a level followed by optical transitions to the other band ("thermo-optical transitions");

b) due to double optical transitions.

We shall write down the criteria for unipolar conductivity in these two cases.

A.   Thermo-Optical Transitions (Fig. 128a)

Assume that light transfers electrons from the M levels to the conduction band and that consequently the conduction band acquires Δn electrons in the steady state, and the number of electrons at the M levels decreases by Δm. The new state corresponds to the appearance of an additional hole density Δp in the valence band due to thermal transitions of electrons to the M levels.

The condition for unipolar conduction is obviously

$$\frac{\Delta n}{\Delta p} \gg 1. \tag{43.1}$$

We shall now find the relationship between Δn and Δp, neglecting direct interband recombination.

In the steady-state case the rate of capture of holes by the M levels is equal to the rate of transition of holes to the valence band, i.e.,

$$(p_0 + \Delta p)(m_0 + \Delta m) = (M - m_0 - \Delta m) P_{vM}. \tag{43.2}$$

Before illumination

$$p_0 m_0 = (M - m_0) P_{vM}, \tag{43.3}$$

and according to the neutrality condition

$$\Delta m + \Delta n = \Delta p, \tag{43.4}$$

and therefore after some transformations we find that the relationship between Δn and Δp for weak excitation has the form

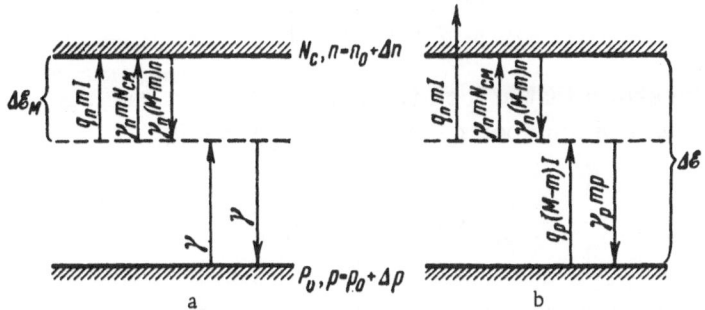

Fig. 128. Two-step transitions: a) Thermo-optical transitions; b) double optical transitions.

$$\frac{\Delta n}{\Delta p} = 1 + \frac{m_0^2}{MP_{vM}}. \qquad (43.5)$$

Hence the condition for unipolar conduction is

$$\frac{m_0^2}{MP_{vM}} \gg 1. \qquad (43.6)$$

Thus to have unipolar conduction it is essential that:

1) the levels should lie as far as possible from that band which gives rise to ambipolar thermal carrier exchange (i.e., P_{vM} should be small);

2) the levels should be strongly populated (i.e., $m_0/M \approx 1$).

If the levels are completely filled, i.e., $m_0 = M$, then the condition for unipolar conduction reduces to $M \gg P_{vM}$.

We shall now evaluate this criterion for the M levels in germanium, located 0.2 eV from the conduction band (i.e., about 0.5 eV from the valence band), at ~170°K (kT ≈ 0.018 eV):

$$P_{vM} \approx 10^{18} \cdot e^{-\frac{0.5}{0.018}} \approx 10^6 \ cm^{-3}.$$

Thus unipolar conduction is obtained beginning from a very low concentration of the impurities: $M \gg 10^6 \ cm^{-3}$.

It can easily be shown that the criterion for unipolar conduction can be written in several ways:

$$\frac{m_0^2}{MP_{vM}} = \frac{n_0^2}{n_i^2} \frac{MN_{cM}}{(N_{cM} + n_0)^2} = \frac{n_0^2}{n_i^2} \alpha = p_0^2 \alpha \gg 1. \qquad (43.7)$$

B. Double Optical Transitions (Fig. 128b)

Double optical transitions are possible in the case of impurity excitation if

$$\left.\begin{array}{l} h\nu \gg \Delta\mathcal{E}_M, \\ h\nu \gg \Delta\mathcal{E} - \Delta\mathcal{E}_M. \end{array}\right\} \qquad (43.8)$$

It can be shown that at high trap concentrations and high excitation levels the extent of ambipolar conduction is given by

$$\frac{\Delta n}{\Delta p} = \frac{q_n \gamma_p}{q_p \gamma_n} \frac{1}{\left(\dfrac{M}{m_0} - 1\right)^2}, \qquad (43.9)$$

where q_n and q_p are the cross sections for the capture of photons in the case of optical transitions to the conduction and valence bands respectively, γ_n and γ_p are the recombination coefficients for the corresponding return transitions.

Thus the conditions for unipolar conduction are

$$\frac{q_n \gamma_p}{q_p \gamma_n} \frac{m_0^2}{(M - m_0)^2} \gg 1 \qquad (43.10)$$

for $\Delta n \gg \Delta p$, or

$$\frac{q_n \gamma_p}{q_p \gamma_n} \frac{m_0^2}{(M - m_0)^2} \ll 1 \qquad (43.11)$$

for $\Delta p \ll \Delta n$.

Chapter X

SOME EFFECTS OF COMBINED EXCITATION

As well as investigating fundamental or impurity photoconductivity separately, in some cases one has to deal with nonequilibrium processes produced by mixed excitation (simultaneous or consecutive) by means of "fundamental" and "impurity" light, or by light and heat. Combined excitation allows us to observe a number of special effects and provides us with new methods of determining properties of substances.

We shall consider briefly some of the effects of combined excitation.

§44. INDUCED IMPURITY PHOTOCONDUCTIVITY

In some photoconductors a special effect known as "induced impurity photoconductivity" is observed: in a semiconductor in which impurity photoconductivity is normally absent or slight, preliminary illumination in the fundamental band produces or strongly enhances the impurity photoconductivity. Thus Fig. 129, which gives the photoconductivity spectra of some materials at the temperature of liquid nitrogen, shows that these materials are not normally photosensitive in the impurity region. However, after preliminary illumination in the fundamental-band region, an induced impurity maximum of the photoconductivity appears in the spectrum [2, 8, 9].

The explanation of this effect is that the "fundamental" excitation fills the empty impurity centers and therefore gives rise to the impurity photosensitivity. Obviously such a photosensitivity due to "nonequilibrium" population of the impurity centers should disappear in time. However, it is found that at low temperatures this photosensitivity is retained for a very long time, reaching perhaps many hours.

We shall consider the problem of the induced photoconductivity kinetics assuming the system of transitions shown in Fig. 130 [8].

Let us assume that there is in the forbidden band of the crystal an equilibrium concentration M of empty electron-trapping levels which become filled with electrons after excitation with light from the fundamental absorption band transfers electrons from the valence to the con-

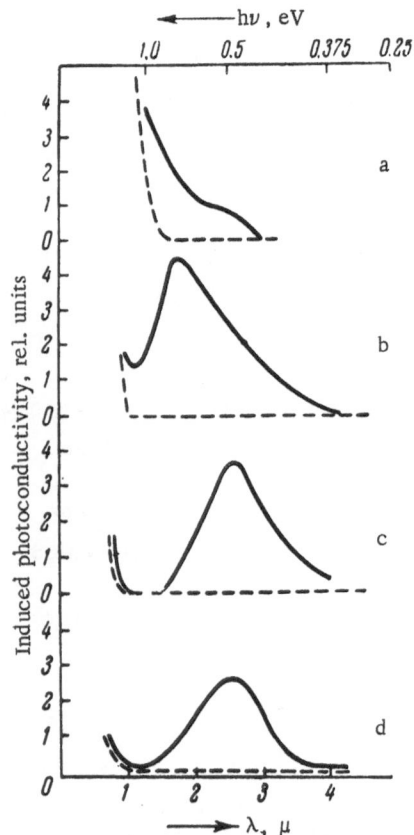

Fig. 129. Photoconductivity spectra of CdSe, CdTe, Sb₂Se₃, and CdS without preliminary illumination (dashed curves) and after preliminary "fundamental" illumination (continuous curves) [8, 9, 11]. a) Sb₂Se₃; b) CdTe; c) CdSe; d) CdS.

duction band (we shall denote the trapped electron density by m). The holes which are then formed in the valence band are captured by recombination centers S which are filled at equilibrium.

Using this model we can predict qualitatively the nature of the photocurrent relaxation. We shall assume that as a result of the preliminary fundamental-band illumination the M levels are mainly filled with electrons and an equal number of holes is localized at the S levels.

If the temperature is sufficiently low, then the electrons at the trapping levels may remain there permanently. *

We shall now consider the nature of the relaxation process when a semiconductor in the "frozen" nonequilibrium state begins to be illuminated with the "impurity" light.

On absorption of this light electrons are transferred from the M levels to the conduction band and induced impurity photoconductivity appears. However, this conductivity should decay with time (Fig. 131) for the following reason: initially the electrons which have reached the conduction band will have a higher probability of recombining with holes at the S levels than of retrapping by the M levels because the hole density at the S levels is high and the number of unfilled M centers is small. Consequently the electrons from the M levels will gradually be pumped to the S levels via the conduction band and this disturbs the equilibrium. The electron density at the M levels, and consequently the rate of impurity transitions, will

_____

* Because thermal transitions to the conduction band are absent and indirect electron transitions from the M levels to the holes at the S levels are impossible since the M and S levels are spatially separated.

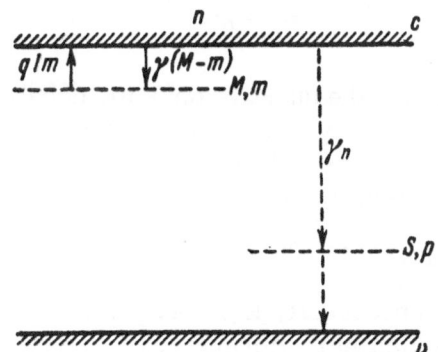

Fig. 130. Electron transitions; dashed arrows represent transitions absent in the "quasi-equilibrium" excited state.

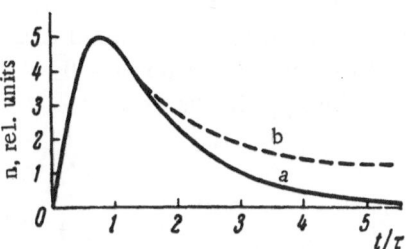

Fig. 131. Photocurrent "flash" (theoretical curves): a) For constant recombination and absence of re-trapping; b) for recombination decreasing and re-trapping increasing with time.

decrease. However as the M levels are emptied and the S levels filled, the probability of electron retrapping by the M levels increases and conversely the probability of their recombination with holes at the S levels decreases. Consequently after a certain time the exchange with the recombination levels may stop and the semiconductor may reach a state of "quasi-equilibrium" (curve b) in which it is photosensitive in the impurity region for a long time even during illumination. *

We shall confirm the above conclusions by the following approximate quantitative treatment.

The transport processes in the case considered (Fig. 130) may, in general, be described by the following system of differential equations:

$$\frac{dn}{dt} = qmI + \gamma_m m N_{cM}$$
$$- \gamma_{n}n(M - m) - \gamma_S n(n + m), \quad (44.1)$$

$$\frac{dm}{dt} = \gamma_m n(M - m) - qmI - \gamma_m m N_{cM}. \quad (44.2)$$

where γ_S is the recombination coefficient.†

We shall consider the special case of low temperatures ($\gamma_m m N_{cM} \approx 0$) and of the M levels completely filled with electrons by the preliminary illumination. We can then neglect retrapping and the system (44.1)-(44.2) may be written as follows:

$$\frac{dn}{dt} = qmI - \gamma_S n(n + m), \quad (44.3)$$

* The sensitivity is retained without illumination because at sufficiently low temperatures the captured carriers remain indefinitely at the trapping levels. The sensitivity is retained during illumination only if the probability of retrapping is much greater than the probability of recombination.

† In fact γ_S represents the coefficient of electron capture by the S levels. Since it is assumed that all holes in the valence band are rapidly captured by the S levels, γ_S governs the recombination process.

$$\frac{dm}{dt} = -\,qmI. \qquad (44.4)$$

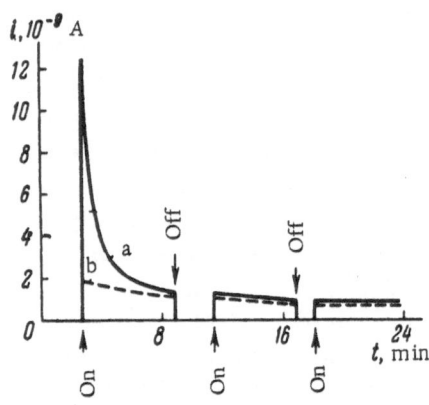

Since the quantity (m + n), equal to the density of holes p which are trapped at the S levels, at first varies only slightly, we shall assume that

$$\frac{1}{\gamma(n+m)} = \tau = \text{const} \quad \text{approximately}$$

and consequently Eq. (44.3) may be rewritten in the form

$$\frac{dn}{dt} = qmI - \frac{n}{\tau}. \qquad (44.5)$$

Fig. 132. Dependence of the induced photocurrent on time for the impurity region of CdS: a) High intensity of preliminary illumination in the fundamental region; b) low intensity of preliminary fundamental illumination.

Integration of the system (44.4)–(44.5) for the case of illumination, i.e., when the initial conditions are n = 0 and m = m_0 ≈ M when t = 0, gives the following expression for the electron density in the conduction band:

$$n = \frac{qIM}{\dfrac{1}{\tau} - qI}\left(e^{-qIt} - e^{-\frac{t}{\tau}}\right). \qquad (44.6)$$

Figure 131 gives the dependence n(t) plotted from Eq. (44.6)(curve a); a "flash" occurs which decreases to zero at high values of t. Numerical analysis shows that the general nature of the curve is similar to that occurring under normal conditions when during relaxation the value of τ does not remain constant but increases with decrease of the hole density, and the trapping of electrons becomes more and more likely. In the latter case, however, the curve at high values of t does not decrease to zero but approaches a practically constant value representing the quasi-equilibrium excited state of the crystal (curve b in Fig. 131). This curve is in qualitative agreement with experimental results [8].

Figure 132 shows that the photocurrent relaxation at the beginning of impurity illumination (after preliminary fundamental illumination) can be divided into two stages:

1) the initial "flash" rise with the photocurrent maximum followed by a relatively fast decay;

2) a very slow decay with the photocurrent remaining practically constant for a long time (the time constant for this stage may amount to hours).

Ignoring the initial "flash" region, we may assume that the preliminary illumination produces a long-lived quasi-equilibrium excited state with an impurity sensitivity which, in the first approximation, does not vary with time. *

Obviously under these conditions of quasi-constant sensitivity the induced impurity photoconductivity may be studied by the same methods as used for normal impurity conductivity. In particular, by studying the induced photoconductivity kinetics we can (§ 42) determine such parameters of the impurity centers as the cross sections for the capture of electrons (holes) or photons by the centers, the concentration of the centers and their population after the preliminary fundamental illumination.

Experiments have shown that, as expected, the kinetics of the induced impurity photoconductivity investigated under the quasi-equilibrium conditions exhibit the same features as the kinetics of the normal impurity photoconductivity.

Experimental curves (Figs.133a and 133b) are in good agreement with the theoretical dependences of Figs.126a and 126b for the linear and nonlinear impurity photoconductivity.

By way of example, we list below the data on the photon capture cross sections for some levels in CdTe and CdS, obtained in a study of the induced impurity photoconductivity:

$$CdS \quad Level \quad \mathscr{E}_c - 0.3\ eV \quad q = 3 \cdot 10^{-15}\ cm^2,$$
$$CdTe \quad " \quad \mathscr{E}_c - 0.33\ eV \quad q = 2 \cdot 10^{-15}\ cm^2,$$
$$CdTe \quad " \quad \mathscr{E}_v + 0.3\ eV \quad q = 2 \cdot 10^{-14}\ cm^2.$$

§ 45. OPTICAL CHARGE EXCHANGE BETWEEN IMPURITY CENTERS

Illumination of a semiconductor not only changes the densities of free electrons and holes in the bands but also changes the population of the local impurity levels in the forbidden band. This very common phenomenon, present in the nonlinear photoconductivity case and appearing directly in such effects as induced impurity photoconductivity (§ 44), can be investigated particularly easily in the case of optical charge exchange in multicharged impurity centers in semiconductors.

To illustrate this, we shall consider qualitatively the phenomena which may be expected to occur during optical excitation of a semiconductor with a multicharged impurity.

Let us assume that this impurity has three acceptor levels in the forbidden band (for example, copper in germanium) and that at equilibrium, because of compensation, the two lower levels are completely filled with electrons and the upper level is empty (Fig. 134a). † When

---

* In this state the photocurrent rise and decay times on application and removal of the impurity illumination are relatively short: from several milliseconds to several seconds (cf. Fig. 132).
† Here, as well as in Chap. VII, we are using one of several possible methods of representing the energy spectrum of multicharged centers with strongly interacting charges. The method is quite simple and "economical." The true energy levels for a given charge state of the centers (the "active" levels) are denoted by thick dashes; electrons at these levels are denoted by black dots. Thin dashes represent the "inactive" levels (which are potentially active in other charged states). Open circles (electrons) at the inactive levels allow us to calculate the total charge of the center.

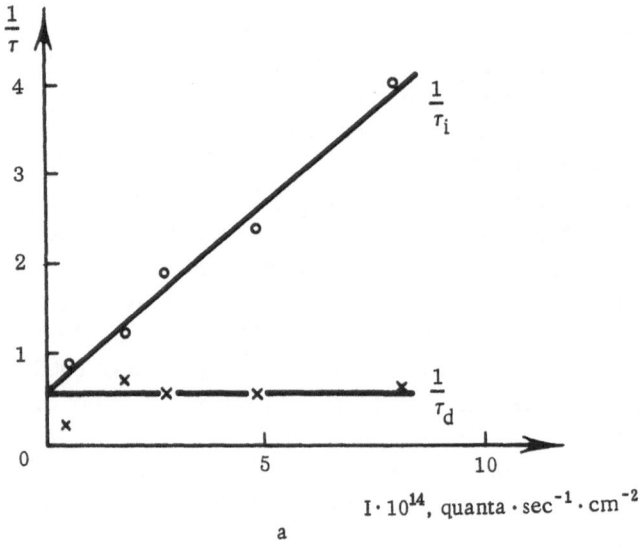

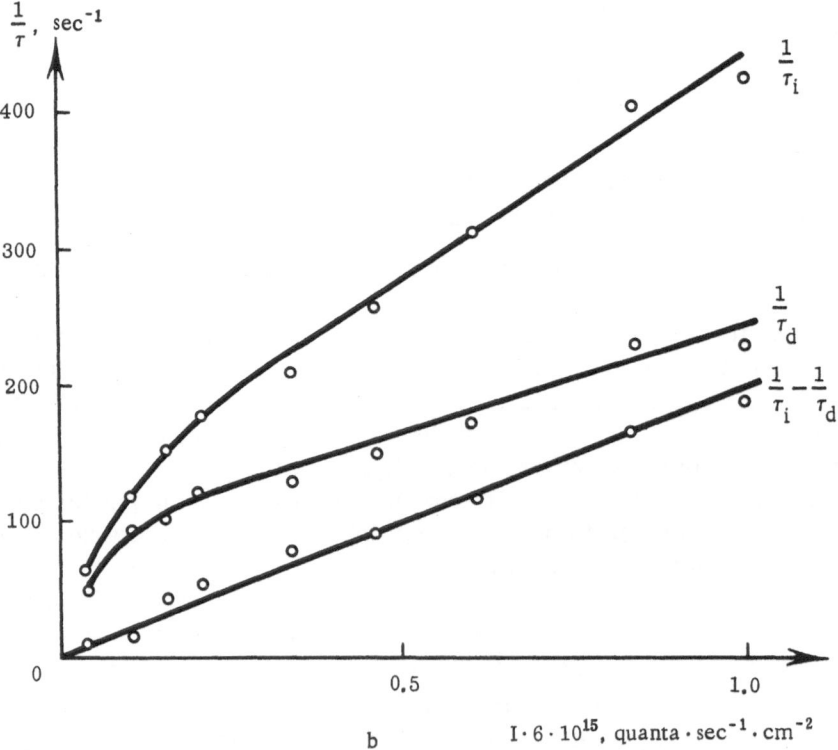

Fig. 133.. Dependence on the intensity of illumination of the reciprocal values of the rise τ_i and decay τ_d time constants of the induced impurity photoconductivity of CdTe in the quasi-equilibrium state [11]: a) Low population of the levels; b) high population of the levels.

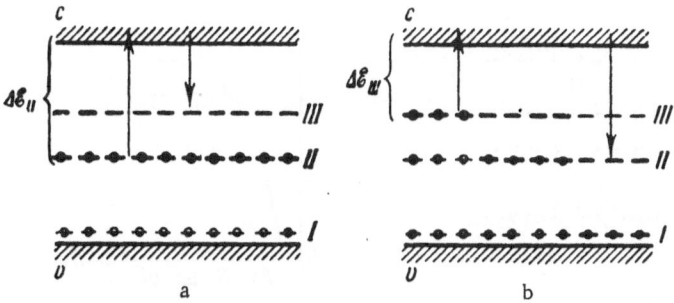

Fig. 134. Semiconductor with a multicharged impurity, in which charge-exchange processes occur. a) Semiconductor at equilibrium. Illumination with light of $h\nu \gtrsim \Delta\mathcal{E}_{II}$ gives rise to electron transitions from II levels to III levels via the conduction band ("forward charge exchange"). b) Nonequilibrium conditions. Illumination with light of $h\nu \gtrsim \Delta\mathcal{E}_{III}$ transfers electrons from III levels to II levels via the conduction band ("reverse charge exchange," equilibrium reestablished).

such a semiconductor is illuminated with quanta of $h\nu \gtrsim \Delta\mathcal{E}_{II}$ the electrons from level II are transferred to the conduction band and then they can be captured either by III levels or by II levels (i.e., they may be recaptured). Since initially there are more empty III levels than II levels, capture occurs primarily at the former.

Thus as a result of illumination, the electrons from II levels are transferred to III levels, i.e., we have here a process of "charge exchange between centers," which makes some of the centers singly-charged by loss of electrons to other centers which themselves become triply-charged (Fig. 134b). *

If the temperature is sufficiently low, the nonequilibrium state (Fig. 134b) may exist for an indefinite time after illumination because the equilibrium can be reestablished only through the conduction band, i.e., there is a barrier to be overcome. † However, "reverse charge exchange" can easily be produced by illumination with light whose quanta have $h\nu \gtrsim \Delta\mathcal{E}_{III}$ (Fig. 134b). Since the processes of forward and reverse charge exchange proceed through the conduction band, they are accompanied by photoconductivity with very special kinetics.

---

* The same charge-exchange effect can be obtained by fundamental excitation ($h\nu > \Delta\mathcal{E}$). The electrons and holes are then captured by different levels (holes by the filled II levels, electrons by the empty III levels), which is equivalent to electron transitions from some copper centers to others (i.e., equivalent to charge exchange).

† The situation when a barrier must be overcome to destroy the nonequilibrium charge-exchange state is called "quasi-stable charge exchange."

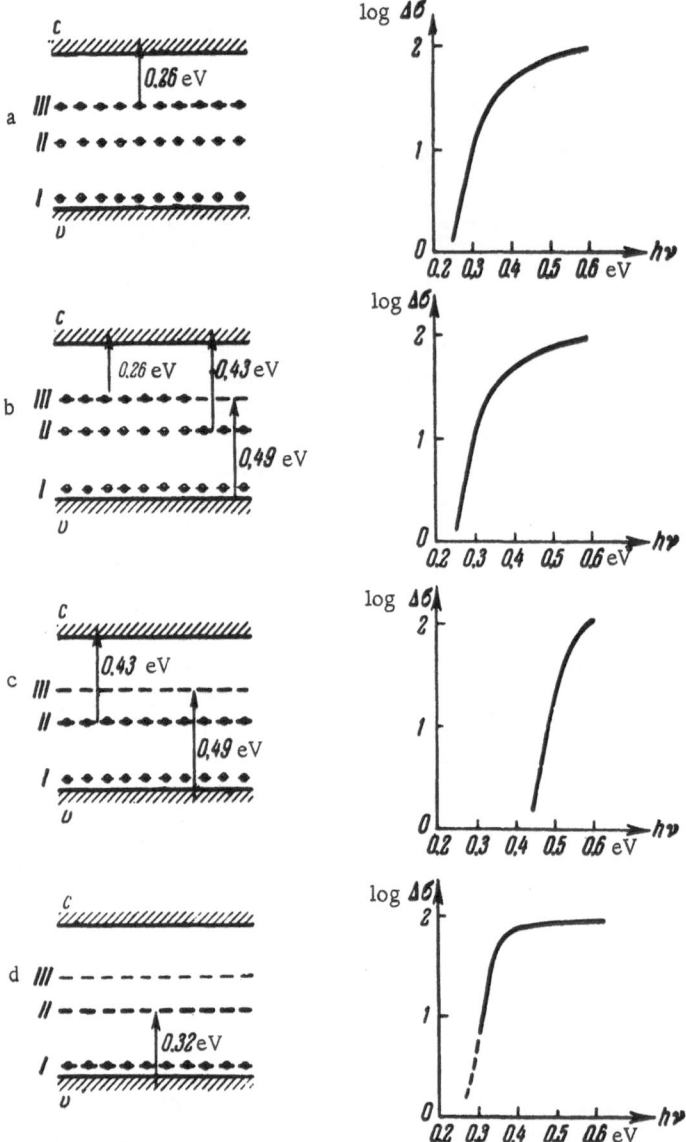

Fig. 135. Energy structure and photoconductivity spectra of four
groups of germanium samples compensated to different extents.

Up to now we have taken "charge exchange" to mean changes of the
impurity charge due to electron (hole) transitions between two identical
multicharged impurity centers. Later we shall find it convenient to extend the term "charge exchange" to mean charge transitions between any
impurities, not necessarily identical and not necessarily multicharged.
From this point of view charge exchange governs such well-known phenomena as induced impurity photoconductivity (§ 44), thermally stimulated
conductivity (§ 46), etc.

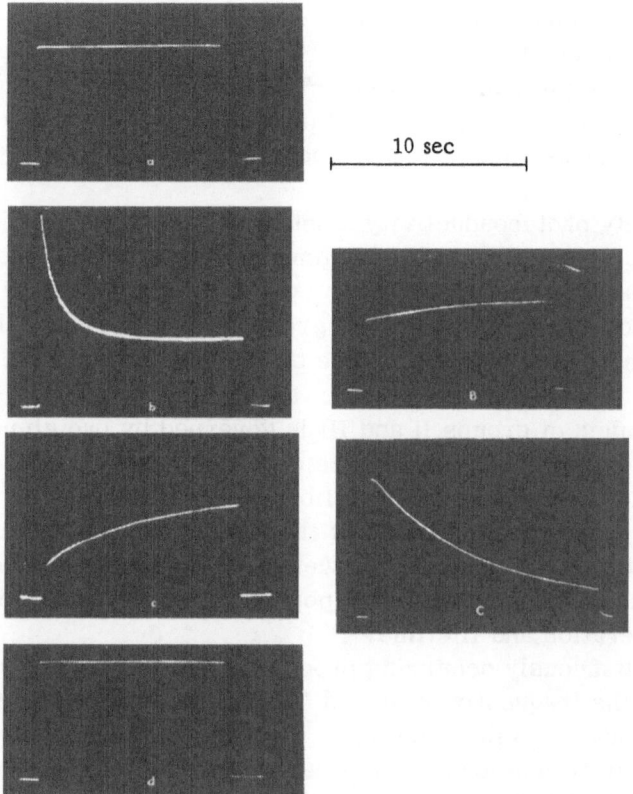

Fig. 136.   Oscillograms of impurity photoconductivity relaxation for the
four groups of samples at $T = 77°K$.

Optical charge exchange should be important in many photoconductors, not only as regards the induced photoconductivity but also in its dominant influence on the kinetics of photoelectric processes.   Charge exchange also makes it possible to observe several special effects, such as induced impurity breakdown, hopping photoconductivity, etc.

By way of example, we shall consider below (in subsections A and B) the results of an experimental study of the photoconductivity kinetics in germanium with copper impurity, and we shall interpret these results on the basis of charge-exchange effects [13].   Subsection C describes briefly the phenomena of induced breakdown and of the related current oscillations, and subsection D deals with the phenomenon of inter-impurity recombination under conditions of "hopping photoconductivity. "

A.   Influence of Charge Exchange on the Kinetics of Impurity Photoconductivity

Different degrees of electron population of the copper levels in germanium may be obtained by compensating the initial donor levels and the acceptor levels of copper; the electron population depends on the concentrations of the donors and copper.

We shall consider the data for four groups of samples which cover all possible degrees of compensation. The energy band structures are given for these groups in the left-hand part of Fig. 135. The right-hand part of the same figure shows the photoconductivity spectra in the impurity region.

The impurity photoconductivity kinetics is illustrated by the curves on the left-hand side of Fig. 136. In general, these curves show "fast" and "slow" relaxation regions. The curves for groups I and IV show only the "fast" component. The curves for groups II and III show both the "fast" and "slow" components. The presence of two components with different time constants is explained as follows: the impurity photoconductivity kinetics in samples of groups II and III is governed by two simultaneous processes. The illumination transfers electrons (or holes) into the allowed bands and gives rise to nonequilibrium conductivity, the magnitude of which is governed by the product of the generation rate and the lifetime of nonequilibrium carriers. However, this process is accompanied by simultaneous changes in the level population which lead to changes in the rate of generation and lifetime.

Such simultaneously occurring processes are, in general, the obvious cause of the frequently observed "nonlinear" photoconductivity. However, in contrast to the usual case when the time constants of the photoconductivity relaxation and generation-rate relaxation are of the same order as the lifetimes, under the conditions considered here these time constants are very different in the case of samples of groups II and III. As a result of this, the impurity photoconductivity kinetics has "fast" and "slow" components.

The steady-state photoconductivity is established rapidly (in one lifetime, which amounts to a fraction of a millisecond). The photoconductivity then varies slowly (over hundreds of seconds) because of the slow change in the lifetime and generation rate due to charge exchange between the centers. Obviously, in the simplest case of impurity photoconductivity which involves one type of level, $\Delta n \equiv \Delta m$ and the time constants of both processes are identical (this is the case in samples of groups I and IV). A difference may occur only when, in addition to the principal "channel" along which the transitions between the levels and the band occur, we can also have electron or hole "leakage" along an additional "channel" to some other levels. In germanium this leakage is governed by the slow charge exchange between copper centers. Bearing this in mind, the relaxation curves of Fig. 136 can be explained as follows.

Illumination of samples of groups I and IV in the impurity region produces electron exchange between one level (level III for group I and level II for group IV) and the corresponding band. Other copper levels do not take part in this process* and the relaxation curves have only the "fast" components, which are governed by the time for the establishment of equilibrium between the band and the levels.

In a sample of group II the action of light with quanta for which $0.43\,\text{eV} > h\nu > 0.26\,\text{eV}$ transfers electrons from level III to the conduction band and these electrons may return only to the same level III. Under these conditions, naturally the relaxation again has only the "fast" component (similar to the curve in Fig. 136a).

When samples of group II are illuminated with light of energy $0.49\,\text{eV} > h\nu > 0.43\,\text{eV}$, the liberated electrons may "recombine" with II and III levels. At the beginning of the process, when there are few empty II levels, recombination occurs mainly at III levels, whose capture cross section is small because of the Coulomb repulsion of the doubly-charged negative center. However, as levels II with the larger capture cross section are emptied, their role in the recombination increases and the electron lifetime decreases. Consequently, the relaxation curve in Fig. 136b shows slow decay. This decay is due to charge exchange between copper centers which reduces the number of doubly-charged centers and increases the number of singly-charged ones.† A sample can retain this "charge-exchange" state for a very long time (many hours) at the test temperature of 77°K.

If the sample is now illuminated with light of energy $0.43\,\text{eV} > h\nu > 0.26\,\text{eV}$, reverse charge exchange takes place, producing an increase in the lifetime. Consequently, the photoconductivity increases slowly with time (Fig. 136B).

The most interesting case is the relaxation of samples of group III (Figs. 135c and 136c). In samples of this group there are no transitions from the upper level to the conduction band. On illumination with quanta of energy $0.49\,\text{eV} > h\nu > 0.43\,\text{eV}$, transitions are produced from II levels to the conduction band and, as in the case of samples of group II, initially the electrons from the conduction band are captured mainly by III levels as long as these are free. Consequently, an intense charge exchange occurs. In principle, this process may alter both the generation rate and the lifetime. Direct measurements during the initial stages of the photoconductivity rise and decay have shown that, in contrast to samples of group II, charge exchange in samples of group III mainly changes the generation rate. The generation rate increases when electrons are

---

*In fact, samples of group IV exhibit the effect of hole trapping at levels I, but this does not affect greatly the nature of the observed effects and we shall not consider it here.

† Since each elementary charge-exchange act reduces by 2 the number of the empty III levels, and the number of II levels increases only by 1, the reduction of the electron lifetime in the conduction band is possible only when the cross section of the electron capture by II levels is not less than double the cross section for III levels.

transferred from II levels to III levels. Since each such transition represents a reduction in the number of absorbing centers per unit volume, obviously a rise of the generation rate is possible only if for a given wavelength the cross section for photon capture by an electron at level III is more than double the cross section for level II. Because of the increase in the generation rate, the relaxation curve of Fig. 136c has a slowly rising component.

Group III samples, in the same way as group II samples, can retain the nonequilibrium "charge-exchange" state for a very long time at low temperatures. If a sample of group III is illuminated with light of energy $0.43 \, eV > h\nu > 0.26 \, eV$, electrons from III levels are transferred through the conduction band to II levels and the system returns to the equilibrium state after a time. Thus the reverse charge-exchange process should appear in the form of a photoconductivity flash. Figure 136C shows the experimentally obtained oscillogram of such a flash.

In conclusion, we note that when $h\nu > 0.49 \, eV$ we can, in principle, have transitions from the valence band to III levels and the appearance of holes in the case of samples of groups II and III. However, the lifetime of the holes is small because of the large cross section for hole capture by the negatively charged copper centers and, consequently, the hole component of the photocurrent is very small compared with the electron component.

B. "Limiting" Optical Charge Exchange Between Impurity Centers

When samples of group III are illuminated with light of energy $0.49 \, eV > h\nu > 0.43 \, eV$, electrons from II levels are transferred to III levels via the conduction band. After illumination, the nonequilibrium "charge-exchange" state may exist in darkness at low temperatures for a very long time. The process of reverse charge exchange, which occurs on illumination with light of energy $0.32 \, eV > h\nu > 0.26 \, eV$, is recorded in the form of a photoconductivity flash (Fig. 136C), the amplitude of which represents the concentration of the nonequilibrium triply-charged centers.

Figure 137a shows that when the preliminary illumination is of sufficiently long duration, the amplitude of the reverse flash and, consequently, the concentration of the centers in the "charge-exchange" state, tend to approach a limit. The higher the intensity of the preliminary illumination, the earlier is this limit reached (the process closely satisfies the reciprocity law, i.e., the rate of the forward charge exchange is proportional to the product of the intensity of light and time). The

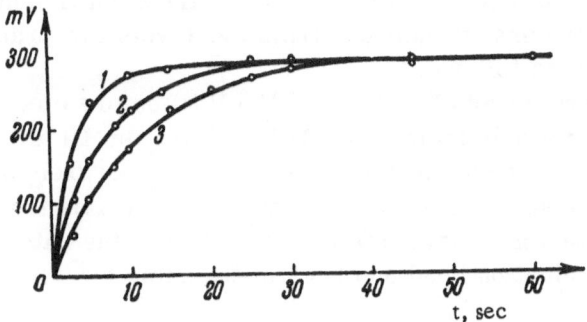

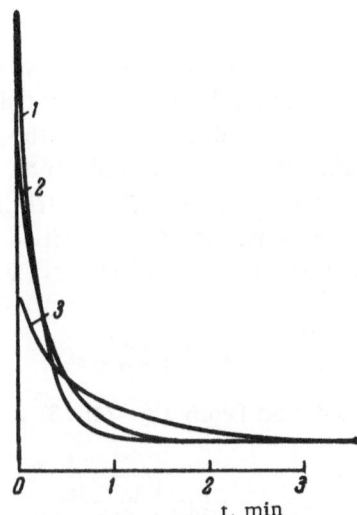

Fig. 137. Limiting charge-exchange characteristics. a) Dependence of the amplitude of the reverse charge-exchange flash on the duration of illumination which produced the forward charge exchange. The intensity of illumination with quanta having $0.49\ eV > h\nu > 0.43\ eV$, which produced the forward charge exchange, was 100% for curve 1, 45% for curve 2 and 25% for curve 3. The intensity of illumination with quanta having $0.32\ eV > h\nu > 0.26\ eV$, which produced the reverse charge exchange, was constant. b) Reverse charge-exchange flashes at different intensities of the $0.32\ eV > h\nu > 0.26\ eV$ illumination. The illumination was 100% for curve 1, 60% for curve 2 and 20% for curve 3. The front of the flash was distorted because of the inertia of the recording instrument.

"limiting charge exchange" effect, which occurs at sufficiently low temperatures when thermal transitions from the levels are practically absent,
may be explained as follows.

Light of energy 0.49 eV > $h\nu$ > 0.43 eV transfers electrons to the
conduction band not only from II levels but also from III levels. The rate
of this "reverse" process should increase with the degree of electron
population of III levels. Obviously, after a certain time from the beginning of illumination a steady state is reached when the rate of optical transitions to the conduction band and back is the same for each of the levels
separately:

$$q_3 m_3 I = \gamma_3 n (m_2 - m_3),\tag{45.1}$$

$$q_2 (m_2 - m_3) I = \gamma_2 n (N - m_2);\tag{45.2}$$

here q_2, q_3 and γ_2, γ_3 are, respectively, the cross sections for the
capture of photons and electrons by levels II and III; I is the intensity of
light of energy 0.49 eV > $h\nu$ > 0.43 eV; n is the density of electrons in
the conduction band (which is zero in the absence of illumination); m_2 and
m_3 are the electron densities at levels II and III, respectively; N is the
concentration of copper centers (cf. Chap. VII).

Eliminating n from Eqs. (45.1) and (45.2), we obtain

$$\frac{q_2 (m_2 - m_3)}{\gamma_2 (N - m_2)} = \frac{q_3 m_3}{\gamma_3 (m_2 - m_3)} ;\tag{45.3}$$

since $m_2 + m_3 + n = N$, we find from Eq. (45.3) that

$$m_3 = N \frac{2 q_2 \gamma_3 - \sqrt{q_2 q_3 \gamma_2 \gamma_3}}{4 q_2 \gamma_3 - q_3 \gamma_2} .\tag{45.4}$$

The above expression is obtained on the assumption that the illumination intensity is sufficiently low for us to neglect the electron density n
in the conduction band compared with the electron densities at the levels
II and III. Such conditions can be realized in practice.

Equation (45.4) shows that the electron density at the III levels is independent of the illumination intensity, i.e., we have here the experimentally observed limiting charge exchange (Fig. 137a) which is governed
only by the relationships between photon and electron capture cross sections. *

---

* In experiments, we frequently have the condition when $q_2 \gamma_3 / q_3 \gamma_2 \ll 1$. It is easily seen that
in this case the expression (45.4) simplifies considerably, becoming

$$\frac{m_3}{N} = \sqrt{\frac{q_2 \gamma_3}{q_3 \gamma_2}} .\tag{45.5}$$

The cross-section ratios may be determined by use of the limiting charge exchange and other methods. The relaxation curves in the reverse charge-exchange case should obviously depend on the intensity of light with energy 0.32 eV $> h\nu > 0.26$ eV. Figure 137b gives the experimental curves. They show that, as expected, the higher the illumination intensity, the shorter and "brighter" the reverse charge-exchange flash.

In conclusion, we note that the "reversible" and relatively simply realized optical charge exchange will undoubtedly extend the methods of studying the structure of impurity centers in various charged states (for example, paramagnetic resonance, etc., may be used).

C. Current Oscillations Related to Induced Impurity Breakdown

It is known that germanium and silicon containing impurities of group V or III elements (we shall consider a donor impurity to make the case definite) exhibit impurity breakdown in relatively weak fields (5-10 V/cm) due to the impact ionization of shallow impurities [14-16]; this happens at sufficiently low temperatures when all the electrons are concentrated at the impurities. However, if the donor impurity is compensated by an acceptor (Fig. 138a) impurity breakdown becomes impossible because all the electrons move from the shallow donors to the deep acceptors. However, conditions favoring breakdown may be reestablished by illuminating the compensated semiconductor using, for example, light from the fundamental absorption band.

In a sample in which the initial donors are completely compensated the electrons and holes generated by light from the fundamental band are first captured by various centers (Fig. 138b). Electrons are captured by vacant donor levels and holes by the compensating acceptor levels, which are filled with electrons at equilibrium. Thus illumination fills the donor levels with nonequilibrium electrons (Fig. 138c). If the temperature is sufficiently low, then after illumination electrons may remain for a very long time at the donor levels since direct recombination with holes at acceptors is impossible and thermal transitions to the conduction band are few (quasi-stable charge exchange). If a sufficiently strong electric field is applied to a sample under these conditions, we should observe the breakdown of donor levels which, in contrast to breakdown involving equilibrium carriers, is called "induced impurity breakdown" (IPB) [17, 18].

In contrast to normal breakdown, IPB kinetics should have special features. In normal impurity breakdown of levels filled with equilibrium charges the current reaches a constant steady-state value, while in the case of IPB a current "flash" should be observed on application of the field (i.e., the current should rise and then decay to zero) (Fig. 139).

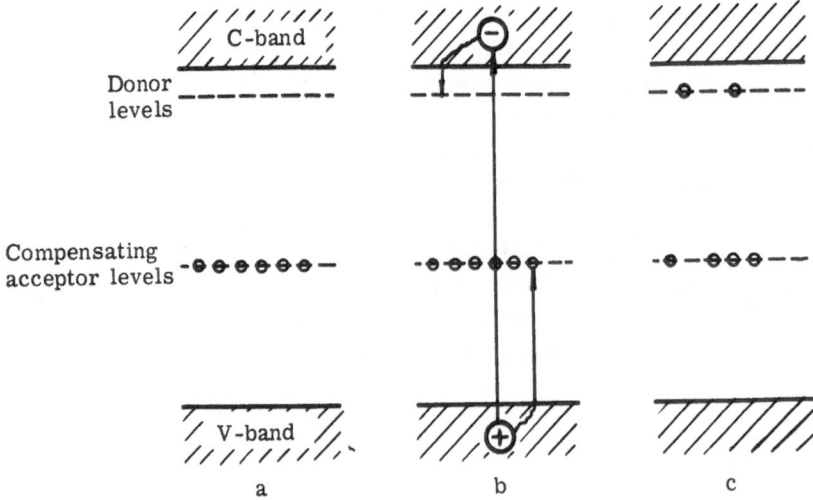

Fig. 138. Band structures illustrating the process of filling (charge exchange) of donor levels with nonequilibrium electrons during illumination.

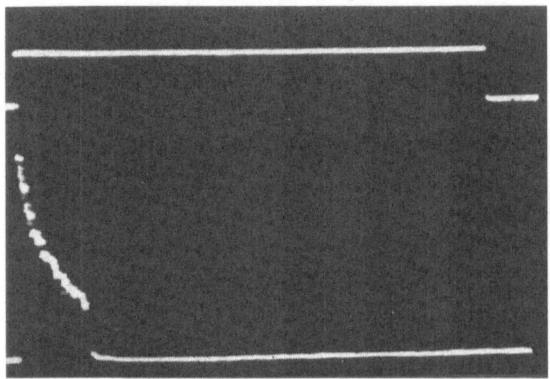

Fig. 139. Oscillograms of induced breakdown. The voltage pulse is shown above, and current pulse below.

This induced-breakdown current kinetics follows from the fact that nonequilibrium electrons, transferred by impact ionization from donor levels to the conduction band, can recombine with holes at the compensating acceptors. As a result of this the electron density at the donors and in the band (as well as the current) gradually decreases and finally the breakdown stops.

It is easy to see that the phenomenon of induced impurity breakdown is analogous to induced impurity photoconductivity, except that the optical impurity excitation is replaced by the excitation due to impact ionization.

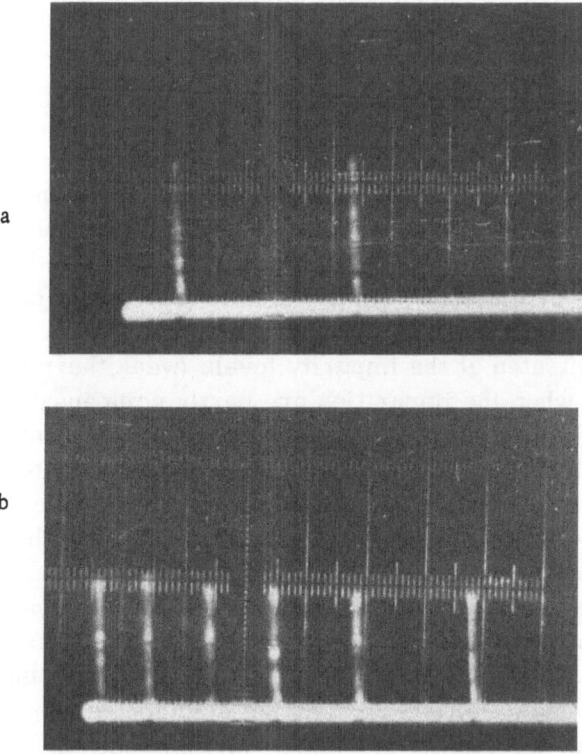

Fig. 140. Oscillations of the induced breakdown current
during continuous illumination in a constant electric field;
the illumination intensity was higher for (b) than for (a).

Induced impurity breakdown may, under certain conditions, be accompanied by current oscillations. If a field is applied to a compensated
sample and it is then subjected to continuous illumination, breakdown will
occur only after a certain time from the beginning of illumination when
the required degree of donor population with nonequilibrium electrons is
reached. On reaching this degree of population the breakdown and related recombination reduces the nonequilibrium electron density and this
stops the breakdown.

However, since the illumination is continuous, after a "flash" the
donors are gradually filled again until breakdown recurs, etc.*

Thus, under conditions of continuous illumination and constant field
we should have current oscillations. The oscillation frequency should be
governed by the intensity of illumination (and the strength of the electric
field). The higher the intensity of illumination, the shorter the time re-

---

*It must be stressed that the breakdown "striking" voltage in compensated samples is greater
than the voltage necessary for the maintenance of breakdown [19]. A detailed analysis [18]
shows the importance of this point in the explanation of the origin of oscillations.

quired to fill the donors with nonequilibrium electrons (i.e., the shorter
the interval between flashes) (Fig. 140).

D. Hopping Photoconductivity and Interimpurity Recombination

a. Hopping Conduction. It is known* that in germanium and silicon
with sufficient concentrations of impurities which form shallow levels (for
example, elements of groups III and V), we can have conduction along
these impurities by electron "jumps" from one impurity center to another.

This "hopping conduction" is observed at low temperatures when carriers are concentrated at the impurity levels (weak thermal ionization
in the band) and when the impurities are partly compensated, permitting
charged jumps from occupied impurity levels to vacant neighbours.

It is obvious that the magnitude of this conduction should depend on
the ratio of the levels that are occupied and free (due to compensation).

We shall consider this dependence qualitatively for the case of hopping
conduction between donors of group V.

In the absence of electrons at the donor levels (complete compensation) there is no hopping conduction. As the electron density at the donors
increases (degree of compensation decreases), the hopping conduction should increase. However, when the electron density becomes greater than about half the density of donors, further increase of the electron
density should reduce the conductivity because then the hopping conduction
is restricted by the decreasing number of vacant levels.

b. Hopping Photoconductivity. Illumination of a semiconductor
should affect the magnitude of its hopping conductivity [21, 22]. Illumination with light from the fundamental band produces charge exchange,
i.e., holes are captured mainly by electron-filled compensating acceptors,
while electrons are captured by donors and change the value of the hopping
conductivity. On illumination the hopping conductivity should increase
when the degree of compensation is more than half, and it should decrease
when the degree of compensation is less than half.

In the latter case we have "negative" photoconductivity [22].

c. Hopping Photoconductivity Kinetics. Nonequilibrium electrons,
located at donors and responsible for the hopping photoconductivity, should
recombine with holes captured by the compensating acceptors. It would
seem that the only way in which this recombination can occur is by thermal transitions of electrons to the conduction band, followed by recombination of these free electrons with holes at the acceptor levels. However, it has been found [23] that even at quite low temperatures when
there are practically no thermal transitions ("quasi-stable charge exchange"), the hopping conductivity disappears quite rapidly after illumina-

---

*See, for example, Mott and Twose's review [20].

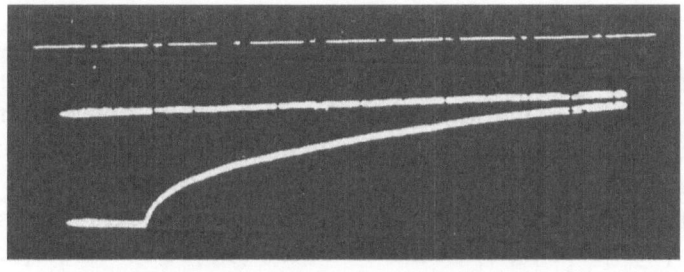

a

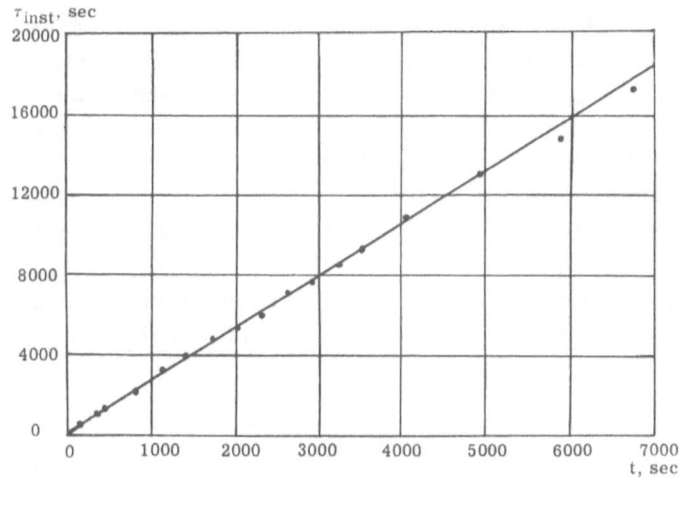

b

Fig. 141. a) Relaxation curve of hopping (negative) photoconductivity of germanium after illumination. The curve extends over two periods of the oscillograph scan (15 sec). The chain curve represents dark conductivity (zero line). T = 4.2°K. b) Dependence of the instantaneous life-

time ($\tau_{inst} = \dfrac{-\Delta\sigma}{d(\Delta\sigma)/dt}$) of the hopping photoconductivity on time from

the end of illumination.

tion. This forces us to assume that under the hopping photoconductivity conditions we have an essentially new type of recombination transition known as "interimpurity recombination" [23].

Hopping conduction is possible because at sufficiently high impurity concentrations the radius of the wave function of an electron located at a shallow impurity center is close in value to the distance between the impurities. Under these conditions we can also have considerable overlap of the wave functions of electrons at donors and of holes at acceptors. Consequently it is highly probable that an electron (at one impurity) will recombine directly with a hole (at another impurity), thus bypassing the conduction band.

Such interimpurity recombination should have characteristic photo-conductivity relaxation curves. The acceptor and donor centers are distributed at random in the lattice but the probability of an electron transition from a donor to an acceptor should depend strongly on their distance of separation.

Thus the holes captured by acceptor centers should be divided into various "types" depending on the distances between the acceptors and the nearest donor centers. After illumination the holes captured by acceptors which lie close to donors will recombine first. Next, the holes located further away from donors will recombine, and so on. Thus, the relaxation curves should, as found by experiment (Fig. 141), be strongly non-exponential with a very fast initial decay followed by a strong deceleration of the rate of decay at the end of relaxation. * Such "nonlinear" decay occurs at low excitation levels, while under normal conditions the decay is exponential. This is the characteristic feature of the relaxation in the impurity recombination case.

We have discussed here only the hopping photoconductivity and inter-impurity recombination conditions. In some cases the interimpurity internal photoeffect [24] (direct transition of an electron from an impurity center to a neighboring center on absorption of a light quantum) can be observed.

§46. THERMALLY STIMULATED CONDUCTIVITY

In §44 we considered the induced photoconductivity arising from the optical excitation of carriers trapped (as a result of a preliminary "fundamental" excitation) at impurity levels. However, as well as optical excitation, we can also have thermal excitation of trapped carriers, which should give rise to thermally-induced conductivity.

The best method of observing this effect (by analogy with the induced photoconductivity) would be as follows: first the trapping levels are filled with carriers by illumination with "fundamental" light at a very low temperature, at which there are practically no thermal transitions from the trapping levels to the appropriate band†; next the sample is rapidly ("instantaneously") heated to a certain temperature, which is equivalent to the application of long wavelength illumination in the case of induced photoconductivity; thermally-induced conductivity is then observed with kinetics, i.e., there should be a "flash" at the initial stage. The only dif-

* Under these conditions the lux-ampere characteristics are, as a rule, sublinear.
† "Quasi-stable charge exchange."

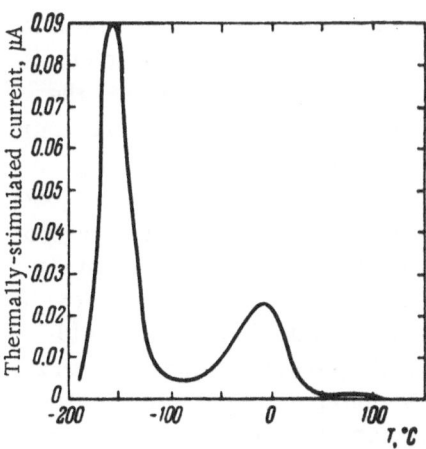

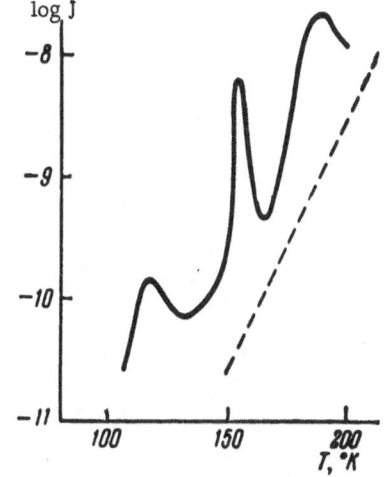

Fig. 142. Temperature dependence of the thermally-stimulated current in CdS samples [5]. The rate of heating was 0.77 deg/sec.

Fig. 143. Thermally-stimulated current for Sb_2Se_3 [6]. The rate of heating was 0.1 deg/sec (the dashed curve represents the variation in the dark current).

ference between thermally-stimulated conductivity and induced photo-conductivity is that the optical excitation, with intensity proportional to m and equal to qmI, is replaced by thermal excitation, which is also proportional to m and equal to γmN_cM.

Thus at a constant illumination intensity and constant temperature (within governs N_cM), the phenomena of induced photoconductivity and thermally-stimulated conductivity are identical. The relaxation curve is again given by an expression of the type of Eq. (44.6) where, however, qI should be replaced by $\gamma_m N_cM$.

Unfortunately, it is experimentally difficult to heat a sample sufficiently rapidly to produce a square excitation step; therefore, the phenomenon of thermally-induced conductivity is rarely investigated in the way indicated above.

More frequently, the phenomenon is studied differently: by measuring the "thermally-stimulated currents."

In this method a sample, illuminated first with "fundamental" light at low temperatures, is heated slowly. The carriers trapped at a given level are transferred more and more rapidly to an allowed band and the conductivity increases with time (temperature). However, since the levels are gradually emptied, there should be a moment (temperature) at which the thermally-stimulated conductivity begins to fall to zero. Thus the dependence of the thermally-stimulated conductivity on time (or temperature) should be a curve with a maximum.

The position and shape of the maximum give information on the trapping-level parameters. The disadvantage of using thermally-stimulated conductivity to study these parameters is the relative complexity of the

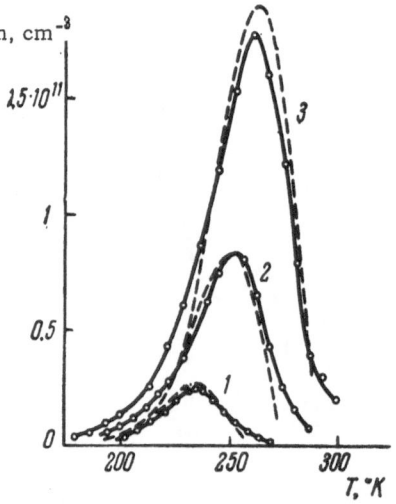

Fig. 144. Comparison of the theoretical (dashed) curves for the thermally-stimulated conductivity (with the experimental (continuous) curves for CdS samples doped with gold [7]. Heating rate: 1) 0.19 deg/sec; 2) 0.58 deg/sec; 3) 1.43 deg/sec.

process in which both time and temperature vary, and the consequent complex relationship between the characteristics of the observed curve and the level parameters.

However, the method described has the very important advantage of allowing us to investigate complex spectra with several types of levels lying at various depths by a relatively simple experimental method.

Obviously, during slow heating of the sample from low temperatures we shall first empty the shallowest level, then the next deeper one, and so on. Thus a curve representing the thermally-stimulated current will have several maxima, each of which represents a certain level. Naturally, we can resolve the maxima only when the levels are located at relatively large energy intervals.

Figures 142 and 143 give the experimental curves of the thermally-stimulated current for CdS and Sb_2Se_3.

Qualitative considerations of the position of a thermally-stimulated current maximum allow us to determine the depth of the corresponding level in the following way.

If heating is sufficiently slow, we may assume that at any moment the given level and the band are in quasi-equilibrium, i.e., the density of carriers in the band and at the level may be represented by a single Fermi quasi-level. At the beginning of the heating run, all the levels are filled with electrons (to make the case definite, we shall consider electron trapping) and, consequently, the Fermi quasilevel lies between these levels and the conduction band. As the temperature increases, the levels considered are emptied and the Fermi quasilevel moves downwards, approaching the trapping level. When the Fermi level coincides with the trapping levels, the population of the trapping levels is only one-half. Further depopulation of the trapping levels makes the Fermi level move further down.

Obviously, the trapping levels are emptied and the corresponding maximum in the thermally-stimulated conductivity will appear in that region of temperatures at which the Fermi quasi-level is close to the trapping levels.

Assuming approximately that the maximum corresponds to the situation in which the Fermi level coincides with the trapping levels, and bearing in mind that $n = N_C \exp{(F/kT)}$, we obtain

$$\Delta\mathscr{E}_M \cong F = kT_{max} \ln \frac{N_c}{n_{max}} . \tag{46.1}$$

Thus, knowing T and n at the maximum, we can determine the depth of the trapping levels.

A more accurate determination of $\Delta\mathscr{E}_M$ on the basis of the available rigorous theory of the phenomenon requires allowance for several factors, including the dependence of τ on time and temperature, dependence of the mobility on temperature, degree of the initial population of the trapping levels, rate of heating, etc.

Figure 144 shows that there is good agreement between the theory and experiment.

§47. SIGN OF THE PHOTOCONDUCTIVITY OF HIGH-RESISTIVITY SEMICONDUCTORS

There is a large group of semiconducting materials which exhibit very low conductivity without being subjected first to special purification processes (such as those used to purify germanium and silicon).

Since there is every reason to assume that the number of accidentally present impurities in such materials is high, * it is obvious that the low conductivity of these materials may be related to the weak ionization of their impurities, i.e., the fact that the levels of these impurities lie quite far from the edges of the allowed bands.

Thus the energy-band structure of a low-conductivity semiconductor may have a large number of deep levels, and because of the weak ionization the total concentration of these levels (there may be i types of these levels) is considerably greater than the density of free carriers

$$\sum_i M_i \gg n + p.$$

It is easily seen that under these conditions the electron and hole lifetimes are independent (§ 23, Chap. V) and, in general, unequal. Electrons are captured by empty impurity levels and holes by occupied levels, and the concentration of the two types of level is governed by equilibrium conditions (position of the Fermi level); it is very unlikely that these conditions would be such that the electron and hole lifetimes are of the same order.

---

* For example, "pure" chemical reagents used to prepare such substances contain 0.1% impurities. If we assume that a considerable fraction (say, one-tenth) of these impurities is dissolved in the host lattice, this corresponds to the enormous impurity concentration of 10^{18} cm^{-3}.

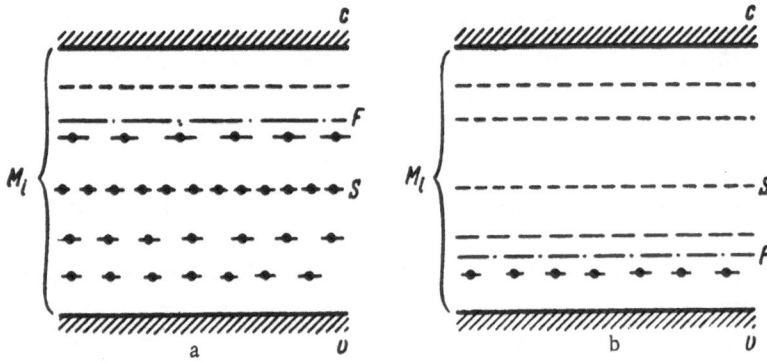

Fig. 145. Explanation of the correlation of the sign of the photoconductivity and of the equilibrium conductivity. Level population in a) an n-type semiconductor; b) a p-type semiconductor.

From general considerations, one may predict the probable dominant type of photoconductivity (n-type or p-type) from the equilibrium conductivity.

Let us assume that the forbidden band of the semiconductor contains a set of different levels M_i (Fig. 145). For simplicity, we shall assume that one of these levels, which will be denoted by S, is present at a density much greater than the densities of all the other levels. * Then if the semiconductor is of n-type and the Fermi level is close to the conduction band, the majority of states at S levels are filled with electrons (Fig. 145a), and they are more likely to capture holes than electrons.

Thus one would expect that, other conditions being equal, the electron lifetime would be longer than the hole lifetime and, consequently, the intrinsic (fundamental) photoconductivity would be predominantly n-type. In the case of a p-type semiconductor, the Fermi level lies closer to the valence band, below the S levels (Fig. 145b), and similar considerations lead us to conclude that the intrinsic (fundamental) photoconductivity is essentially p-type. Thus we have a situation in which the sign of the photoconductivity is the same as the sign of the dark conductivity.

Similar considerations lead us to conclude that the type (sign) of the induced impurity photoconductivity and of the thermally-stimulated current should also be the same as the type (sign) of the equilibrium conductivity.

Induced impurity photosensitivity may appear because the electrons and holes formed by the "fundamental" excitation simultaneously recombine at the S levels and are captured by the M_i levels.

The sign of the induced photoconductivity will be the same as that of the dark conductivity for the following reason [we shall illustrate this for an n-type semiconductor (Fig. 145a)].

* This assumption allows us to neglect some complications in the "flash" effect.

The capture of nonequilibrium holes by the filled levels M_j competes effectively with capture by the filled recombination centers S. At the same time, the nonequilibrium electrons, whose lifetime for capture by the S centers is long, are much more likely to be accumulated above the Fermi level in the free trapping levels. Thus the number of trapped electrons should be greater than the number of trapped holes and, consequently, the induced photoconductivity due to excitation of trapped carriers should be of n-type, i.e., it should have the same sign as the equilibrium conductivity.

The electron component of the induced photoconductivity will exist for a longer time than the hole component. When light from the impurity region excites the M_j centers, which have captured electrons and holes, the holes (the minority carriers) are quickly captured by the S centers, while the electrons (majority carriers), whose lifetime for capture by the recombination centers S is long, exist for a longer time.

Thus the majority carriers trapped as a result of preliminary fundamental excitation govern the magnitude and lifetime of the induced impurity photoconductivity.

For the same reasons, the thermally-stimulated conductivity has the same sign as the equilibrium conductivity.

§ 48. METHOD OF LONG-WAVELENGTH PROBING OF LOCAL LEVELS

The data obtained in studies of photoconductivity kinetics may be satisfactorily interpreted in simple cases only. In the presence of several types of level in the forbidden band and in the case of "nonlinear processes," solution of the transport equations becomes impossible in practice and the selection of a model is, to a considerable extent, arbitrary.

This is because the information which can be obtained by experimental studies of the relaxation curves is insufficient. The best that can be obtained from such studies are the time dependences of the free-carrier densities in the bands [for example, the functions n(t) and dn(t)/dt], but we cannot measure the densities of the carrier densities in the bound state which fill the local levels in the forbidden band. It is clear that extension of the measurements to the bound carriers [i.e., the functions m(t) and dm(t)/dt] would have allowed us to reduce the mathematical difficulties radically by transforming differential equations into algebraic equations and by reducing the number of unknowns.

The method of long-wavelength probing of levels [10, 12] provides us with the possibility of such an extension. A sample, illuminated with sufficiently long square pulses of the fundamental excitation which give rise to the relaxation process under investigation, is also illuminated by "probing" pulses of long-wavelength light from the impurity region. The

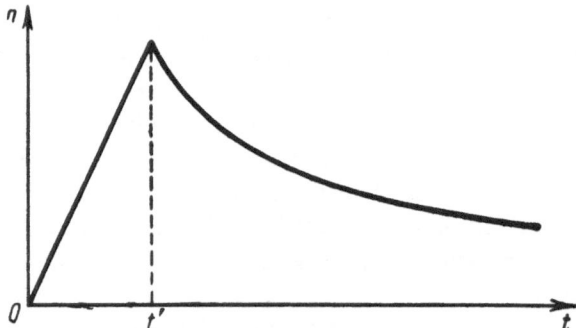

Fig. 146. Photocurrent signal produced by a short square
probing pulse (t' ≪ τ).

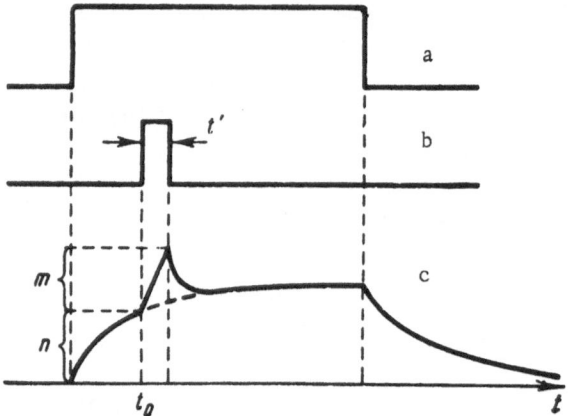

Fig. 147. Explanation of the probing method. a) Shape
of the fundamental light pulse; b) long-wavelength probing pulse; c) photoresponse signal. At a time ~t_0, we
can determine both the carrier density in the band (n)
and the density of the carriers at the probed level (m).

wavelength of the probing pulses is such that the levels investigated are
ionized optically. The relaxation curve observed on the oscillograph
screen is related to the fundamental excitation and obviously governs the
variation with time of the free-carrier density in the bands [for example,
$n(t)$ and $dn(t)/dt$].

The variation with time of the additional signals (peaks) produced by
the probing pulses can be used to determine the relaxation of the bound
carriers at the level being probed [i.e., $m(t)$ and $dm(t)/dt$]. Obviously,
the intensity of impurity excitation is directly proportional to the carrier
density at the level being probed:

$$n' = mqI \qquad (48.1)*$$

* The quantity qI is constant and can be determined, for example, by the method described
in §42.

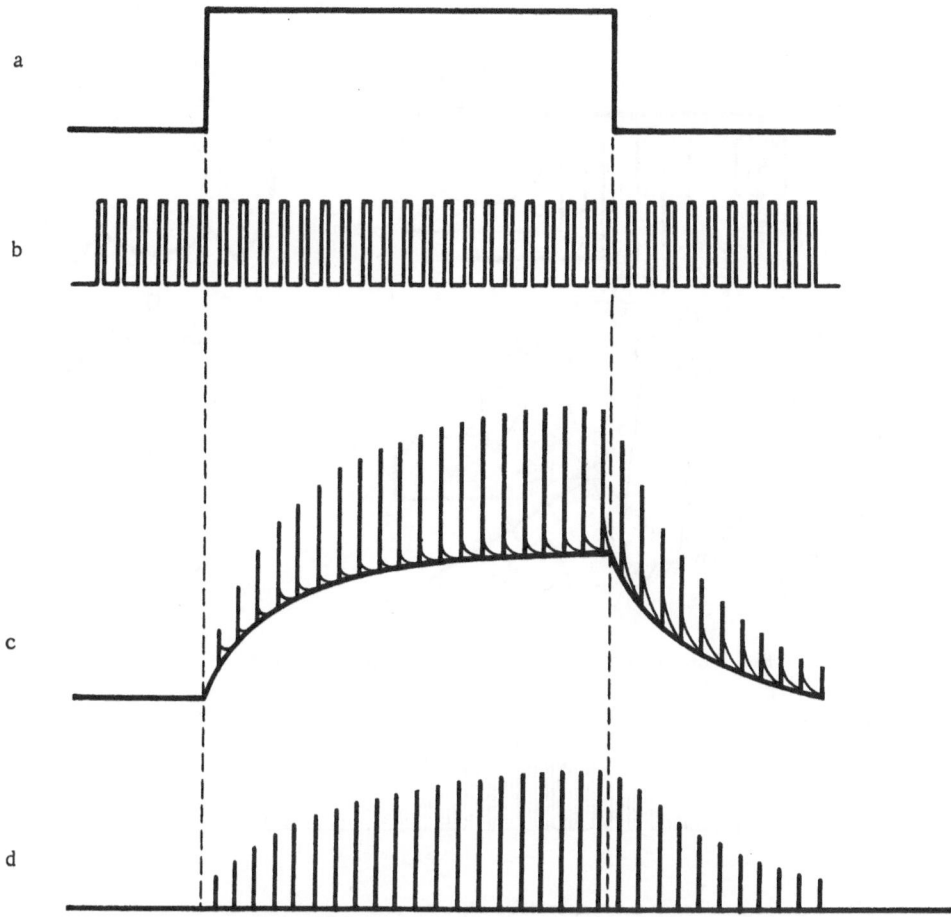

Fig. 148. Continuous probing ("sliding-pulse") method) of the relaxation process. a) Light pulse which produces the process being studied; b) long-wavelength probing pulses; c) oscillogram of the complete photoresponse. The continuous curve in Fig. 148c represents the variation with time of the free-carrier density, while the envelope of the peaks due to the probing pulses represents the relaxation of the carrier density at the level being investigated (d).

If we use sufficiently short single pulses of the impurity excitation, the signal due to this excitation should have the form shown in Fig. 146. The peak value of the signal is $mqIt'$, i.e., it is proportional to m.

Thus simultaneous illumination with impurity (probing) and fundamental light should give on the oscillograph screen a curve of the type shown in Fig. 147c. The oscillogram can then be used to determine the values of both n and m at a time t_0. By displacing the probing pulse in time with respect to the main pulse, we can obtain the value of m at various times and, consequently, we can determine the dependence of m on time throughout the whole relaxation process ("sliding-pulse" method).

Very frequently, the time constant of the main process of the relaxation n is considerably longer than the time constant of the probing signal.

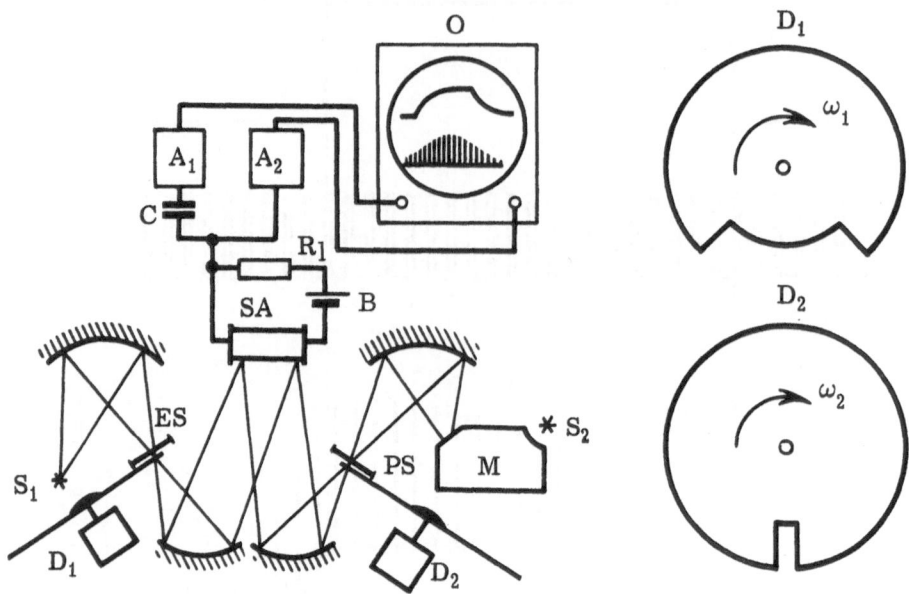

Fig. 149. Schematic representation of the apparatus.

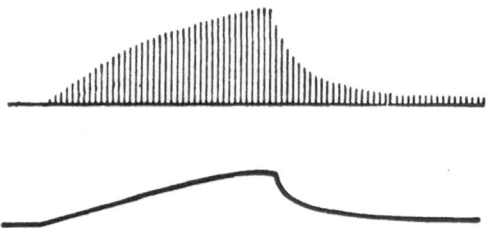

Fig. 150. Probing oscillograms for the level
$\mathscr{E}_c$ −0.4 eV in CdS during photoconductivity
rise and decay, produced by illumination in the
fundamental absorption region at 100°K.

The problem then simplifies considerably: during the main pulse we can
apply a number of probing pulses (Fig. 148b). The oscillograph screen
then shows the main relaxation curve (Fig. 148c) for n, on which a
"comb" of probing pulses is superimposed. The envelope of this "comb"
obviously represents the relaxation of m. Thus the screen shows simultaneously the dependences n(t) and m(t) ("continuous-probing" method).

     Figure 149 shows the apparatus with which we can probe the local
levels and also investigate the usual photoconductivity relaxation. Light
is directed on to a sample (SA) from a source S_1. The probing infrared
light from a source S_2 is interrupted by a disc D_2. To obtain single probing pulses, we can use a photographic shutter PS, in addition to the disc
D_2. The fundamental illumination is switched on and off by an electro-

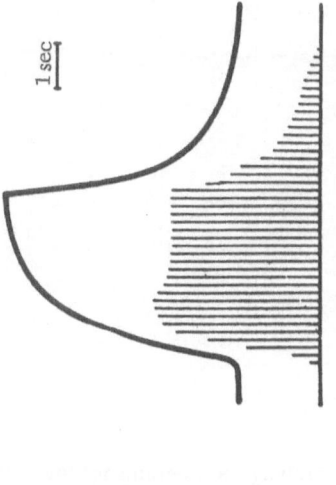

Fig. 152. Probing oscillogram for the level $\mathscr{E}_v + 0.35$ eV in Si doped with Au during relaxation of the fundamental photoconductivity at 100°K.

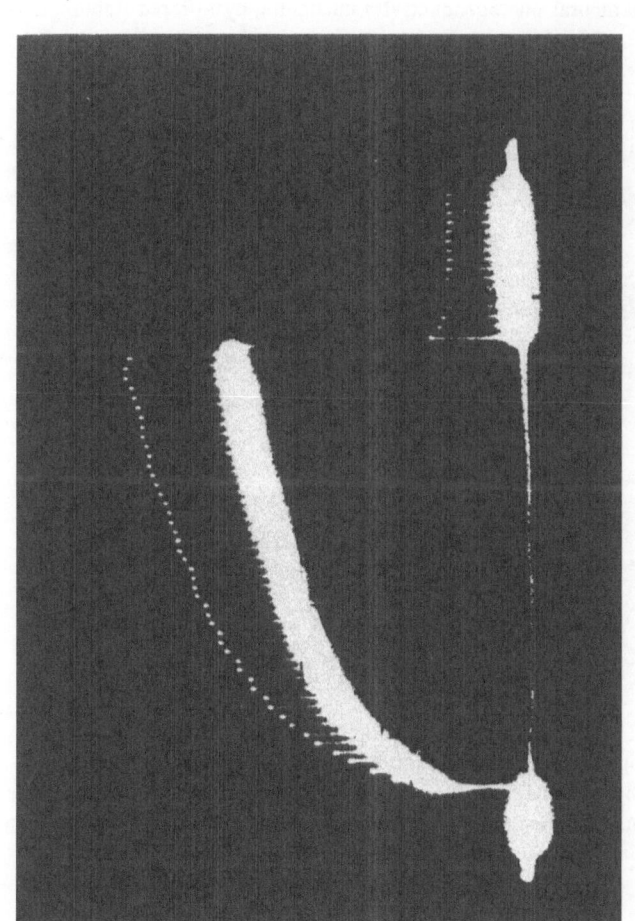

Fig. 151. Oscillogram obtained by probing the level $\mathscr{E}_c - 0.26$ eV in Cu-doped Ge during the rise and decay of the impurity photoconductivity produced by electron excitation from the level $\mathscr{E}_c - 0.43$ eV ($\lambda = 2.7$ μ) at 77°K.

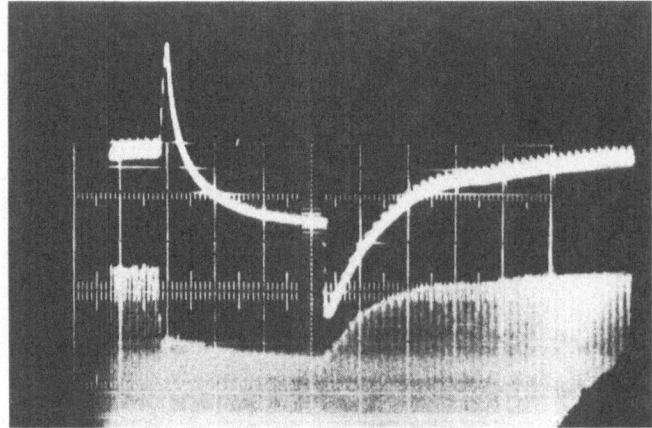

Fig. 153. Probing oscillograms for the level $\mathcal{E}_c - 0.4$ eV in CdS during fundamental photoconductivity quenching by infrared light of $\lambda = 0.9$ μ.

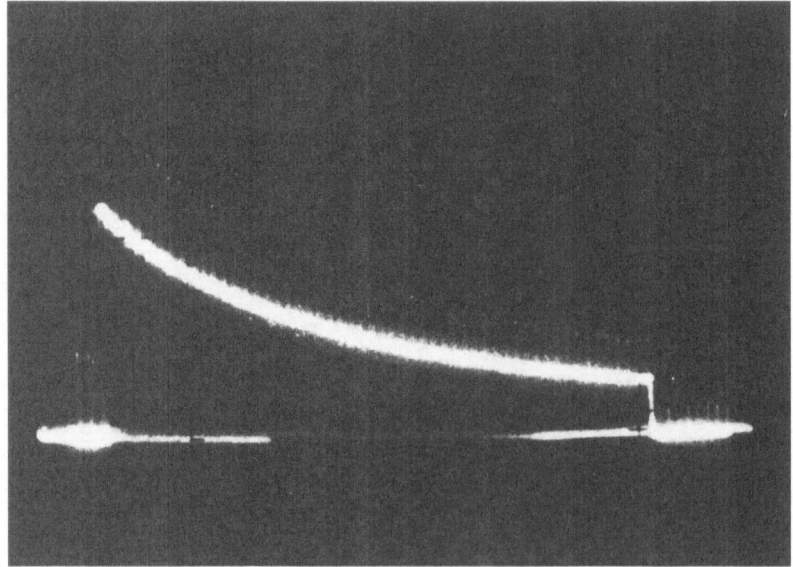

Fig. 154. Probing of the "flash" relaxation of the induced impurity photoconductivity from the copper level $\mathcal{E}_c - 0.26$ eV in germanium (the wavelength of the probing pulses and of the light producing the "flash" was 4 μ). The sample was illuminated first with light of energy 0.43 eV > hν > 0.26 eV (§ 46).

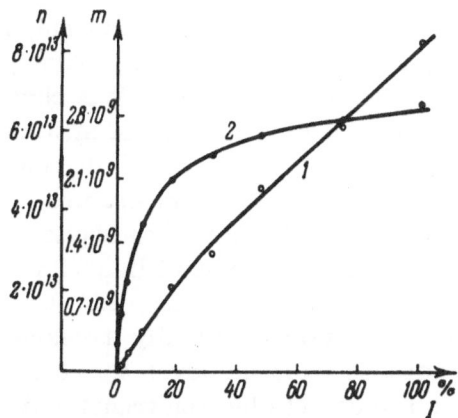

Fig. 155. Dependence of the steady-state
electron density in the conduction band
(curve 1) and at the $\mathcal{E}_c - 0.4$ eV level in
CdS (curve 2) on the illumination intensity
($\lambda = 0.52\,\mu$) at 100°K.

dynamic shutter ES (the disc D_2 is then switched off). The photoconductivity signals are recorded by photographing the screen of a double-beam
oscillograph O. The total signal from the load resistance R_l is divided
according to its frequency: the main low-frequency signal which represents the relaxation n(t), is amplified by a dc amplifier A_1 and is fed
to channel I of the oscillograph; the high-frequency signal, produced by
the probing pulses and representing m(t), is amplified by a hf amplifier
A_2 and fed to channel II. This ensures that the amplification of the two ·
signals is independent.

When the "sliding-pulse" method is used, light from the source S_1
is interrupted by the disc D_1, which is cut to ensure a square pulse. The
disc D_2 which produces the probing pulses, is cut similarly. The discs
D_1 and D_2 are rotated at slightly different speeds. Thus if the disc D_1 has
an angular frequency ω_1 and the disc D_2 a frequency ω_2, for every revolution the probing pulse is displaced from the main pulse by a time

$$\Delta t = 2\pi \left(\frac{1}{\omega_1} - \frac{1}{\omega_2} \right).$$

Thus the probing pulse is automatically displaced during the process
being studied. By synchronizing the display and the camera, we can
photograph a series of oscillograms with displaced probing pulses on one
plate (Fig. 156b). *

Figures 150-155 show oscillograms of the probing pulses for several
typical relaxation processes. The lower curve in Fig. 150 gives the de-

* We can also use a spark discharge in air as the source of probing pulses. It gives pulses of
down to 1.5×10^{-6} sec duration and quite stable amplitude.

pendence of the fundamental photoconductivity on time, i.e., it gives n(t). The upper curve of this oscillogram shows the probing peaks as a function of time, i.e., m(t). The level $\mathcal{E}_c - 0.4$ eV is obviously an effective recombination center, since the rate of decay of m(t) after switching off the fundamental excitation is quite rapid (return thermal transitions back to the band are relatively few at the test temperature).

The oscillogram in Fig. 151 shows that on application of light with $\lambda = 2.7\,\mu$, the level $\mathcal{E}_c -0.26$ eV becomes filled with electrons. When this light is switched off, the electrons remain for quite a long time at the level (the probing pulses retain practically the same amplitude). This is because thermal transitions from the level $\mathcal{E}_c -0.26$ eV to the band are few at low temperatures and direct electron transfer from the level -0.26 eV to the level $\mathcal{E}_c -0.43$ eV is impossible.

Probing high-resistivity silicon, compensated with gold, shows (Fig. 152) that the fundamental photoconductivity curve is monotonic while the time-dependence of the carrier density at the probed level shows a flash. Such a dependence of m(t) indicates that the conditions for electron and hole capture by this level vary during the relaxation process.

The behavior of the electron density at the level $\mathcal{E}_c -0.4$ eV in CdS during infrared quenching of the fundamental photoconductivity is shown in Fig. 153. The dependence n(t), given by the upper curve, shows a photocurrent dip on switching-on the quenching infrared radiation, which is usually accounted for by partial removal of electrons from the local levels to the conduction band simultaneously with the formation of holes in the valence band. The electron density at the level decreases at first (lower curve).

The probing technique allows us to investigate directly the variation in the electron (hole) density at the probed level in the case of induced impurity photoconductivity (Fig. 154). This figure shows that the photoconductivity flash is accompanied by a reduction in the probing-pulse amplitude, i.e., a reduction in the electron density at the level $\mathcal{E}_c -0.26$ eV.

The method of local-level probing may also be used to study the steady-state characteristics. Figure 155 gives the dependence of the steady-state fundamental photoconductivity (more precisely, the electron density in the conduction band) and the electron density (m) at the $\mathcal{E}_c -0.4$ eV level, found by probing, on the illumination intensity ($\lambda = 0.52\,\mu$) for the CdS sample for which the kinetics of the relevant processes was considered above (Fig. 150). The value of n varies approximately linearly with intensity but m tends to saturation.

It should be noted that the probing technique using the "sliding-pulse" method can be used not only for long wavelengths but also to solve other problems related to photoconductivity kinetics.

Figure 156 gives oscillograms illustrating the photoactivation of the phenomenological quantum yield for the fundamental photoconductivity of

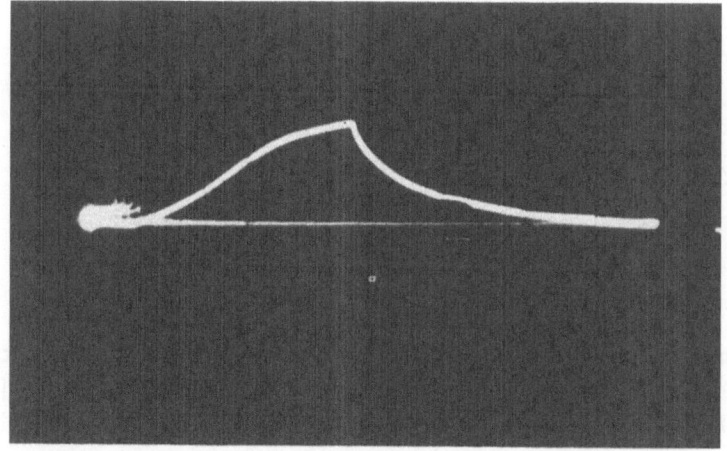

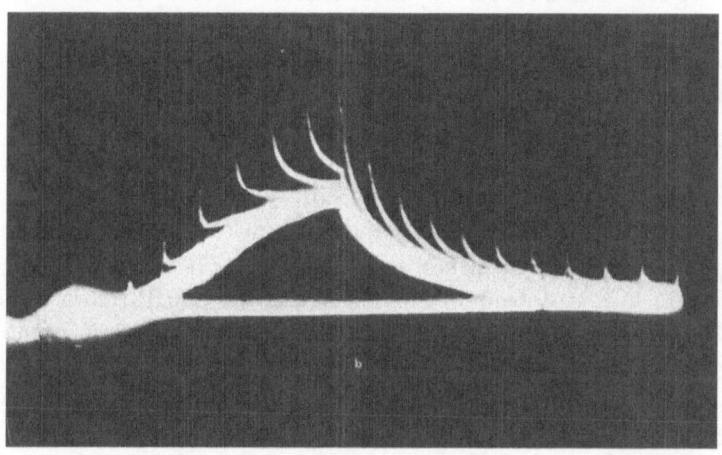

Fig. 156. Photoactivation of the phenomenological quantum yield in CdSe.
a) Relaxation curve of the fundamental photoconductivity in CdSe ($\lambda \approx 0.8\,\mu$);
b) oscillogram of the photocurrent produced by the fundamental illumination
and single short fundamental light pulses ($\lambda \approx 0.8\,\mu$) moving along the main
relaxation curve (the "sliding-pulse" method). The photoresponse to single
pulses was proportional to the phenomenological quantum yield.

a CdSe single crystal during relaxation of a photocurrent produced by illumination in the fundamental region.

These examples show that the long-wavelength probing method extends the range of information which can be obtained from experimental
measurements.

Again, in conclusion, it should be noted that this method can also be
used to probe several types of level, but the method is somewhat more
complicated.

Chapter XI

MEANING OF THE CONCEPT "LIFETIME"

Extensive studies of nonequilibrium processes in semiconductors have led to a variety of interpretations of the concept "lifetime" [1, 3].
We shall analyze below meaning of some of these interpretations.

§49. LIFETIME OF CARRIERS IN THE BANDS

The following quantities are used to describe nonequilibrium carriers in the conduction and valence bands *:
1) τ_n the lifetime of electrons in the conduction band;
2) τ_p the lifetime of holes in the valence band;
3) τ_{np} the lifetime of an electron-hole pair;
4) τ_{neq} the lifetime of the nonequilibrium conduction state;
5) τ_{mc} the lifetime of minority carriers.

In general, a complete description of the nonequilibrium carriers in the bands is given by the first two quantities: τ_n and τ_p. In some cases, however, one uses the concept of the pair lifetime τ_{np} and on other occasions it is useful to use the lifetime for the nonequilibrium conduction state, τ_{neq}.

The concept of the pair lifetime can be used only in those cases where $\Delta n = \Delta p$, i.e., where we are dealing only with "complete" electron-hole pairs (this occurs, for example, in the case of low-trapping concentration considered in detail in §23); then $\tau_n = \tau_p = \tau_{np}$. If $\Delta n \neq \Delta p$ (for example, when the number of traps is large, as in Chap. V, §23B), there are complete pairs of free electrons and holes and also a certain number of carriers of one sign. Under these conditions, use of the concept "pair lifetime" is hardly justified although in some cases it is used to denote the lifetime of the complete pairs (the pair is regarded as annihilated even if only one of its components is captured by a trap). Obviously then, τ_{np} is identical with the shorter of the lifetimes of the carriers of different sign.

---

* In general, these quantites should be regarded as the instantaneous values of the corresponding lifetimes.

The concept of the nonequilibrium conduction lifetime may be defined by the following expression, analogously to Eq. (5.16):

$$\tau_{neq} = -\frac{\Delta\sigma}{\dfrac{d}{dt}\Delta\sigma} = \frac{\mu_n\Delta n + \mu_p\Delta p}{\mu_n\dfrac{\Delta n}{\tau_n} + \mu_p\dfrac{\Delta p}{\tau_p}}. \tag{49.1}$$

If the electron and hole mobilities are similar, $\mu_n \approx \mu_p$, then

$$\tau_{neq} \cong \frac{\Delta n + \Delta p}{\dfrac{\Delta n}{\tau_n} + \dfrac{\Delta p}{\tau_p}}. \tag{49.2}$$

Analysis of Eq. (49.2) shows that if the electron and hole densities are very different, except during a short initial period, the instantaneous value of τ_{neq} (as well as the total duration of the relaxation) is governed by the lifetime of those carriers whose density is higher (in this sense, τ_{neq} behaves in the opposite way to τ_{np}). However, during the initial stage and under steady-state conditions, the value of τ_{neq} is only half the lifetime of the majority carriers. Under steady-state conditions, when $\Delta n = \Delta n_{st} = \beta kI\tau_n$, $\Delta p = \Delta p_{st} = \beta kI\tau_p$, Eq. (49.1) transforms into

$$\tau_{neq}^{st} = \frac{\tau_n\mu_n + \tau_p\mu_p}{\mu_n + \mu_p}. \tag{49.3}$$

The minority-carrier lifetime τ_{mc} has a meaning when the electron and hole densities are very different and, by definition, it is identical with the lifetime of those carriers whose density is lower.

As well as the concepts listed above which apply to the band electrons and holes, we can similarily define the electron or hole lifetime for local levels, the lifetime for a given transition, etc. However, quantities of this kind are essentially auxiliary and do not define directly the experimentally measured properties of carriers in the bands under the usual conditions.

§ 50. MICROSCOPIC AND RELAXATION LIFETIMES

Usually, the lifetime is understood to be the reciprocal of the coefficient standing in front of the nonequilibrium carrier density in the transport equation. For example, for electrons

$$\frac{d}{dt}\Delta n = U - \frac{\Delta n}{\tau}. \tag{50.1}$$

Here, U is the rate of the nonequilibrium process of electron generation, i.e., the number of nonthermal transitions of electrons to a band per unit time per unit volume,* and τ is the electron lifetime.

* The quantity U may represent the action of light (internal photoconductivity) or of other radiations, injection, etc.

From Eq. (50.1) it follows that (cf. §5C)

$$\tau = \frac{\Delta n}{U - \dfrac{d}{dt}\Delta n} \cdot$$

(50.2)

Under steady-state conditions

$$\tau^{st} = \frac{\Delta n}{U} \cdot$$

(50.3)

This is the lifetime appropriate to recombination through traps (Chaps. V and VII) and to interband recombination (Chap. VIII). *

It is usually understood that the quantity τ, defined by Eq. (50.2) or (50.3), represents the average duration of the existence of an electron in a band, i.e., it represents the time between the electron's appearance and departure from the band. In fact, the quantity τ has this meaning only in certain special cases while, in general, the average duration of the time spent by an electron in a band is not given at all by Eq. (50.2) or (50.3). Let us consider the dynamics of a process involving transitions of electrons to the conduction band and their departure from this band (due to capture by holes in the valence band or by local levels) [4].

Transitions to a band are, in general, due to thermal transitions from various levels (the overall rate of these transitions will be denoted by Q)† and to nonequilibrium excitation, the intensity of which has been denoted by U. The total electron density in the band n due to both types of excitation (Q and U) may usually be represented in the form

$$n = n_0 + \Delta n,$$

where n_0 is the electron density at equilibrium, i.e., when U = 0.

The strong interaction with the lattice makes the nonequilibrium carriers practically indistinguishable from the equilibrium carriers and, therefore, the average lifetimes of carriers in the band are the same for the two different types of carrier. This true microscopic lifetime of equilibrium and nonequilibrium carriers is denoted by τ_{micr}.

The quantity τ_{micr} is understood to be the time which an electron spends on the average between the moment of its excitation (irrespective of the levels from which it is excited and the form of energy used for this purpose) to the moment of its capture (irrespective of the type of level by which it is captured). In other words, had we been able to follow the fate of a single electron and determine its average length of stay in the band, this quantity would be found to be τ_{micr}.

* The various lifetimes listed in §49 are also calculated by means of expressions such as Eqs. (50.2) or (50.3).

† $Q = \sum_k Q_k$, where Q_k is the rate of transitions from the kth level.

The time a free electron stays in the conduction band is limited by its capture by holes at the local levels or in the valence band. Obviously, the average time between two "collisions" with holes (and each such "collision" captures an electron and therefore removes it from the band), has the meaning of the true average microscopic lifetime of electrons in the band, and may be written in the form

$$\tau_{\text{micr}} = \frac{1}{\sum_k v_{nk} q_{nk} p_k} = \frac{1}{\sum_k \gamma_{nk} p_k},$$ (50.4)

where p_k is the density of holes of the kth type, q_{nk} is the cross section for the capture of electrons by these holes, v_{nk} is the relative velocity of motion of electrons and holes of the kth type, γ_{nk} is the capture coefficient. * Formula (50.4) defines the microscopic lifetime in general. In the particular case of the steady state, the quantity p_k is understood to be the steady-state hole density at the appropriate level. Comparing Eqs. (50.4) and (50.3) in the steady state, we can easily see that they are not equivalent. This is particularly clear when Eq. (50.4) is written in a different way. Under steady-state conditions, the total rate of all generation processes $Q + U$ should be equal to the total rate of capture processes

$$Q + U = n \sum_k \gamma_{nk} p_k.$$ (50.5)

Comparing the above expression with Eq. (50.4), we find that

$$\tau_{\text{micr}} = \frac{n}{U + Q} = \frac{n_0 + \Delta n_{\text{st}}}{U + Q}.$$ (50.6)†

Obviously, Eqs. (50.3) and (50.6) are not, in general, equivalent, i.e., $\tau \ne \tau_{\text{micr}}$. Only when the excitation level is high, i.e., $\Delta n \gg n_0$ and, consequently, $U \gg Q$, are these quantities equal. Thus the difference between τ_{micr} and τ is related to the equilibrium (thermal) excitation and is absent when this excitation is sufficiently high.

In contrast to the microscopic lifetime τ_{micr}, we shall call the quantity τ the relaxation lifetime and denote it by τ_{rel}. ‡ The quantities τ_{rel} and τ_{micr} may differ by many orders of magnitude. Below, we give a

---

* It is understood that the quantities $q_{nk} v_{nk} = \gamma_{nk}$ are averaged out for all electrons in the conduction band (§ 4).

† For the steady-state conditions, we shall use Eq. (50.6), and not Eq. (50.4), to define τ_{micr} because Eq. (50.6) allows us to analyze more simply the limiting cases of high and low excitation levels.

‡ The meaning of τ_{rel} will be considered in greater detail in subsection C below.

Conduction band

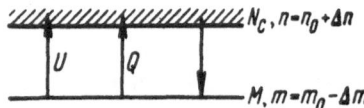

Valence band

Fig. 157.   System of transitions in the
impurity photoconductivity case.

comparison of the quantities τ_{micr} and τ_{rel} for two important special cases:

    1) impurity photoconductivity;

    2) fundamental photoconductivity in the case of recombination through traps.

A. Impurity Photoconductivity

Let us assume that transitions to the conduction band (both thermal and and optical) proceed from the impurity levels M (Fig. 157).

We shall consider only the case of low illumination intensity when Δn and Δm are small compared with the corresponding equilibrium densities n_0 and m_0. Using the equation

$$\frac{dn}{dt} = Q + U - \gamma n (M - m) \tag{50.7}$$

and the neutrality condition

$$\Delta n = \Delta m, \tag{50.8}$$

we then find that in the steady-state case (dn/dt = 0)

$$n = \frac{Q + U}{\gamma (M - m)}, \tag{50.9}$$

and, consequently, the microscopic lifetime follows from Eq. (50.6):

$$\tau_{micr} = \frac{n}{Q + U} = \frac{1}{\gamma (M - m)} \approx \frac{1}{\gamma (M - m_0)}. \tag{50.10}$$

To calculate τ_{rel}^{st}, we must calculate Δn_{st} in accordance with Eq. (50.3). Using Eqs. (50.7) and (50.8), we obtain, after simple transformations,

$$\Delta n_{st} = \frac{U}{\gamma (M - m_0) \left(1 + \frac{n_0}{M - m_0}\right)}, \tag{50.11}$$

and hence

$$\tau_{rel} = \frac{\Delta n_{st}}{U} = \frac{1}{\gamma (M - m_0) \left(1 + \frac{n_0}{M - m_0}\right)}. \tag{50.12}$$

Comparing Eqs. (50.12) and (50.10), we find that

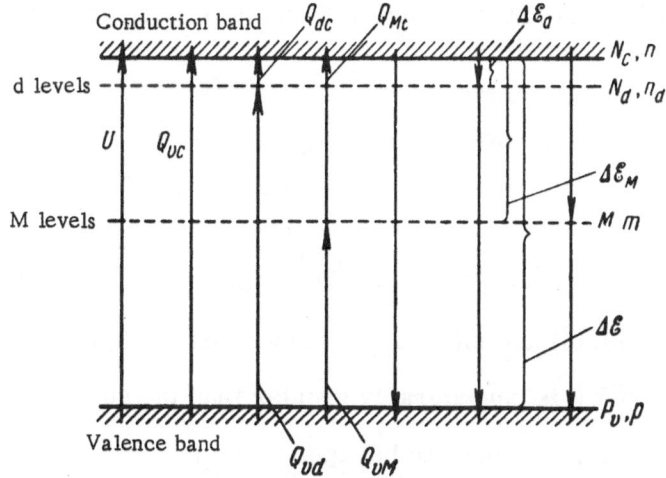

Fig. 158. System of transitions for fundamental photoconductivity and recombination through traps.

$$\tau_{rel} = \tau_{micr} \frac{1}{1 + \dfrac{n_0}{M - m_0}}. \qquad (50.13)$$

Thus τ_{rel}, usually regarded as the "nonequilibrium carrier lifetime," is not the true microscopic lifetime of carriers in a band.

In the case considered, we have found that $\tau_{rel} \leq \tau_{micr}$, but the difference between them is not very great. * However, in other cases, such as that considered below in subsection B, these quantities may differ by many orders of magnitude.

B.   Fundamental Photoconductivity.   Recombination Through Traps

We shall consider an n-type semiconductor such as germanium (Fig. 158). Illumination produces electron-hole pairs which recombine through traps. The system of transitions is the same as in § 23 except that thermal transitions between the conduction band and the donor levels d and between the conduction and valence bands, are now explicitly included because these are necessary to calculate τ_{micr} whereas they were not required in the calculation of τ_{rel}.

In § 23, we calculated τ_{rel} for this system and obtained Eq. (23.11) for a low trap concentration in the absence of band degeneracy. We shall now calculate τ_{micr} for electrons and holes and compare it with the value

---

* For example, if all the M levels are completely filled with electrons at absolute zero and, consequently, $n_0 = M - m_0$, then $\tau_{rel} = \tau_{micr}/2$.

of τ_{rel} given by Eq. (23.11). Figure 158 shows that, in accordance with Eq. (50.6), we have for electrons

$$\tau_{micr}^n = \frac{n_0 + \Delta n}{U + Q_{dv} + Q_{Mc} + Q_{vc}}, \qquad (50.14)$$

and for holes

$$\tau_{micr}^p = \frac{p_0 + \Delta n}{U + Q_{vd} + Q_{vM} + Q_{vc}}. \qquad (50.15)$$

From Eqs. (50.14) and (50.15), it follows that at high injection levels when $\Delta n \begin{cases} \gg n_0 \\ \gg p_0 \end{cases}$ and U is considerably greater than the sum of the thermal transitions, τ_{micr} is found to be equal to τ_{rel}

$$\tau_{micr}^n = \tau_{micr}^p = \frac{\Delta n}{U} = \tau_{rel}. \qquad (50.16)$$

However, in the case of moderate and low excitation levels, this is not true.

We shall consider below the case of low excitation levels when the microscopic lifetime is equal to the equilibrium carrier lifetime τ_{eq}.

Dropping Δn from the numerators and U from the denominators of Eqs. (50.14) and (50.15), we obtain

$$\tau_{micr}^n = \frac{n_0}{Q_{dc} + Q_{Mc} + Q_{vc}} = \tau_{eq}^n, \qquad (50.17)$$

$$\tau_{micr}^p = \frac{p_0}{Q_{vd} + Q_{vM} + Q_{vc}} = \tau_{eq}^p, \qquad (50.18)$$

or, using Eq. (50.4),

$$\tau_{micr}^n = \tau_{eq}^n = \frac{1}{\gamma_{dc}p_d + \gamma_{Mc}p_M + \gamma_{vc}p_0}, \qquad (50.19)$$

$$\tau_{micr}^p = \tau_{eq}^p = \frac{1}{\gamma_{vd}n_d + \gamma_{vM}n_M + \gamma_{vc}n_0}. \qquad (50.20)$$

A more complete comparison of τ_{micr} and τ_{rel} requires the plotting of the temperature dependences of these quantities (Fig. 159). Region 1 in Fig. 159a represents the density rise with temperature in the impurity region; region 2 represents the total ionization of impurities; and region 3 corresponds to intrinsic conduction.

Figure 159b shows the temperature dependence of the Fermi level (the band structure is given in the background), and Fig. 159c shows the well-known curve for the temperature dependence of the nonequilibrium carrier lifetime τ_{rel} (Fig. 69).

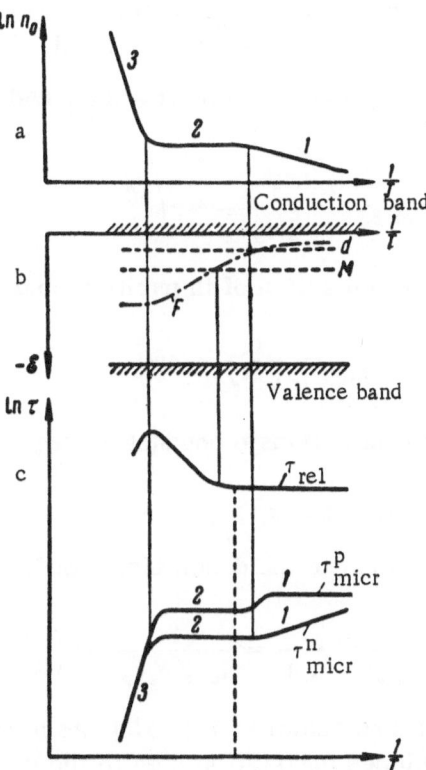

Fig. 159. Temperature dependence of the majority
carrier density (a), Fermi level (b), and lifetimes
τ_{micr} and τ_{rel} (c).

Finally, the same figure (159c) illustrates the temperature dependences of the true microscopic lifetimes of electrons and holes plotted
from Eqs. (50.19) and (50.20).

In the case of low trap concentration, we can neglect the second
terms in the denominators of Eqs. (50.19) and (50.20). Then

$$\tau_{micr}^{n} = \frac{1}{\gamma_{dc}p_d + \gamma_{vc}p_0}, \tag{50.21}$$

$$\tau_{micr}^{p} = \frac{1}{\gamma_{vd}n_d + \gamma_{vc}n_0}. \tag{50.22}$$

It follows from Eq. (50.21) that at low temperatures in the impurity
conduction regions, when $p_d = n_0 \gg p_0$,

$$\tau_{micr}^{n} = \frac{1}{\gamma_{dc}p_d} = \frac{1}{\gamma_{dc}n_0}. \tag{50.23}$$

The temperature dependence of τ_{micr}^{n} (Fig. 159c) is the converse of the temperature dependence of n_0 (Fig. 159a) and, therefore, also has three regions. In region 1, $n_0 = \sqrt{N_c N_d} \exp\left(-\frac{\Delta\mathscr{E}_d}{2kT}\right)$ and, consequently,

$$\tau_{micr}^{n} = \frac{1}{\gamma_{dc}\sqrt{N_c N_d}} \exp\left(\frac{\Delta\mathscr{E}_d}{2kT}\right). \tag{50.24}$$

In region 2, $n_0 = N_d$ (total ionization of impurities) and, consequently,

$$\tau_{micr}^{n} = \frac{1}{\gamma_{cd} N_d} = \text{const.} \tag{50.25}$$

At high temperatures, in the intrinsic conduction region (region 3), we have

$$p_0 = n_0 = n_i \gg p_d = N_d$$

and, neglecting the first term in the denominator of Eq. (50.21), we find

$$\tau_{micr}^{n} = \frac{1}{\gamma_{vc} p_0} = \frac{1}{\gamma_{vc} n_i} = \frac{1}{\gamma_{vc}\sqrt{N_c P_v}} \exp\left(\frac{\Delta\mathscr{E}}{2kT}\right). \tag{50.26}$$

These three regions are shown in Fig. 159c as $\ln \tau = f(1/T)$: they are straight lines with slopes opposite in sign to the slopes for the carrier density n_0.

The temperature dependence of τ_{micr}^{p} (Fig. 159c) can also be divided into three regions.

At low temperatures, when the electron density at the donor levels n_d is considerably greater than the density of electrons in the band n_0 (region 1): $n_0 \ll n_d \approx N_d$,

$$\tau_{micr}^{p} = \frac{1}{\gamma_{vd} N_d} = \text{const.} \tag{50.27}$$

In the region where impurities are totally ionized, $n_d \approx 0$ and $n_0 \approx N_d$ (region 2),

$$\tau_{micr}^{p} = \frac{1}{\gamma_{vc} n_0} = \frac{1}{\gamma_{vc} N_d} = \text{const;} \tag{50.28}$$

and in the intrinsic region (region 3), where $n_0 = n_i$,

$$\tau_{micr}^{p} = \frac{1}{\gamma_{vc} n_i} = \frac{1}{\gamma_{vc}\sqrt{N_c P_v}} \exp\left(\frac{\Delta\mathscr{E}}{2kT}\right) = \tau_{micr}^{n}. \tag{50.29}$$

C.   The Meaning of Relaxation Lifetime

The cases just considered show that the true microscopic lifetime of carriers in the bands is not identical with τ_{rel}; it can be either larger or smaller than the latter. The meaning of τ_{rel} can be understood from Eq. (50.1), which introduces it.

During the relaxation of the nonequilibrium carrier density, given by Eq. (50.1), the quantity τ_{rel} behaves as a time constant of the process of relaxation of the excess carrier density. This period is not necessarily equal to the true lifetime of carriers in the band. Strictly speaking, τ_{rel} is not a lifetime but an effective time during which a system of particles (carriers) regains its equilibrium with the crystal, i.e., it is the "total relaxation time."

In contrast to the relaxation time τ, which occurs in the expression for the mobility $\mu = \frac{e}{m} \langle \tau \rangle$ and gives the time for the dissipation of the excess momentum, τ_{rel} represents the time for the dissipation of all the excess energy. The quantity τ_{rel} appears in the majority of experimentally observed phenomena, e.g., τ_{rel} is equivalent to a time constant of the nonequilibrium conductivity, etc.; in contrast, the quantity τ_{micr} does not usually appear explicitly in the experiments.

However, there are, at least in principle, phenomena in which τ_{micr} may appear as the important parameter. Thus, in many cases (for example, when there is a large number of shallow trapping levels) τ_{micr} may be shorter than the time which a carrier with high kinetic energy (for example, a carrier liberated by ionization with hard radiation) needs to transfer this energy to the lattice.

In this case, the process of energy transfer may occur in a way different from that usually assumed, i.e., not as a series of successive collisions of a carrier with phonons but as a result of a single capture by a trapping level, followed by reemission to a band with normal or close-to-normal energy.

We can also have situations when τ_{micr} is comparable with the relaxation time (the time taken by a carrier to travel a distance equal to its mean free path) and, consequently, it can have some influence on the mobility. In other words, carrier trapping followed by thermal reemission may act as a completely independent mechanism of carrier scattering. At low temperatures, when scattering is solely due to impurities, the relative role of this mechanism is determined by the ratio of the effective cross sections for the capture (or trapping) of carriers by impurities and the scattering by impurities via other mechanisms.

Chapter XII

DIFFUSION AND DRIFT OF NONEQUILIBRIUM CARRIERS
(UNIPOLAR CASE)

In experiments with nonequilibrium carriers we frequently meet with nonuniform carrier distribution in the volume of a semiconductor. Such distributions may be the result of the nonuniform generation of the carriers or of the inhomogeneity of the semiconductor itself (for example, enhanced recombination near the surface, variation in the equilibrium conductivity with coordinates, etc.). The diffusion processes occurring under nonuniform distribution conditions are important in many extensively studied physical problems and in processes which govern the operation of various semiconducting devices.

Moreover, a study of some of the effects connected with diffusion allows us to measure the parameters of nonequilibrium carriers (mobility, lifetime, etc.) and, therefore, these effects form the bases of the methods used for measuring the nonequilibrium parameters.

Frequently, phenomena to which we are referring here are investigated in the presence of external electric fields when carrier drift is superimposed on diffusion. Furthermore, we have the additional possibility of measuring semiconductor parameters by studying diffusion in a magnetic field.

In the present chapter, we shall consider the diffusion and drift of nonequilibrium carriers in the unipolar conduction case. Ambipolar diffusion is dealt with in Chap. XIII.

§51. GENERAL CONSIDERATIONS AND FUNDAMENTAL EQUATIONS

Let us assume that we have a completely homogeneous semiconductor whose unipolar equilibrium conductivity is, say, n-type and whose band structure is that shown in Fig. 160. The valence band is omitted from the figure because it is assumed that the density of mobile holes is zero and that mobile electrons appear in the conduction band by transfer from the M levels. If nonequilibrium electrons are generated in such a semiconductor (for example, by illumination) and the generation rate de-

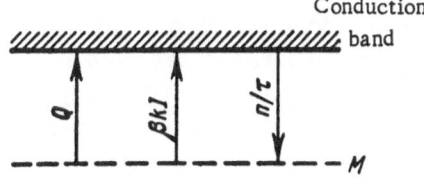

Fig. 160. System of transitions.

pends on the coordinates (for example, as a result of nonuniform distribution of the illumination intensity along the sample surface), the density of free carriers is different at different points of the sample. The resultant diffusion of charged particles (electrons) disturbs the electrical neutrality which had existed in a homogeneous semiconductor before it was subjected to nonuniform illumination. A space charge ρ appears and, consequently, an electric field E is set up.

In the region where excess electrons are accumulated by diffusion a negative space charge appears with a corresponding positive charge in the electron-depleted regions. These charges obviously establish a field which prevents further motion of electrons in the original direction, i.e., which prevents diffusion.

These considerations can be formulated quantitatively as follows.

A density gradient (grad n) produces a diffusion current $j_D = eD \, \mathrm{grad} \, n$ (here D is the diffusion coefficient) which gives rise to an electric field E and, consequently, a conduction current $j_E = \sigma E = e\mu nE$ (which is, in-general, opposite to the diffusion current). Consequently, the total current j at any point in a semiconductor at any time is given by

$$j = j_D + j_E = e \, (D \, \mathrm{grad} \, n + \mu nE). \qquad (51.1)$$

In an isolated semiconductor under steady-state conditions the total current should be zero; in other words, at each point the conduction current should compensate the diffusion current. This steady-state condition corresponds to a definite spatial distribution of carriers, space charge and electric field (and, consequently, of the potential difference between any two points of the semiconductor, which, in the case of nonuniform illumination, is known as the photo-emf).

Our problem is to calculate these distributions for some of the simplest cases. For this purpose, we use, in addition to Eq. (51.1), two other equations: Poisson's equation

$$\mathrm{div} \, E = \frac{4\pi\rho}{\varepsilon}, \qquad (51.2)$$

and the equation of carrier conservation (continuity equation). The latter equation can be written down from the following considerations. Any change in the carrier density in a certain volume of a semiconductor is the result of:

1) Carrier generation (for example, thermal and optical transitions from the M levels to the conduction band, Fig. 160);

2) Carrier recombination (for example, with holes at the M levels);

3) The difference between the number of carriers entering a given volume and the number leaving it. The first two processes are represented by electron transitions in the energy space, and the third is represented by electron motion in the coordinate space. The latter process should obviously depend on the rate of variation of the current with coordinates. This is described quantitatively by div j which, in Cartesian coordinates, is

$$\operatorname{div} j = \frac{\partial}{\partial x} j_x + \frac{\partial}{\partial y} j_y + \frac{\partial}{\partial z} j_z.$$

It follows that the equation of continuity for the conduction band electrons may be written in the form

$$\frac{\partial n}{\partial t} = Q + \beta kI - \frac{n}{\tau} + \frac{1}{e} \operatorname{div} j, \qquad (51.3)$$

where Q is the thermal ionization rate, βkI is the optical ionization rate, and n/τ is the recombination rate.

The system of equations (51.1)-(51.3) allows us, in principle, to find the distribution of the carrier density, charge, and field in discussing very general problems.

We shall restrict ourselves to a detailed consideration of the simplest one-dimensional problem of diffusion in the case of a sharp boundary between the illuminated and dark parts of a semiconductor. Other simplifications – to be brought in next – will allow us to introduce into the solution an important characteristic known as the "screening length" which is the effective length over which diffusion is active in the unipolar case.

§52. CARRIER DENSITY, CHARGE, AND FIELD DISTRIBUTIONS DURING DIFFUSION UNDER WEAK EXCITATION CONDITIONS. DEBYE SCREENING LENGTH

Let us assume that we have a sufficiently long semiconductor sample (Fig. 161a), in one part of which, at x < 0 nonequilibrium electrons are generated uniformly by the action of light of intensity I.

In the region x > 0, there is no illumination. At the point x = 0, there is a sharp boundary between darkness and illumination. The energy band structure of this semiconductor is the same as in Fig. 160.*

---

*It is assumed that in darkness at absolute zero the M levels are completely filled with electrons and the conduction band is empty.

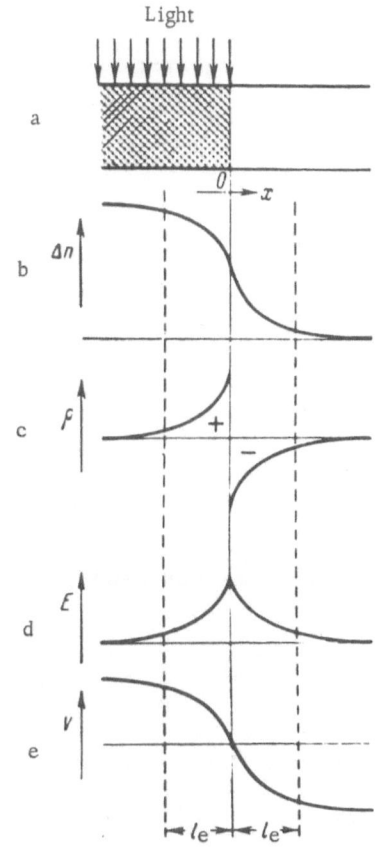

Fig. 161. The distribution of the density (b), space charge (c), electric field (d) and potential (e) during partial illumination of a semiconductor (a).

The illumination generates nonequilibrium electrons which diffuse into the dark part and raise the carrier density there.

Eqs. (51.3), (51.1), and (51.2) may be rewritten, therefore, in the form

$$\frac{\partial n}{\partial t} = \gamma m N_{cM} + \beta kI - \gamma np + \frac{1}{e}\frac{\partial j}{\partial x},$$

$$(52.1)$$

$$j = \mu kT \frac{\partial n}{\partial x} + e\mu nE, \qquad (52.2)*$$

$$\frac{\partial E}{\partial x} = \frac{4\pi e}{\varepsilon}(n - p). \qquad (52.3)$$

In Eq. (52.1), the thermal transition rate is given as $Q = \gamma m N_{cM}$. The recombination term is written in its general form: γnp, where p is the density of holes at the M levels ($p = M - m$).

In the absence of diffusion, n = p at every point and there is no space charge. In the case of diffusion, n differs, in general, from p and, consequently, a space charge $\rho = e(p - n)$ appears, which should be allowed for in Poisson's equation (52.3).

We shall solve the system (52.1)-(52.3) in the steady-state case. Then $\partial n/\partial t = 0$, j = 0, div j = 0, and Eq. (52.1) becomes

$$\gamma m N_{cM} + \beta kI - \gamma pn = 0. \qquad (52.4)$$

We shall assume also that the illumination intensity is so low that in the expressions for the total electron and hole densities

$$n = n_0 + \Delta n,$$
$$p = p_0 + \Delta p, \qquad (52.5)$$

the nonequilibrium densities Δn and Δp are negligibly small compared with the equilibrium densities n_0, p_0

---

*In accordance with the Einstein relationship, we assume in Eq. (52.2) that $D = \mu kT/e$.

$$\Delta n \ll n_0,$$
$$\Delta p \ll p_0,$$
(52.6)

and that the generation rate (β kI) is constant. Substituting Eq. (52.5) into Eq. (52.4) and bearing in mind that at equilibrium

$$\gamma m_0 N_{cM} - \gamma n_0 p_0 = 0,$$
(52.7)

we obtain (neglecting the small term $\gamma \Delta n \Delta p$)

$$\beta kI - \gamma \Delta p (N_{cM} + n_0) - \gamma p_0 \Delta n = 0.$$
(52.8)

Hence, noting that $n_0 = p_0$, we find

$$\Delta p = \frac{\dfrac{\beta kI}{\gamma} - n_0 \Delta n}{n_0 + N_{cM}}$$
(52.9)

(this relationship is needed later). Using Eq. (52.6), we can rewrite Eqs. (52.2) and (52.3) in the form

$$E = -\frac{kT}{en_0} \frac{d}{dx} \Delta n,$$
(52.10)

$$\frac{dE}{dx} = \frac{4\pi e}{\varepsilon} (\Delta p - \Delta n).$$
(52.11)

Differentiating Eq. (52.10), we obtain

$$\frac{dE}{dx} = -\frac{kT}{en_0} \frac{d^2}{dx^2} \Delta n.$$
(52.12)

Equating the right-hand parts of Eqs. (52.11) and (52.12), using Eq. (52.9) and carrying out simple transformations, we obtain a differential equation for the distribution of the nonequilibrium carrier density Δn:

$$\frac{d^2}{dx^2} \Delta n - \frac{4\pi e^2 n_0}{\varepsilon kT} \left(\frac{N_{cM} + 2n_0}{N_{cM} + n_0} \right) \Delta n + \frac{4\pi e^2 n_0 \beta kI}{\varepsilon kT \gamma (n_0 + N_{cM})} = 0.$$
(52.13)

The quantity in parentheses in the second term can vary within the narrow limits of 1 to 2. We shall assume that $n_0 \gg N_{cM}$ so that this quantity is equal to 2, and also that in the third term of the equation we can neglect N_{cM} compared with n_0. * Then introducing the notation

---

* It is obvious that the condition $n_0 \gg N_{cM}$ represents weak ionization of impurities (because the Fermi level lies above the impurity levels); the case $n_0 \ll N_{cM}$ represents total ionization of the impurities. Since the photoeffect is obviously impossible in the total ionization case, we are not interested in it.

Table 4

σ_0, $\Omega^{-1}\cdot cm^{-1}$	n_0, cm^{-3}	l_e, cm	l_e, cm for $k = 10^{12}$ cm^{-3}	$\theta = \varepsilon/4\pi\sigma$, sec
10^3	$0.63 \cdot 10^{20}$	$5 \cdot 10^{-8}$	$5 \cdot 10^{-8}$	10^{-15}
1	$0.63 \cdot 10^{17}$	$1.58 \cdot 10^{-6}$	$1.58 \cdot 10^{-6}$	10^{-12}
10^{-3}	$0.63 \cdot 10^{14}$	$5 \cdot 10^{-5}$	$5 \cdot 10^{-5}$	10^{-9}
10^{-5}	$0.63 \cdot 10^{12}$	$5 \cdot 10^{-4}$	$5 \cdot 10^{-4}$	10^{-7}
10^{-6}	$0.63 \cdot 10^{11}$	$1.58 \cdot 10^{-3}$	$5 \cdot 10^{-4}$	10^{-6}
10^{-7}	$0.63 \cdot 10^{10}$	$5 \cdot 10^{-3}$	$5 \cdot 10^{-4}$	10^{-5}
10^{-8}	$0.63 \cdot 10^{9}$	$1.58 \cdot 10^{-2}$	$5 \cdot 10^{-4}$	10^{-4}
10^{-9}	$0.63 \cdot 10^{8}$	$5 \cdot 10^{-2}$	$5 \cdot 10^{-4}$	10^{-3}
10^{-12}	$0.63 \cdot 10^{5}$	1.58	$5 \cdot 10^{-4}$	1
10^{-15}	$0.63 \cdot 10^{2}$	$5 \cdot 10$	$5 \cdot 10^{-4}$	10^3

$$l_e = \sqrt{\frac{\varepsilon kT}{8\pi e^2 n_0}},\qquad (52.14)$$

we can rewrite Eq. (52.13) in its final form

$$\frac{d^2}{dx^2}\Delta n - \frac{\Delta n}{l_e^2} = \frac{1}{l_e^2}\frac{\beta kI}{2\gamma n_0}.\qquad (52.15)$$

The general solution of the above equation without its right-hand part (which represents the dark region where I = 0) has the form

$$\Delta n = C_1 e^{k_1 x} + C_2 e^{k_2 x}.\qquad (52.16)$$

Here C_1 and C_2 are constants which are found from the boundary conditions; k_1 and k_2 are the roots of the characteristic equation:

$$k^2 - \frac{1}{l_e^2} = 0,\qquad (52.17)$$

i.e.,

$$k_1 = \frac{1}{l_e};\quad k_2 = -\frac{1}{l_e}\ \text{and}\ \Delta n = C_1 e^{\frac{x}{l_e}} + C_2 e^{-\frac{x}{l_e}}.\qquad (52.18)$$

In the dark region, only the term with the negative power exponent in Eq. (52.18) has any meaning because the term with the positive exponent tends to infinity as x increases, contrary to the obvious observation that $\Delta n \to 0$ as $x \to \infty$.

Therefore, in the dark region

$$\Delta n = C_2 e^{-\frac{x}{l_e}},$$

(52.19)

i.e., the nonequilibrium carrier density decreases exponentially away from the boundary between light and darkness. The distance l_e at which the density is $1/e$ of its initial value, is known as the screening radius or the Debye radius [2, 3].

It is evident from Eq. (52.14) that the screening radius depends strongly on the equilibrium density n_0. Consequently, depending on the conductivity of the semiconductor, the value of the screening radius may vary within very wide limits. Table 4 lists the values of l_e (column 3) at room temperature (300°K) for several values of n_0 (column 2). In addition to n_0, Table 4 lists the electrical conductivity (column 1) for a mobility of $\mu = 100 \text{ cm}^2 \cdot \text{V}^{-1} \cdot \text{sec}^{-1}$.

It follows from Table 4 that at densities n_0 typical of germanium or silicon the screening radius is very small (10^{-4}-10^{-6} cm) and, consequently carriers diffuse in them (in the unipolar case) to negligible distances.

To determine Δn in the illuminated region, it is necessary to find the solution of Eq. (52.15) when the right-hand part is included. It is easily seen that this solution is

$$\Delta n = C_3 e^{\frac{x}{l_e}} + C_4 e^{-\frac{x}{l_e}} + \frac{\beta k I}{2\gamma n_0}.$$

(52.20)

To determine C_3 and C_4 we shall use the condition of continuity of Δn and dn/dx at the point $x = 0$. As a result, we finally obtain for the dark and illuminated regions, respectively:

$$(\Delta n)_{x>0} = \frac{\beta k I}{4\gamma n_0} e^{-\frac{x}{l_e}},$$

(52.21)

$$(\Delta n)_{x<0} = \frac{\beta k I}{4\gamma n_0} \left(2 - e^{\frac{x}{l_e}}\right).$$

(52.22)

In the illuminated region, far from its boundary, we have $(\Delta n)_{x=-\infty} = \frac{\beta k I}{2\gamma n_0}$.

In the dark region, again far from its boundary, we have $(\Delta n)_{x=\infty} = 0$.

At the boundary between light and darkness, Δn is half its maximum value

$$(\Delta n)_{x=0} = \frac{\beta k I}{4\gamma n_0}.$$

The resultant distribution of Δn is shown in Fig. 161b. Next we shall determine the distribution of the space charge ρ:

$$\rho = e\,(\Delta p - \Delta n). \tag{52.23}$$

Substituting into Eq. (52.33) the value of Δp from Eq. (52.9) and using the fact that $n_0 \gg N_{cM}$, we obtain

$$\rho = e\left(\frac{\beta kI}{\gamma n_0} - 2\Delta n\right). \tag{52.24}$$

Substituting the values of Δn for the dark and illuminated regions, we obtain, respectively

$$(\rho)_{x>0} = -\frac{e\beta kI}{2\gamma n_0}\,e^{-\frac{x}{l_e}}, \tag{52.25}$$

$$(\rho)_{x<0} = +\frac{e\beta kI}{2\gamma n_0}\,e^{\frac{x}{l_e}}. \tag{52.26}$$

The charge distribution is shown in Fig. 161c.

To obtain the electric field distribution E, we shall substitute Eqs. (52.21) and (52.22) into Eq. (52.10). Then we obtain (Fig. 161d)

$$(E)_{x>0} = \frac{kT}{e}\,\frac{\beta kI}{4\gamma n_0^2 l_e}\,e^{-\frac{x}{l_e}}, \tag{52.27}$$

$$(E)_{x<0} = \frac{kT}{e}\,\frac{\beta kI}{4\gamma n_0^2 l_e}\,e^{\frac{x}{l_e}}. \tag{52.28}$$

The potential distribution $V = \int E\,dx$ is shown in Fig. 161e. The total potential difference between the illuminated and dark regions is

$$V = \frac{kT}{e}\,\frac{\beta kI}{2\gamma n_0}\,\frac{1}{n_0} = \frac{kT}{e}\,\frac{(\Delta n)_{x=-\infty}}{n_0}, \tag{52.29}$$

i.e., it is governed by the ratio of the densities in the interiors of the illuminated and dark regions. This potential difference is frequently called the Dember emf.*

We have just considered the relatively simple case of a sharp boundary between light and darkness and a low excitation level. This discussion has led to the introduction of an important constant which represents the diffusion in the unipolar case: the screening length, which can be

---

* In contrast to the low excitation level discussed here, the Dember emf is, in general, given by (§ 63)

$$V = \frac{kT}{e}\,\ln\left(1 + \frac{\Delta n}{n_0}\right).$$

used in quantitative discussions of more complex problems. Among them is the case of a high excitation level. It can be shown that in the weak excitation case ($n_0 \gg \Delta n$) the screening length is governed by the equilibrium distribution n_0 but in the general case of an arbitrary excitation level we must replace n_0 in the expression for l_e with the total density $n = n_0 + \Delta n$. If the total density varies with the coordinates (which is frequently the case in diffusion), the quantity $l_e = f(n)$ has different values at different points, increasing with decrease of the density.

Consequently, the dependence of Δn on the coordinate in the case when $\Delta n \gg n_0$ should be as follows. First, the value of n decreases very sharply on crossing over from the illuminated to the dark region (because n is high in the illuminated region and therefore l_e is small). As one moves further into the dark region, the rate of fall of n slows down and l_e increases more and more. Thus, when $\Delta n \gg n_0$, it becomes difficult to determine the effective region over which diffusion is active. We can only say that the main part of the density fall occurs in a distance l_e whose order of magnitude is given by the carrier density in the interior of the illuminated region, but there is a density "tail," which is due to the considerably lower density in the dark region.

§53. EFFECTIVE TIME FOR THE ESTABLISHMENT OF DIFFUSION EQUILIBRIUM ("MAXWELL TIME CONSTANT")

The general theory of diffusion gives the well-known expression for the effective length L over which the diffusion spreads in time t: $L = \sqrt{Dt}$, where D is the diffusion coefficient.

It is easily seen that the screening length l_e can also be represented in the form of a root of the product of the diffusion coefficient and a time interval representing the effective time for the establishment of the diffusion equilibrium.

Multiplying the numerator and denominator under the root sign in Eq. (52.14) by the mobility μ and bearing in mind that $e\mu n_0 = \sigma_0$, we find, using the Einstein relationship $\mu kT/e = D$,

$$l_e = \sqrt{D \frac{\varepsilon}{8\pi\sigma_0}} = \sqrt{D\theta}. \qquad (53.1)$$

The quantity θ, which has the dimensions of time and is in our case given by

$$\theta = \frac{\varepsilon}{8\pi\sigma_0} \qquad (53.2)*$$

represents the time necessary for diffusion to a distance l_e.

* The value of the numerical coefficient in the denominator of Eq. (53.2) may vary somewhat, taking values of 8, 4, 2, etc., depending on the model of the semiconductor.

The time

$$\theta = \frac{\varepsilon}{4\pi\sigma_0} \tag{53.3}$$

is known as the Maxwell time constant.

The meaning of this quantity can be understood from the following considerations. Imagine a homogeneous semiconductor with conductivity σ_0 placed between two parallel electrodes of area S separated by a distance d.

The resistance R of the semiconductor between the electrodes is then

$$R = \frac{1}{\sigma_0} \frac{d}{S},$$

and its capacitance is

$$C = \frac{\varepsilon S}{4\pi d}.$$

Hence, we find that RC = $\varepsilon/4\pi\sigma_0$, which is identical with Eq. (53.3).

Thus, θ represents nothing else but the "RC" of the semiconducting material, i.e., it is the effective time for the establishment of the diffusion-drift equilibrium. The values of θ for various values of σ_0 when $\varepsilon \approx 4\pi$ are given in § 52 in Table 4 (column 5).*

§ 54. DENSITY DISTRIBUTION IN THE PRESENCE
OF AN EXTERNAL ELECTRIC FIELD

In the absence of a field, the diffusion drop of the carrier density is located (Fig. 161b) symmetrically with respect to the boundary between the illuminated and dark regions, extending into both of them to an effective distance l_e. The density distribution is different in the presence of an electric field.

Assume that a field E_0 – directed along the X axis so that the electrons drift in this field from the illuminated to the dark region – is established in the sample by means of an external source shown in Fig. 161a. Then, obviously, the electrons are dragged by the field into the dark part and, consequently, the effective range of the excess electrons increases in this region. At the same time, the field in the illuminated region "pushes" the electrons toward the boundary between darkness and light and their density in the illuminated region is made to approach the density in the interior.

---

* Strictly speaking, Eqs. (53.3) and (53.4) can be used only when the determined values of θ and l_e are greater than the relaxation time and the carrier mean free path, respectively.

We shall not deal in quantitative detail with this case but we shall quote the final expression for the density distribution in the dark region [4]:

$$\Delta n = \text{const} \, e^{-\frac{x}{l}},\tag{54.1}$$

where

$$l = \frac{1}{\frac{eE_0}{2kT}\left(\sqrt{1 + \left(\frac{2kT}{eE_0 l_e}\right)^2} - 1\right)}.\tag{54.2}$$

It is easily seen that in a weak field ($E_0 \gg 2kT/el_e$)

$$l = l_e = \sqrt{D\theta}.\tag{54.3}$$

In a strong field $\left(E_0 \gg 2\frac{kT}{el_e}\right)$, we obtain, by expanding the root in the denominator of Eq. (54.2) as a series and neglecting small terms:

$$l = \frac{eE_0}{kT}\, l_e^2 = \frac{eE_0}{kT}\, D\theta = \frac{eE_0 D\varepsilon}{4\pi kT\sigma_0} = l_E.\tag{54.4}$$

The quantity l_e, given by Eq. (54.4), is known as the effective drift length for the unipolar case. It is evident from Eq. (54.4) that, like the screening length, the drift length is greater for substances of low conductivity, other conditions being equal.

If we introduce a "diffusion field" $E_{\text{diff}} = kT/el_e$, it follows from Eq. (54.4) that

$$\frac{l_E}{l_e} = \frac{E_0}{E_{\text{diff}}}.\tag{54.5}$$

§55. DEBYE SCREENING LENGTH FOR LOW-CONDUCTIVITY SEMICONDUCTORS AND FOR DIELECTRICS

It follows from Eq. (52.14) that the screening length increases when the carrier density n_0 decreases and it should be very long in insulators. Table 4 in §52 shows that for low-conductivity semiconductors, for example those with $\sigma \approx 10^{-12}\ \Omega^{-1} \cdot \text{cm}^{-1}$, the screening length should reach centimeters. Under these conditions, the illumination of one end of a semiconducting sample of normal dimensions with light which produces photoconductivity should (because of the diffusion of carriers to a distance $\sim l_e$) raise considerably the conductivity at the other end of the sample. Experiments on low-conductivity semiconductors did not show this effect

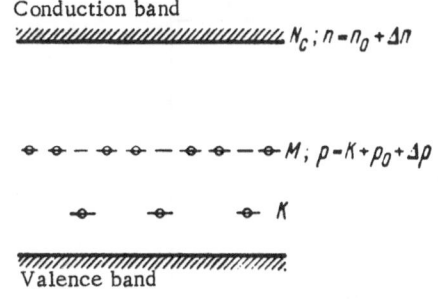

Fig. 162. Energy band structure in the presence of a compensating impurity.

and so for some reasons the effective screening radius is considerably smaller than that given by Eq. (52.14).

It can be shown that a sharp reduction of the screening length may occur in low-conductivity semiconductors in the presence of very small amounts of impurities at which bound charges may accumulate. We shall consider this case, using a "compensating impurity" as an example. *

Let us assume that the band model of a semiconductor differs from that given in §52 (Fig. 160) by the presence, in addition to the principal donor levels whose density is M, of a certain number of compensating acceptor levels with a density K, where K < M (Fig. 162). Then K electrons are transferred from the donors to the lower compensating acceptors and at absolute zero the M levels have M − K electrons and K holes, i.e., they are only partly filled with electrons.

A calculation similar to that given in §52 (we shall not consider it in detail) leads to the following expression for l_e:

$$l_e = \sqrt{\frac{\varepsilon kT}{4\pi e^2 n_0} \frac{(n_0 + N_{cM})}{(2n_0 + N_{cM} + K)}} \cdot \qquad (55.1)$$

In the absence of a compensating impurity (K = 0)

$$l_e = \sqrt{\frac{\varepsilon kT}{4\pi e^2 n_0} \frac{n_0 + N_{cM}}{2n_0 + N_{cM}}}, \qquad (55.2)$$

i.e., the expression is identical with Eq. (52.14), which has been obtained for the case $n_0 \gg N_{cM}$.

We shall estimate the value of the multiplier $\dfrac{n_0 + N_{cM}}{2n_0 + N_{cM} + K}$ under the root sign in Eq. (55.1) in the presence of a compensating impurity. If, for example, the Fermi level lies above the M levels, then $N_{cM} \ll n_0$ and this multiplier reduces to

$$\frac{n_0}{2n_0 + K} \cdot \qquad (55.3)$$

* An acceptor in an n-type semiconductor or a donor in a p-type material is a compensating impurity.

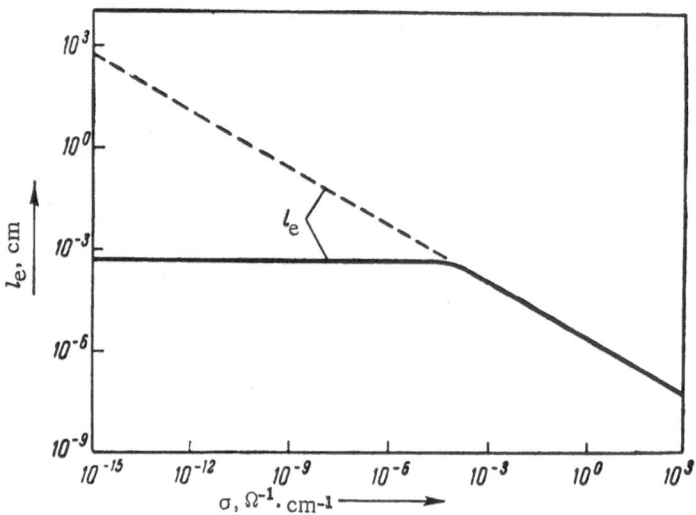

Fig. 163. Dependence of the screening length on the electrical conductivity (carrier density). The continuous curve represents the case when a compensating impurity is present, the dashed curve, the case without such an impurity.

For high-conductivity semiconductors with $n_0 \gg K$, the multiplier (55.3) becomes equal to 1/2 and we find that l_e is given by Eq. (52.14), i.e., a compensating impurity does not affect l_e. In the case of low-conductivity semiconductors with $n_0 \ll K$, the multiplier (55.3) becomes n_0/K and l_e is given by

$$l_e = \sqrt{\frac{\varepsilon kT}{4\pi e^2 K}} \, , \tag{55.4}$$

i.e., K now replaces n_0 in the expression for l_e.

If the quantity K is compared with the concentration of accidentally-present, uncontrolled impurities and if it is assumed that, say,

$$K = 10^{12} \, \text{cm}^{-3^*},$$

we find that (cf. Fig. 163 and Table 4, column 4 in § 52) upon decrease of σ_0 the quantity l_e remains constant, amounting to several microns, in semiconductors with $\sigma_0 < 10^{-5} \, \Omega^{-1} \cdot \text{cm}^{-1}$ ($n_0 < 10^{12} \, \text{cm}^{-3}$). Conse-

---

* 10^{12} cm^{-3} or $10^{-8}\%$ represents the content of accidental impurities in semiconductors such as germanium or silicon for which the methods of purification have been developed to the greatest extent. In other semiconductors, particularly low-conductivity ones (CdS, Sb$_2$S$_3$, etc.), the concentration of uncontrolled impurities and defects is much greater than 10^{12} cm^{-3}.

quently – for example, in the case of semiconductors with $\sigma \approx 10^{-12}$ $\Omega^{-1} \cdot cm^{-1}$ – a compensating impurity present in a negligible concentration of $\sim 10^{12}$ cm^{-3} reduces the screening length by a factor of ~ 1000. *

We have just considered the influence of a compensating impurity on l_e for a special model when the compensating acceptor impurity levels lie below the principal donor levels M. However, it is easily seen that the main results are unaltered if a different model is used. For example, if the K levels lie above the M levels,† then, as before, some of the M levels will be only partially filled and the K trapping levels weakly filled, we can show that the screening radius is [2]

$$l_e = \sqrt{\frac{\epsilon kT}{4\pi e^2 n_0 (1 + K/N_{cK})} \cdot \frac{n_0 + N_{cM}}{2n_0 + N_{cM}}}. \qquad (55.5)$$

This reduction of l_e due to the presence of a compensating impurity is characteristic not only of impurity excitation (impurity photoconductivity). The unipolar fundamental photoconductivity is usually explained by the formation of equal numbers of electrons and holes with one type of carrier being rapidly captured by traps. It is easily seen that this is equivalent to photoionization of traps with the transfer of carriers of the other type to a band, i.e., it is equivalent to the impurity excitation case.

In many (although not in all) low-conductivity semiconductors and in dielectrics, the effect of the screening length reduction due to the presence of compensating impurities plays a very important role.

§ 56. "SECONDARY" AND "CONDUCTION" PHOTOCURRENTS IN SEMICONDUCTORS

A.   Introduction

Serious difficulties in the study of the internal ionization in semiconductors are the absence of a rational terminology and the continued extensive use (even in the most recent papers and monographs) of the obsolete concepts of the so-called "primary" and "secondary" photocurrents.

Investigating the internal photoeffect in insulating crystals, Gudden and Pohl [1] discovered that the photocurrent in these crystals increases with time from a certain initial constant value.

---

* The physical meaning of this influence of a compensating impurity is that the space charge which "balances" the diffusion is partly located at the levels of the compensating impurity. If the density of such levels is much greater than the free-carrier density, the screening length becomes much shorter.

† The K levels act then as trapping levels.

They called the initial photocurrent "primary." The subsequent increase of the photocurrent Gudden and Pohl ascribed to secondary processes, not directly related to the principal effect. The difference, which increased with time, between the measured photocurrent and its initial value was called the "secondary" photocurrent.

This terminology, proposed by Gudden and Pohl in the early 1920's on the basis of some very fruitful and important investigations of the photoelectric properties of insulators, was later mechanically extended to all photoelectric phenomena in every type of substance. The concepts of primary and secondary photocurrents, defined by Gudden and Pohl in a purely phenomenological way, are quoted in textbooks as being universally valid and are widely used to describe new observations, including those observed in n-type semiconductors for which the concept of the secondary photocurrent has no meaning whatsoever, as will be shown below.

Over the years since the publication of Gudden and Pohl's work, the problem of the nature of the secondary photocurrent has been discussed many times. * A large number of possible mechanisms has been suggested to explain this phenomenon but none has gained general acceptance. The position changed only very recently when several workers suggested that the secondary current is related to the injection of carriers from the electrodes into the crystal and, consequently, it is simply the "conduction" photocurrent† which is usually observed in semiconductors.

However, we shall show that this identification of the secondary and conduction photocurrents is erroneous.

In the present section, we shall consider this problem for a semiconductor with unipolar conductivity by analyzing the concepts of the primary and secondary photocurrents and show the irrationality of their use . in the description of photoelectric phenomena in semiconductors, and we shall propose what seems to be a more rational terminology.

B. Photocurrent Kinetics

We shall consider certain general ideas on the photocurrent kinetics [6]. If a semiconductor sample is connected up to the circuit shown in Fig. 164 and illuminated from a certain moment with light of constant intensity, then the relaxation to new steady-state values of the carrier density and of the current is, in general, governed by two processes:

---

*Gudden and Pohl related the secondary photocurrent to electrolytic processes. Later, however, this term was used to describe electron processes.

†A "conduction" ("through") current flows when diffusion equilibrium is established between a crystal and the electrodes and when carriers from the electrodes may enter the crystal. A more detailed definition of the term "conduction" photocurrent is given in subsection B.

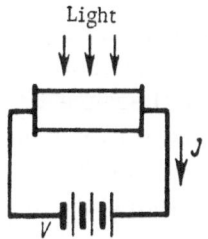

Fig. 164. Circuit for connecting up a sample.

1) changes in the density due to electron transitions in the energy space, i.e., due to ionization and recombination; *

2) changes of the state due to the motion of carriers in the coordinate space, i.e., due to diffusion. and drift in an electric field. †

Both processes occur simultaneously and are interrelated. However, an estimate of the influence of each of them on the effective relaxation time of the overall process cannot be obtained from the time constants for each process separately.

The time for the establishment of a recombination equilibrium (i.e., the equilibrium in the ionization-recombination process) τ, known as the carrier relaxation time, is related to the rate of recombination. The time for the establishment a diffusion-drift equilibrium is governed by the electrical conductivity σ and the permittivity ε; in the simplest case, it is expressed in terms of these quantities in the form (cf. §53)

$$\theta = \frac{\varepsilon}{4\pi\sigma}. \tag{56.1}$$

If one of these times, θ or τ, is considerably larger than the other, then the larger time constant governs the total relaxation time of the whole process. ‡ Thus, for example, in low-conductivity substances the time for the establishment of the diffusion-drift equilibrium, θ, is usually greater than the lifetime τ and, consequently, the former governs the total duration of the relaxation process.

In this case, the steady-state current after an effective time τ is the same at each point in the circuit and over each cross section of the sample; therefore, it is reasonable to call it the "conduction" current. We note that in the case considered, when $\theta \gg \tau$, the concept of the conduction current is identical with the concept of the steady-state current. However, this is not true in other cases.

For example, if the lifetime τ is much greater than the time θ, then the total duration of the relaxation process is governed by τ but at any moment at any cross section of the sample, or any point in the circuit, the current is the same.

---

* The term recombination as used here also includes trapping.

† When the whole sample between the electrodes is uniformly illuminated, the carriers diffuse toward the electrodes.

‡ We are speaking here of the total relaxation time; in some cases, when the main component of the photocurrent relaxes in a time corresponding to the shorter of the relaxation times, the residual photocurrent may relax during a time governed by the longer of the relaxation times.

When $\tau \gg \theta$, at any moment of the process of the slow establishment (the value of τ is large) of equilibrium in the ionization-recombination process, the conduction current is the same at all points because its value is established rapidly (θ is small). This conduction current varies with time and reaches its steady-state value after a time τ. It is therefore clear that the steady-state current is always the conduction current but the conduction current need not be a steady-state one.

It also follows that the conduction current may be defined as that current which at any given moment is the same at any point in the circuit. The time for the establishment of the diffusion-drift equilibrium θ (which is also the time for the establishment of the conduction current) represents the process of the establishment of equilibrium in the coordinate space in the same way as the lifetime represents this process in the energy states.

The process of the establishment of the diffusion-drift equilibrium or of the conduction current has a simple meaning and, for example, in the case when $\theta \gg \tau$ it is in many ways analogous to the processes occurring in electric circuits when the circuit parameters (resistance, capacitance, voltage) are suddenly altered. Such processes have been extensively investigated by electronic engineers and are known as transient processes.

Therefore, it seemed reasonable to call the processes and the photocurrent, observed before the establishment of the diffusion-drift equilibrium, the "transient" processes and the "transient" photocurrent, respectively. *

On the establishment of equilibrium in the coordinate space, the transient current becomes the conduction current.

C. "Primary" Photocurrent

It is now obvious that the transient current (or more precisely its initial value) corresponds to Gudden and Pohl's "primary" current.

The principal assumption made by Gudden and Pohl to describe the primary current was the assumption that carriers from the electrodes do not take part in this current. The primary current in the closed circuit shown in Fig. 164 is solely due to the motion of charges liberated by ionization directly in the sample itself.†

The transient current in the circuit of Fig. 164 is also initially due to the motion of carriers liberated directly by ionization. This can be seen from Fig. 165.

---

* During the transient conditions, the currents across different cross sections of the semiconductor sample are not the same. By the transient current, we shall understand the current in the metal conductors of the circuit external to the sample (Fig. 164).

† This is fully analogous to the case with a crystal replaced by an ionized gas contained between cold metal electrodes.

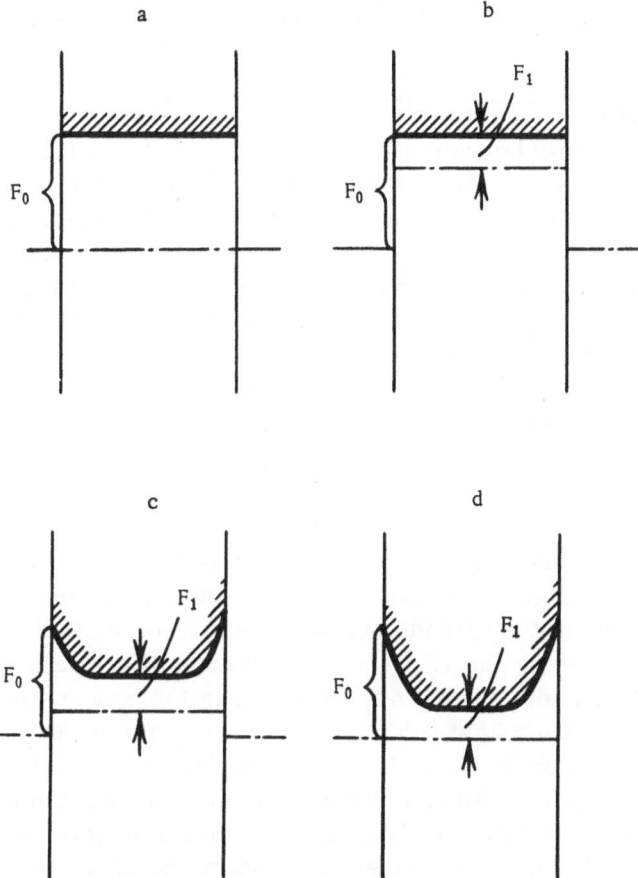

Fig. 165. Changes in the energy band structure during the establishment of the conduction current: a) Before illumination;
b) immediately after illumination — the recombination equilibrium established but not the diffusion equilibrium; c) the
diffusion equilibrium partly established; d) the diffusion equilibrium completely established.

Let us assume that a semiconductor is placed between the metal electrodes and that in darkness there are no blocking or antiblocking layers
(neutral contacts) (Fig. 165a). We shall assume also that the carriers in
darkness and during illumination are electrons* (that is why the valence
band is not shown in Fig. 165). Finally, we shall assume that the lifetime τ is much shorter than θ, the time for the establishment of diffusion-
drift equilibrium.

---

*In darkness, there are few holes, and those appearing due to ionization are captured very ra-
pidly. The conditions formulated here are identical with those assumed by Mott and Gurney [3]
in their treatment of the secondary photocurrent.

Under these conditions, the illumination of the crystal produces very rapidly (τ is short) a higher carrier density in the upper band, i.e., the quasi-Fermi level for electrons is displaced closer to the conduction band (Fig. 165b).

The discontinuity between the Fermi level in the metal electrodes and the quasi-Fermi level in the semiconductor, equivalent to a change in the contact potential difference between the metal and the semiconductor gives rise to the diffusion of electrons from the semiconductor to the metal electrodes. If, moreover, an electric field is applied to the sample, we then have directed motion (current) of photoelectrons from the cathode to the anode. It is easily seen that, before the establishment of the diffusion equilibrium (Fig. 165b), the electrons which move away from the cathode are not replaced by electrons from the cathode itself. This is because the current in the semiconductor is governed by the higher electron density during illumination while the current through the metal-semiconductor contact at the cathode is still governed by the previous conditions which have existed in darkness (the number of electrons in the semiconductor which move away from the cathode is larger than the number entering the semiconductor from the cathode because the former motion involves overcoming a small energy gap (F_1) between the edge of the band and the quasi-Fermi level in the semiconductor, while the latter process involves overcoming a barrier equivalent to the gap (F_0) between the Fermi levels of the semiconductor and in the metal). Consequently, until the diffusion-drift equilibrium is established, the carriers entering from the electrodes do not play any important role and the phenomenon may be described by means of the model used by Gudden and Pohl to interpret the primary current, assuming that there is no carrier exchange between the crystal and the electrodes. *

When the diffusion-drift equilibrium is established, the quasi-Fermi level in the semiconductor drops (Fig. 165c), the relative importance of the carriers entering the crystal from the cathode increases and, moreover, the conditions depart from those postulated for the primary current. †

The primary current is therefore simply the initial value of the transient current. ‡   After an effective time θ, a conduction current is es-

---

* It is easily seen that under these conditions the photocurrent should tend to saturation with increase of the field, which is also typical of the primary photocurrent.

† If a voltage is applied to the sample, the establishment of the diffusion equilibrium between the electrodes and the sample is accompanied by a redistribution of the applied voltage between the middle part and the electrode regions of the sample: the field in the electrode regions rises and in the middle part falls. This process of the establishment of the drift equilibrium finally gives rise to the conduction current. We shall not consider it in greater detail.

‡ The later stages of the development of the transient current are characterized by the increasingly important role of the electrons entering the crystal from the electrodes and therefore these stages cannot be identified with the primary current.

tablished which is equal at all points of the series circuit shown in Fig. 164. Then the number of electrons in the semiconductor moving away from the cathode is exactly equal to the number entering the semiconductor from the cathode (Fig. 165d). In this sense, the conduction and primary currents are completely opposite concepts.

When the equilibrium in the energy space is established rapidly (τ is small), the relaxation process during illumination simply represents the transformation of the transient current into the conduction current. This completes the process of the establishment of the equilibrium in the coordinate space.

D.   "Secondary" Photocurrent

We have mentioned in the foregoing that it is wrong to identify the secondary with the conduction photocurrent.

In order to prove this, we shall consider more carefully the definition of the secondary current given by Gudden and Pohl.

Gudden and Pohl showed experimentally that the charge that passed through zinc sulfide during illumination was directly proportional to time only for illumination of short duration. After longer illumination, the amount of charge which passed through the crystal began to increase faster with time. Hence they concluded that as a result of some changes in the crystal due to the passage of a photocurrent the total dark conductivity of the crystal increased. The part of the current due to the conductivity which increased with time was called the "secondary" photocurrent. Gudden and Pohl found that the time dependence of the charge Q which passed through the crystal could be approximated by the parabolic function

$$Q = at + bt^2, \qquad (56.2)$$

where a and b are constants.

Differentiating Eq. (56.2) with respect to time t, we obtain the relationship for the current

$$J = a + 2bt, \qquad (56.3)$$

where J is the total measured current, $a = J_p$ is the primary photocurrent, $2bt = J_s$ is the secondary photocurrent.

Thus, according to the definition of Gudden and Pohl, the secondary photocurrent J_s increases with time. Consequently, the total measured photocurrent,

$$J = J_p + J_s \qquad (56.4)$$

should also increase with time.

If we identify the secondary photocurrent* with the conduction current, we find that since the secondary photocurrent increases with time, the conduction photocurrent should also become greater than the primary current. This is not true. Without making any special assumption (for example, without considering the barrier layers at contacts, the variation of F_0 during illumination, very strong fields, etc.) we can easily show that the photocurrent appearing immediately after the beginning of illumination can only decrease on approach to the diffusion equilibrium with the electrodes.†

In order to prove this, we shall consider the variation with time of the current in the circuit of Fig. 164 during illumination of the whole area of the sample between the electrodes, which provide neutral contacts in darkness (it is assumed that $\tau \ll \theta$). Thus immediately after the beginning of illumination (actually after time τ, i.e., practically at the time t = 0) a higher carrier density is established in the sample. Consequently, an additional current (photocurrent) will appear in the circuit, which is

$$\Delta J_{ph} = \frac{eN}{L} \mu E, \tag{56.5}$$

where e is the electronic charge, N is the total number of photocarriers during illumination, L is the length of the crystal, μ is the carrier mobility, and E is the electric field (E = V/L). However, the illumination disturbs the equilibrium which existed in darkness at the electrodes.

The higher carrier density in the sample causes carriers to diffuse toward the electrodes (cf. subsection C) and lowers the edge of the conduction band in the interior of the sample (cf. Fig. 165). This process was described in detail by Mott and Gurney. However, in spite of the possible impression that might be gained from Mott and Gurney's treatment, this can lead only to a reduction of the photocurrent compared with its initial value [cf. Eq. (56.5)].

The "blocking" layers, which appear near the electrodes due to the diffusion of electrons toward the electrodes and due to the lowering of the conduction band edge in the interior of the sample, reduce the initial photocurrent.‡

After the establishment of the diffusion equilibrium (Fig. 165d), the sample consists of three regions connected in series: two blocking layers

---

* Here and later, the secondary photocurrent will be understood to be the total photocurrent, which, strictly speaking, is valid only when $J_s \gg J_p$.

† The case $\tau \ll \theta$ is considered, as in the experiments of Gudden and Pohl. The photocurrent appearing immediately after illumination is understood to be the photocurrent after the establishment of the recombination equilibrium, i.e., after a time τ from switching on the illumination.

‡ The thickness of these blocking layers is usually sufficient to neglect the possible tunnel effect and to represent them as resistances connected in series with the middle part of the sample.

whose conductivity is approximately equal to the conductivity before illumination (we shall denote the effective blocking-layer thickness by l_e) and the middle portion of the sample of length $L - 2l_e$ whose conductivity represents the illuminated state.

Obviously, because of the presence of the blocking layers, the additional current (the conduction photocurrent) is smaller than the current rise immediately after illumination, which is given by Eq. (56.5).

It follows, therefore, that it is wrong to identify the "secondary" photocurrent observed by Gudden and Pohl in ZnS, which (and this is its most important characteristic) is greater than the primary current, with the conduction photocurrent which is smaller than the primary photocurrent. *

In fact, the term "secondary" photocurrent should not be used in the description of typical processes in semiconductors. This must be stressed particularly as it is frequently stated that in semiconductors the photocurrent exhibits its "secondary" nature.

E. Some Terminological Problems

It is now necessary to depart from the traditional terminology. The question is whether one should introduce new concepts which represent correctly the photoelectric phenomena on the basis of our present knowledge, or to alter the meaning and definitions of the earlier terms: "primary" and "secondary" photocurrents.

The latter solution is unacceptable because the term "secondary" photocurrent has been used to represent many various phenomena and its retention would lead to confusion. If we accept the former solution, we must make it possible to describe separately the kinetics of processes in the coordinate space and in the energy space, i.e., we must distinguish between the processes of the establishment of the diffusion-drift equilibrium and of the recombination equilibrium.

In A (see above), we suggested the terminology for the description of the kinetics of the establishment of the diffusion-drift equilibrium. This terminology includes the concepts of transient photocurrent, representing the absence of the diffusion-drift equilibrium, and of the conduction photocurrent, flowing when such equilibrium is reached.

To describe the photocurrent kinetics related to the establishment of the recombination equilibrium, we are proposing to use the terms "equilibrium" photocurrent and "nonequilibrium" photocurrent. The

---

* We must stress that this conclusion is drawn from consideration of the same model (Fig. 165) which is used by the workers who identify the conduction with the secondary current. There are models which may give rise to a conduction current larger than the primary current but they do not invalidate the above conclusions.

photocurrent under the conditions when the recombination equilibrium is established should be called the equilibrium photocurrent, and in the absence of such equilibrium we have the nonequilibrium photocurrent.

It follows that we can have four cases which can be described by means of the proposed terminology. For example, if the illumination of a sample begins at a time $t = 0$, then depending on the relationship between the time t and the time constants for the establishment of the recombination and diffusion-drift equilibria, τ and θ, we have:

1) transient nonequilibrium photocurrent (when $\tau \gtrless \theta$ and $t < \tau$ and $t < \theta$);

2) transient equilibrium photocurrent (when $\tau \ll \theta$ and $\tau < t < \theta$);

3) conduction nonequilibrium photocurrent (when $\tau \gg \theta$ and $\tau > t > \theta$);

4) conduction equilibrium photocurrent, or in other words the steady-state photocurrent (when $\tau \gtrless \theta$ and $t > \tau$ and $t > \theta$).

In the case of semiconductors of relatively high conductivity and, consequently, small θ, one usually has the conduction nonequilibrium photocurrent during the relaxation process, * while in the case of low-conductivity semiconductors and insulators, we usually have the transient equilibrium photocurrent.

---

* This allows us to determine the photocarrier lifetime and other important parameters of semiconductors using the methods described in Chap. III for studying photoconductivity.

Chapter XIII

DIFFUSION AND DRIFT OF NONEQUILIBRIUM CARRIERS
(AMBIPOLAR CASE)

In conductors containing free charges any disturbance of the neutrality, i.e., the appearance of uncompensated space charge and electric fields, generates currents directed in such a way as to reestablish the neutrality.

In the case of a nonuniformly illuminated semiconductor with unipolar conduction (Chap. XII), the diffusion current disturbs the neutrality and gives rise to a space charge and a field. This field is directed so that it opposes diffusion and further departure from the neutrality. Consequently, the neutrality is disturbed only in a small region defined by the screening length. The diffusion also extends in this small region. Thus the carriers cannot be displaced far during diffusion in the unipolar case because of the electrostatic forces of attraction of the fixed charges of opposite sign.

The conditions are different when there are carriers of two signs. Then the diffusion of carriers of one sign sets in motion the carriers of the opposite sign (which were fixed in the unipolar case) in such directions as to compensate the space charge and reestablish the neutrality.

In this way, the diffusing carriers cause diffusion of carriers of the opposite sign (which retarded diffusion in the unipolar case).

These conditions are known as the ambipolar case and here the diffusion may spread over considerable distances, governed by the lifetimes of free carriers and their mobilities but not by the screening length. The present chapter deals with the diffusion and drift under these conditions.

§ 57. DIFFUSION AND DRIFT OF MINORITY CARRIERS

The diffusion and drift of minority carriers is of special interest for reasons which will emerge from the forthcoming treatment.

Let us consider the following example. Assume that electrons are introduced in some way into a p-type region of a semiconductor (here electrons are the minority carriers). The resultant departure from the neutrality in this region may be eliminated in two ways: 1) by the "expulsion" of electrons from the region considered, or 2) by the "drawing in" of holes. Since there are more holes than electrons, the neutrality is obviously reestablished by the second method. It follows that the diffusion or drift of the minority carriers occurs under conditions such that the charge of the minority carriers is rapidly compensated by a redistribution of the charge of the majority carriers which are drawn into the region where there are excess minority carriers. Obviously, the number of such majority carriers necessary to compensate completely for the presence of the minority carriers is equal to the number of minority carriers. If the majority carrier density is sufficiently high, this redistribution and the reestablishment of neutrality occur very easily and in a very short time. * The diffusion of minority carriers is practically unimpeded by this redistribution and the appearance of the opposing electric fields, i.e., it proceeds as if the minority carriers were uncharged but still had the same mobility and diffusion coefficient (§ 62).

In an electric field, the minority carriers behave in a special way: they drift in the field as charged particles but they do not produce a space charge in those regions where they appear (because their charge is rapidly compensated by a redistribution of the majority carriers) and, consequently, they do not affect the nature of the electric field, which is determined completely by external conditions [10].

Minority carriers may be generated not only by internal ionization (for example, by illumination) but also by injection from a rectifying contact connected up in the forward direction. Such injection generates equal numbers of minority and majority carriers, which ensures neutrality. Thus the conductivity of a sample increases as the result of injection.

When a current is passed through a sample with nonrectifying contacts, which brings majority carriers into the sample, the total density of carriers in the sample does not increase. This is because, to conserve neutrality, the same number of majority carriers leaves the sample. Thus, in contrast to the injection of minority carriers, the introduction of majority carriers cannot increase the conductivity of the sample.

Characteristic features of the behavior of minority carriers can be used to study, by means of their diffusion and drift, the lifetime and mobility of carriers in a very exact way.

We shall consider quantitatively the problem of minority-carrier diffusion and drift using an example similar to that employed in the unipolar case (§ 52).

We shall assume that we have a sufficiently long semiconductor sample (Fig. 166), part of which from $x = 0$ to $x = -L$ is uniformly illuminated

* This time is given by $\theta = \varepsilon/4\pi\sigma$ (§ 53).

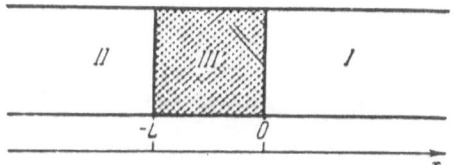

Fig. 166. Illumination of the sample: the hatched region III is illuminated, while regions I and II are in darkness.

with light, which increases the minority carrier density. * We shall also assume that the minority carrier density Δn is small compared with the density of majority carriers and that the minority carrier lifetime τ is everywhere the same.

Minority carriers can diffuse and drift in the dark portions I and II along the x axis (we shall consider only the one-dimensional case). The minority carrier density Δn can be found (as in §52) by solving the equations

$$\frac{\partial}{\partial t}\Delta n = k\beta I - \frac{\Delta n}{\tau} + \frac{1}{e}\frac{\partial j}{\partial x}, \qquad (57.1)$$

$$j = eD\frac{\partial}{\partial x}\Delta n + e\mu n E. \qquad (57.2)$$

Here, the term $k\beta I$ determines the number of minority carriers liberated by illumination, the intensity of which is I (k is the absorption coefficient and β the internal photoeffect yield); the term $\Delta n / \tau$ allows for the recombination of minority carriers, and the term $\frac{1}{e}\frac{\partial j}{\partial x}$ is the divergence of the minority carrier current (j is the density of the minority carrier current).

rier current (j is the density of the minority carrier current).

Since we are dealing with the diffusion and drift of minority carriers, the space charge of which is compensated by a suitable redistribution of the majority carrier density, the field E is constant and equal to the external field.

We shall find the distribution of the minority carrier density along the x axis in the steady-state case $\left(\frac{\partial \Delta n}{\partial t} = 0\right)$ Differentiating Eq. (57.2) with respect to x and substituting $\partial j/\partial x$ into Eq. (57.1), we obtain

$$D\frac{d^2}{dx^2}\Delta n + \mu E\frac{d}{dx}\Delta n - \frac{\Delta n}{\tau} = -\beta k I. \qquad (57.3)$$

Dividing both sides of the equation by D (where $D \neq 0$) and introducing the notation

$$\left.\begin{array}{l} l_D = \sqrt{D\tau}, \\ l_E = \tau\mu E, \end{array}\right\} \qquad (57.4)$$

* In contrast to the case described in §52, we are assuming here that the sample is illuminated by a light "strip," as is frequently done in the study of diffusion and drift of minority carriers. It is, moreover, assumed that the light is weakly absorbed so that the rate of generation of minority carriers does not in practice vary with depth.

we finally obtain

$$\frac{d^2}{dx^2}\Delta n + \frac{l_E}{l_D^2}\cdot\frac{d}{dx}\Delta n - \frac{\Delta n}{l_D^2} = -\frac{\tau\beta kI}{l_D^2}. \qquad (57.5)$$

(For regions I and II we have $I = 0$.) The above equation differs from Eq.
(52.15) by the presence of a term in its first derivative, which, however,
vanishes when $E = 0$. Solving this equation in the same way as Eq. (52.15)
and using the boundary conditions (the minority carrier density must
vanish at $x \to +\infty$ and $x \to -\infty$) as well as the condition that the particular
carrier current must be finite and continuous at the boundaries $x = 0$ and
$x = -L$, we obtain [4, 11]:

for region I ($+\infty \geq x \geq 0$)

$$\Delta n = \tau\beta kI\,\frac{l_1}{l_1+l_2}\left(1 - e^{-\frac{L}{l_1}}\right)e^{-\frac{x}{l_1}} = \text{const}\,e^{-\frac{x}{l_1}}, \qquad (57.6)$$

for region II ($-\infty \leq x \leq -L$)

$$\Delta n = \tau\beta kI\,\frac{l_2}{l_1+l_2}\left(e^{\frac{L}{l_2}} - 1\right)e^{\frac{x}{l_2}} = \text{const}\,e^{\frac{x}{l_2}}, \qquad (57.7)$$

for region III ($-L \leq x \leq 0$)

$$\Delta n = \tau\beta kI\,\frac{1}{l_1+l_2}\left[l_1\left(1 - e^{-\frac{L+x}{l_1}}\right) + l_2\left(1 - e^{\frac{x}{l_2}}\right)\right], \qquad (57.8)$$

where

$$l_1 = \frac{2l_D^2}{\sqrt{l_E^2 + 4l_D^2} - l_E}, \qquad (57.9)$$

$$l_2 = \frac{2l_D^2}{\sqrt{l_E^2 + 4l_D^2} + l_E}. \qquad (57.10)$$

Thus, in the dark regions (I and II), the minority carrier density decreases exponentially with distance and the decrease is by a factor of e
over the distances l_1 and l_2, respectively. The quantities l_1 and l_2 are
the "displacement constants" when the diffusion occurs in the presence
of an electric field, i.e., when the diffusion and drift occur simultaneously.

In the absence of a field (i.e., when $l_E = \mu\tau E = 0$) there is only diffusion and we find from Eqs. (57.9) and (57.10)

$$l_1 = l_2 = l_D = \sqrt{D\tau}. \qquad (57.11)$$

Thus the quantity $l_D = \sqrt{D\tau}$ introduced above represents the effective
diffusion length, i.e., the average distance travelled by a diffusing carrier during its lifetime.

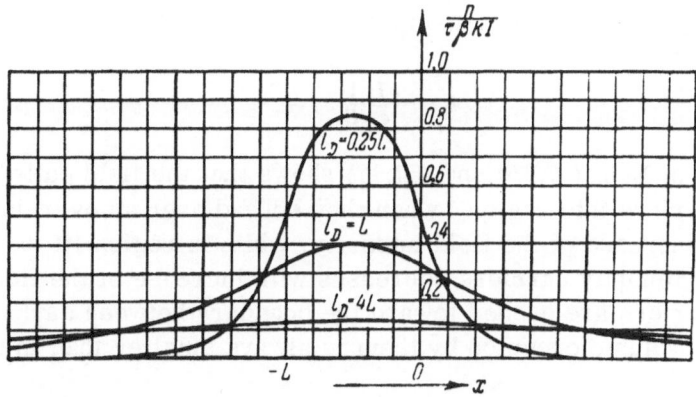

Fig. 167. Minority carrier distribution in the absence of a field.

In the ambipolar case, the diffusion length plays to some extent the role of the screening length in the unipolar case, because it defines the effective region in which the diffusion occurs. Assuming that the Einstein relationship $D/\mu = kT/e$ is valid, we can rewrite Eq. (57.11) in the form

$$l_D = \sqrt{\frac{kT}{e}\mu\tau}. \tag{57.12}$$

We shall consider also the opposite case of a strong field such that the motion of carriers is governed primarily by the drift and not by the diffusion, i.e.,

$$l_E \gg l_D. \tag{57.13}$$

If we rewrite Eqs. (57.9) and (57.10) in the form

$$l_1 = \frac{2l_D^2}{l_E}\frac{1}{\sqrt{1 + 4\frac{l_D^2}{l_E^2}} - 1}, \tag{57.14}$$

$$l_2 = \frac{2l_D^2}{l_E}\frac{1}{\sqrt{1 + 4\frac{l_D^2}{l_E^2}} + 1}, \tag{57.15}$$

then, expanding as a series the expression under the square root in the denominator and using Eq. (57.13), we obtain

$$l_1 \cong l_E = \tau\mu E, \tag{57.16}$$

$$l_2 \cong \frac{l_D^2}{l_E} = \frac{D}{\mu} \cdot \frac{1}{E} \tag{57.17}$$

or, using the Einstein relationship

$$l_2 = \frac{kT}{e} \frac{1}{E}.$$

(57.18)

Comparison of Eqs. (57.16) and (57.17) shows that the field pulls in minority carriers into region I when they spread over an over-increasing region with increase of the field intensity, while in region III the volume occupied by minority carriers decreases with increase of the field. Thus minority carriers are pushed toward the boundary between darkness and light and the volume occupied by them is always smaller than in the absence of a field.

Figure 167 shows the distribution of minority carriers in the absence of a field while Figs. 168a and 168b give the distribution in the presence of a field. Figure 167 shows that in the absence of a field the density in the illuminated region approaches its maximum value only when l_D/L is sufficiently small: this value is $\tau\beta kI$ in the center, and $\tau\beta kI/2$ at the ends of the illuminated region. The density maximum in the absence of a field is in the middle of the illuminated region.

It follows from Fig. 168 that with increase of the field intensity the density maximum is displaced more and more toward one of the boundaries of the illuminated region and its value decreases; the direction of the displacement of the maximum corresponds to the direction of the minority carrier drift.

The value of the minority carrier density in region II where the field "compresses" these carriers decreases monotonically with increase of the field, while in region I where the field draws in the carriers there is no such monotonic dependence: for given x, the density first increases with increase of the field but this is followed by a decrease. Consequently, when a field which causes a drift of minority carriers from the illuminated region into the dark region I is applied to the sample, the density of the carriers in that region (especially near the boundary between darkness and light) may even decrease compared with the field-free conditions, although the total number of carriers in the dark region increases because of the drift.

We have considered the distribution of minority carriers on the assumption that their charge is always compensated by a suitable redistribution of majority carriers, which conserves the neutrality. This means that wherever there are nonequilibrium minority carriers there must be an equal number of excess (nonequilibrium) majority carriers. Thus the minority carrier distributions shown in Figs. 167 and 168 may also be considered to be the distributions of nonequilibrium majority carriers.

The characteristic features of the behavior of the minority carriers which diffuse as if they were neutral but drift in an electric field as

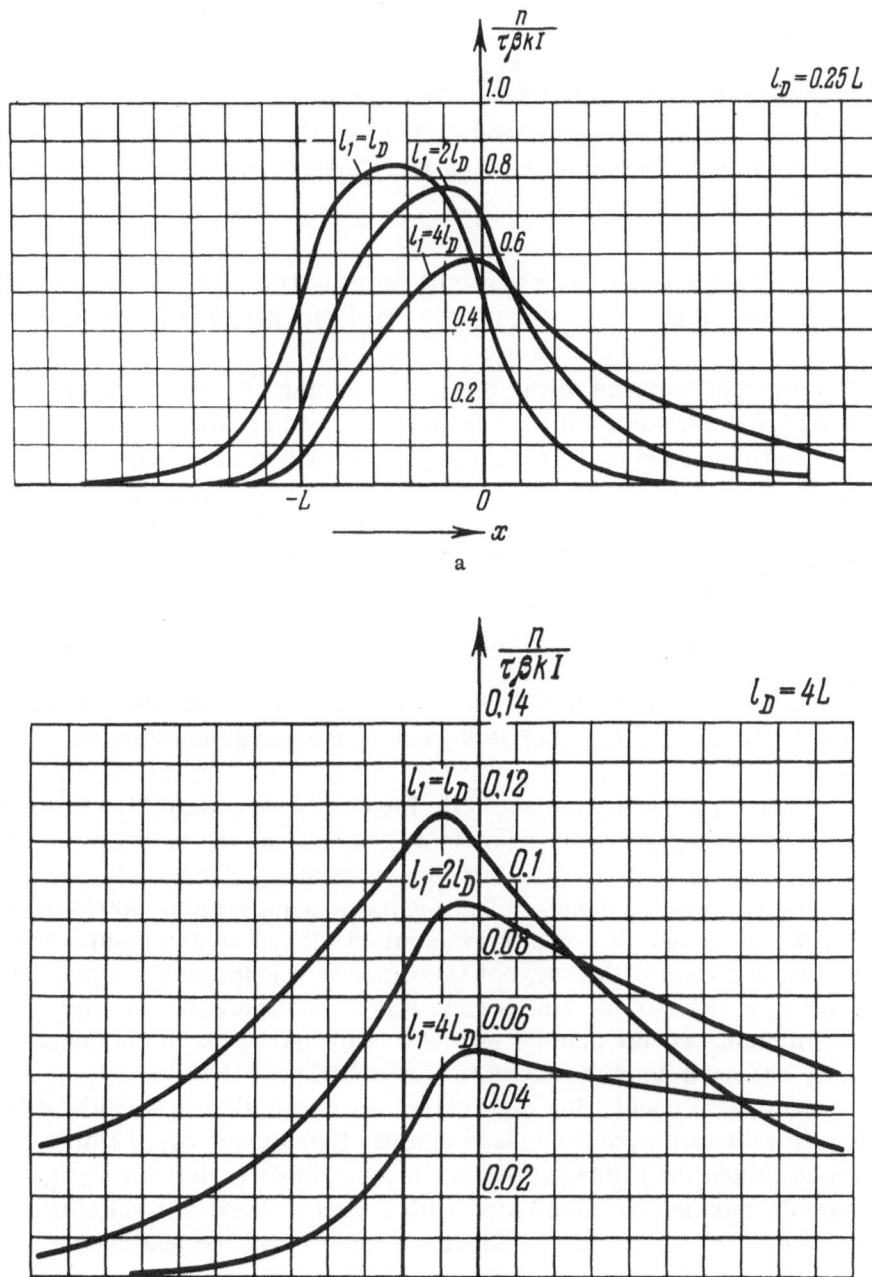

Fig. 168. Minority carrier distribution in the presence of a field.

charged particles (without, however, generating a space charge) allows us to use them in several effective methods of the study of semiconductors and the determination of their parameters.

We shall illustrate this by considering next the methods of measuring the lifetime (diffusion length), and drift mobility, as well as the results obtained in an experimental check of the Einstein relationship for nonequilibrium carriers.

§58. EXPERIMENTAL METHOD OF DETERMINING THE DIFFUSION LENGTH l_D AND LIFETIME τ OF MINORITY CARRIERS

The special features of the minority carrier diffusion allow us to use a method for determining their lifetime which is essentially different from the methods discussed in the foregoing sections (the relaxation curve and frequency dependence methods). The method for minority carriers is based on the fact that, as shown by Eqs. (57.6), (57.7), and (57.11), the steady-state distribution of the density of minority nonequilibrium carriers in the dark is a simple exponential function of the diffusion

length $l_D = \sqrt{D\tau} = \sqrt{\dfrac{kT}{e}\mu\tau}$. Consequently, if their distribution can be

found experimentally it can then be used to obtain the parameter l_D as well as the lifetime τ (if other independent measurements are used to find μ).

The density distribution could be found by determining the dependence of the conductivity on a coordinate, using, for example, point probes.* However, this method is much inferior to another which allows us to determine directly the density of minority nonequilibrium carriers at any given point by means of a rectifying contact placed at this point. It is known that the rise of the current through a p-n junction connected in the blocking direction (or the emf across a p-n junction when the minority nonequilibrium carrier density is low), is proportional to the density of minority carriers near the junction (for details see §70).

In many semiconducting materials, a p-n junction is formed at the contact of a metallic point pressed against the surface of a sample. This allows us to use the following simple arrangement to determine l_D and τ: the sample surface is illuminated with a narrow band of light with sharp edges (Fig. 169). A point (a collector probe) is placed against the surface and a voltage applied in the blocking direction. The current through the probe is proportional to the minority carrier density at the point where it is placed. By moving the probe along the sample, we can record the

---

* It is known that the electrical conductivity measured by means of wire probes is governed by the electrical conductivity of the small regions of the sample in the direct vicinity of the probes.

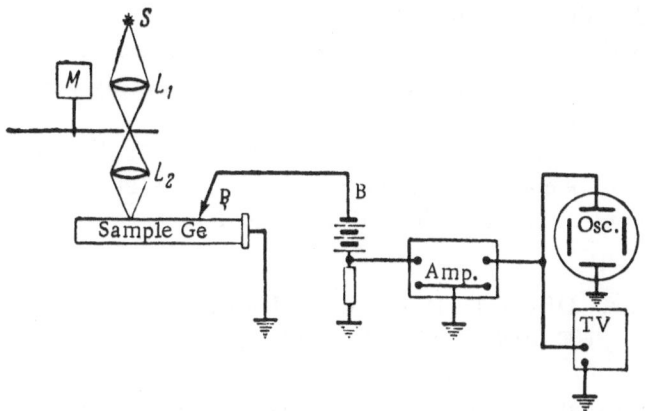

Fig. 169. Apparatus used for measuring the diffusion length. P is the collector probe, B is a battery which provides the bias in the blocking direction, S is the source of light, L_1 and L_2 are lenses, and M is the modulator motor.

dependence of the minority carrier density on the distance from the boundary between light and darkness, denoted by x. However, it is simpler to displace the strip of light and keep the probe fixed. Usually the light beam is modulated by means of a disk with apertures. The signal v obtained in the probe circuit is then alternating and can be easily recorded after amplification with an ac amplifier. This signal is proportional to the nonequilibrium minority carrier density Δn. Since

$$\Delta n = \text{const}\, e^{-\frac{x}{l_D}}$$

therefore

$$\ln v \sim \ln \Delta n = \text{const} - \frac{x}{l_D}.$$

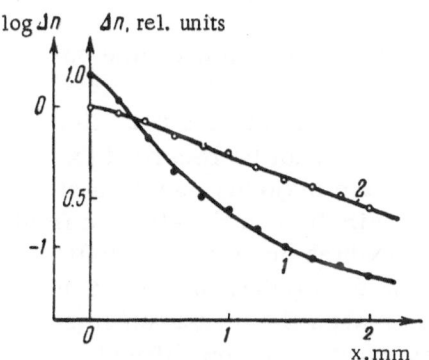

Fig. 170. 1) Dependence of the minority carrier density on the distance from the light strip; 2) the same dependence on a semilogarithmic scale.

Thus, if the results are plotted as $\ln v$ against x, we should obtain a straight line and the cotangent of the angle of its slope should be equal to l_D. Knowing the mobility and sample temperature, we can also determine τ:

$$\tau = \frac{e}{kT}\, \frac{l_D^2}{\mu}. \qquad (58.1)$$

This method of determining τ is widely used for such materials as germanium and silicon, in which the diffusion length is sufficiently large (of the order of 0.1 cm) and blocking layers are relatively easily formed at the pressure contacts of metal points [3, 5, 15].

Figure 170 gives a typical curve (1) for the carrier distribution in germanium. The dependence $\ln \Delta n$ on x (curve 2) is a straight line, indicating an exponential relationship between Δn and x.

§59. DIRECT METHOD OF DETERMINING THE MINORITY-CARRIER MOBILITY

It would seem that the most direct method of measuring the carrier mobility should be the following one.

Assume that excess carriers are introduced in some way at a point a (Fig. 171). They move in an external uniform electric field established in the sample and after a certain time they reach a point b. If we record their arrival at this point then, knowing the distance L between the points a and b, the field E and the time t_0 taken by the carriers to travel the distance from a to b, we can determine the mobility from the formula

$$\mu = \frac{L}{t_0 E} = \frac{L^2}{t_0 V_{ab}}.$$
(59.1)

However, this most direct method cannot be used until we know the behavior of the minority carriers.

The introduction of majority carriers at the contact does not raise their density in the interior and consequently it cannot be recorded experimentally. Even if we raise the density of the majority carriers at the point a by means of internal ionization, their drift in the electric field to the point b will give rise to a space charge which opposes this drift and distorts the field distribution in the sample. All this makes the measurements and their interpretation very difficult.

The only method of determining the carrier mobility during a fairly long time interval is the indirect method of the simultaneous measurement of the conductivity and the Hall coefficient.

Recently, using the special characteristics of the minority carrier motion, it became possible to use the direct method to determine the drift mobility [2, 8]. In practice, one uses sharp probes which

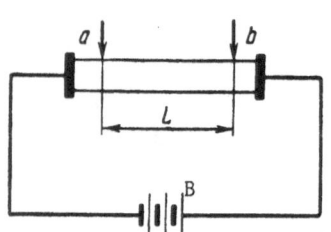

Fig. 171. Circuit for measuring the mobility.

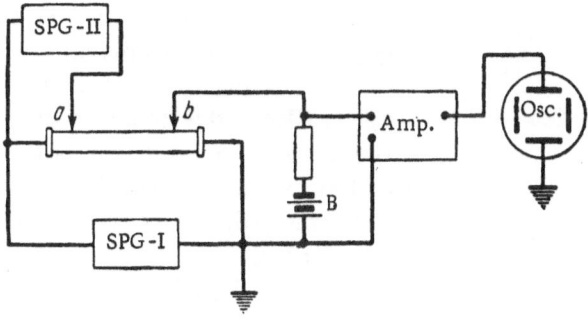

Fig. 172. Circuit for measuring the "drift" mobility.

form rectifying contacts at the points a and b. A voltage is applied in
the forward direction at the point a and this injects minority carriers at
this point. Moving in the field toward the point b the minority carriers do
not produce a space charge and therefore (if there are few of them) they
do not distort the field distribution established by an external source. The
contact b is connected in the blocking direction and registers the minority carriers.

The electrical circuit used in measurements is shown schematically
in Fig. 172. A square-pulse generator (SPG-I) with a small duty factor
(a pulse voltage is used in order to avoid heating the sample) produces a
sweeping field. Another generator of very short pulses (SPG-II), which
follow the sweeping pulse after a short delay (Fig. 173), is used to inject
carriers at the point a. A packet of injected minority carriers moving
in the sample under the action of the sweeping field passes the collector
probe. The form of the signal recorded in the collector circuit is shown
schematically in Fig. 173c.

The time, between the injection pulse and the pulse representing the
arrival of the carrier packet at the collector, t_0, is the value referred
to in Eq. (59.1). It is interesting to note that the collector signal representing the carrier distribution in the packet has the shape of a broad
peak which is sometimes asymmetric. During its drift between the points
a and b, the initially compact packet spreads because of the diffusion of
minority carriers, which broadens the packet symmetrically. The symmetry is disturbed by the modulation of the conductivity. The conductivity
in the region of the packet is higher than in other points of the sample because of the concentration of nonequilibrium minority carriers and an
equal number of equilibrium carriers to ensure neutrality.

The electric field, which has a distribution inversely proportional to
the conductivity, is weaker in the region of the packet than in neighboring
regions. Moreover, the electric field distribution varies within the
packet, having a minimum in the middle and increasing toward the edges
of the packet. Consequently, the center of the packet moves more slowly

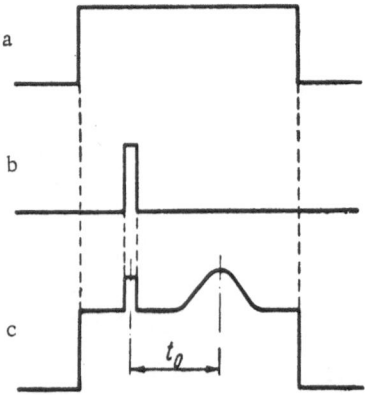

Fig. 173. Nature of the signals in the circuit shown in Fig. 168: a) Signal of the generator SPG-I; b) signal of the generator SPG-II; c) signal in the collector circuit, observed on the oscillograph screen.

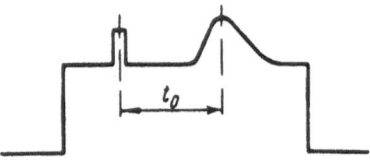

Fig. 174. Distortion of the shape of the signal due to modulation of the conductivity within the packet.

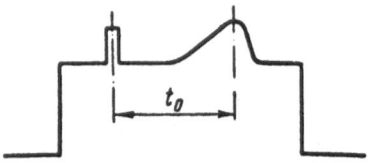

Fig. 175. Distortion of the shape of the signal due to trapping.

than its edges and therefore the leading edge of the packet spreads more and more, moving away from the center, while the trailing edge tends to catch up with the center. This disturbs the symmetry of the carrier packet and of the corresponding pulse observed at the collector (Fig. 174).

The modulation of the conductivity which distorts the pulse shape makes it difficult to measure the mobility because the average field in the packet is not equal to the average field in the sample and the former is difficult to determine. Therefore, to obtain accurate values of the mobility one records the dependence of t_0 on the voltage of the injection pulse (and, consequently, on the carrier density in the packet); this dependence is extrapolated to the zero value of the voltage to obtain t_0.

Trapping levels may distort considerably the drift mobility measurements. When these levels are present the minority carriers are trapped on the way to the collector and then they are returned to the band by thermal motion. Obviously, the time t_0 determined under these conditions will be overestimated because the carrier would have spent some time at the trapping levels. The presence of trapping distorts the symmetry of the carrier packet in the direction opposite to the distortion due to the conductivity modulation. The effect of trapping is illustrated in Fig. 175.

Apart from the method described, there are other ways of determining the drift mobility in which constant injecting fields and pulsed sweeping fields are used. In the latter case, injection by a point may be replaced by injection by a small light spot [6, 13]. The effective dimensions of a region with nonequilibrium carriers under steady-state conditions are then determined by the sum of the diameter of the light "probe" and two diffusion lengths. The pulsed sweeping field displaces the whole packet toward the collector, which is reached in a time t_0.

§60. DETERMINATION OF THE RATIO OF THE MOBILITY TO THE DIFFUSION COEFFICIENT FOR NONEQUILIBRIUM MINORITY CARRIERS

It is known that at equilibrium, charged particles that obey the Boltzmann distribution satisfy a certain relationship between the mobility μ and the diffusion coefficient D, which is known as the Einstein relationship:

$$\frac{D}{\mu} = \frac{kT}{e},$$ (60.1)

where k is the Boltzmann constant, T is the absolute temperature, and e is the electronic charge. This relationship should apply also to carriers in semiconductors liberated by thermal motion because we can assume that in the majority of actual cases the Fermi statistics of electrons in the conduction band and holes in the valence band can be replaced by the Boltzmann statistics (absence of degeneracy).

However, in the case of nonequilibrium carriers in semiconductors (for example, carriers liberated by the internal photoeffect) the relationship (60.1) may not be satisfied. The carriers which have been liberated by quanta of energy considerably greater than that of the forbidden band, have energies immediately after ionization which are considerably greater than the average thermal energy of carriers in the band (Fig. 2).

We shall consider these possible extreme cases.

1. The carriers liberated by light interact mainly with one another and the relaxation time is much shorter than the carrier lifetime. In this case, there is a "quasi-equilibrium" distribution of carriers whose temperature is different from the lattice temperature (like the electrons in plasma whose temperature may be very different from the temperature of the ions); the relationship (60.1) is satisfied if T is understood to be not the lattice temperature but the carrier temperature.

2. The carriers interact mainly with the lattice and the relaxation time is then much shorter than the lifetime; a quasi-equilibrium distribution is established corresponding to a temperature common to the carriers and the lattice; the relationship (60.1) is satisfied and T represents the experimentally measured temperature.

3. The relaxation time of carriers is comparable with or greater than their lifetime. Now we cannot speak of the carrier temperature, and, consequently, the experimentally measured ratio D/μ loses its simple meaning given by the Einstein relationship.

A theoretical consideration of the problem of the relaxation of nonequilibrium carriers liberated by light in semiconductors shows that (a) nonequilibrium carriers interact mainly with the lattice, and (b) the relaxation time for this interaction is usually much shorter than the lifetime of nonequilibrium carriers.

Thus, the theoretical consideration leads to the conclusion that the nonequilibrium carriers in semiconductors become, in a time which is a negligible fraction of their lifetime, quasi-equilibrium ones, i.e., they acquire the lattice temperature T. Such carriers should obey the Einstein relationship. However, this should be confirmed experimentally. For this purpose, we can study the diffusion and drift of the minority nonequilibrium carriers. To estimate the value of the ratio D/μ, it is natural to turn to the diffusion (to determine the diffusion coefficient D) and drift of carriers in an electric field (to determine the mobility μ). However, as mentioned previously the determination of the diffusion coefficient of majority carriers meets with difficulties because of the appearance of space charge.

The ratio D/μ for minority carriers can be found by determining experimentally the diffusion length l_D and the drift length l_E.

According to Eq. (57.4)

$$l_D^2 = D\tau, \tag{60.2}$$

$$l_E = \tau\mu E. \tag{60.3}$$

Hence

$$\frac{D}{\mu} = \frac{l_D^2}{l_E} E. \tag{60.4}$$

The quantities l_D and l_E can be found simply by investigating the steady-state distribution of the minority carrier density in the dark regions. The method for determining l_D is described in § 58. To determine l_E, all that is needed is to apply also a uniform electric field E. When the field is sufficiently strong we can determine experimentally the drift length l_E.

However, since it is not always possible to apply strong fields (they may heat the sample), we can find l_E in another way without using a strong field. In any field, the effective length l_1 (or l_2) is given in the form of Eq. (57.14) [or (57.15)]. It follows from this expression that

$$l_E = l_1 - \frac{l_D^2}{l_1}. \tag{60.5}$$

Thus, measuring l_D in the absence of field and l_1 in the presence of a field, we can use Eq. (60.5) to calculate l_E.

Experiments carried out by this method showed, as expected, that the ratio D/μ is equal to kT/e, i.e., the Einstein relationship is valid for nonequilibrium carriers [11, 7].

§ 61. METHOD OF DETERMINING THE LIFETIME AND MOBILITY OF MINORITY CARRIERS USING A TRAVELING LIGHT SPOT

The method for the simultaneous determination of the mobility and lifetime of nonequilibrium minority carriers at some fixed point, employing a uniformly moving injection region [12], is of considerable interest.

In this method, a light beam reflected from a rotating mirror moves across a sample on which a collector probe is placed. The shape of the signal in the collector circuit is shown in Fig. 176b. In the one-dimensional case, the signal consists essentially of two exponential branches with different time constants. *

The asymmetry of the signal allows us to determine simultaneously the lifetime τ and mobility μ (or the diffusion coefficient D) using formulas the derivation of which is given below:

$$\tau = \theta_1 - \theta_2. \tag{61.1}$$

$$D = c^2 \frac{\theta_1 \theta_2}{\theta_1 - \theta_2}. \tag{61.2}$$

Here, c is the linear velocity of motion of the beam across the sample, and θ_1 and θ_2 are the effective times calculated from the exponential branches of the oscillogram. From these formulas, it follows that the method is based on the use of the asymmetry in the distribution, which appears for the following reasons.

If we imagine a very long sample (a filament) along which the injection region (the light spot) moves at a constant velocity, it is obvious that in a moving system of coordinates fixed to the injection region the distribution of the minority carrier density is independent of time.

We shall consider two extreme cases: very slow and very fast motion of the beam.

As shown in § 57 for a stationary beam, a minority carrier packet has initially the form of a symmetrical "bell" with exponential branches (Fig. 176a). Obviously, during very slow motion this symmetry is not disturbed and the signal at the collector probe, which repeats in time the shape of the packet in space, is also symmetrical.

---

* Clearly, Adam [12] intended initially to reproduce the method of determining the diffusion length of minority carriers from the dependence of their density on the distance from the injection point (Fig. 170), except that instead of carrying out the measurements at separate points, a complete curve was obtained on the oscillograph screen by moving the light probe with respect to a sample provided with a collector point.

However, the motion of the injection region introduced an essentially new element: the asymmetry of the observed curve, the exponential branches of which became different (Fig. 176b) and the difference increased with the velocity of motion of the injection point.

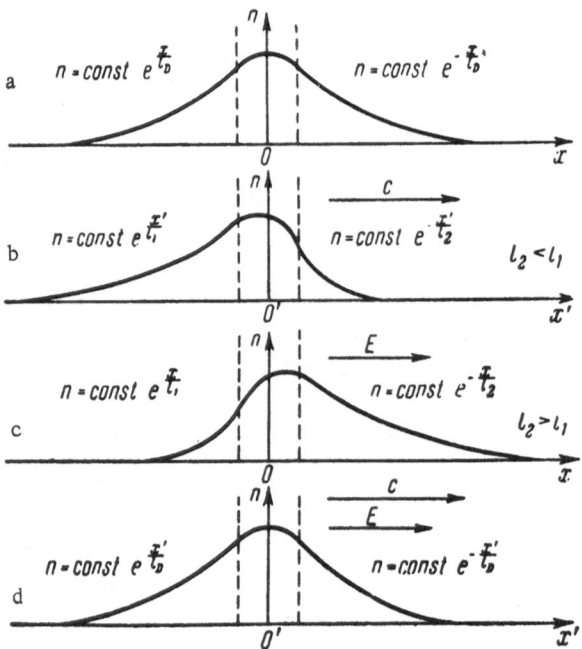

Fig. 176. Distribution of the density of minority carriers. a)
Injection region stationary, field absent, signal symmetrical;
b) injection region moving at a velocity c, field absent, signal
asymmetrical; c) field present, injection region stationary,
signal asymmetrical; d) injection region moving at a velocity
c, "compensating" field applied, signal symmetrical.

In the opposite case of very fast motion of the beam, the situation
reduces to "instantaneous" ionization, as a result of which the whole
sample is uniformly filled with minority carriers, the density of which
will then decrease due to recombination in time τ. Consequently, the
front of the signal is very short (infinitely short in the limit) and the end
of the signal extends over a time τ. The resulting signal has then maximum asymmetry.

At intermediate values of the light beam velocity, the degree of asymmetry of the signal will also be intermediate.

The appearance of the asymmetry may be explained also in a different way: when the injection region moves, the velocity of motion of this
region should be added (for one direction) or subtracted (for the opposite
direction) to the velocity of diffusion of minority carriers from the injection region (described by the quantity $\sim l_D/\tau$). This makes the curve
asymmetrical.

Figure 177 shows a series of oscillograms which illustrate the increase of the degree of asymmetry with increase of the velocity of motion
of the injection region.

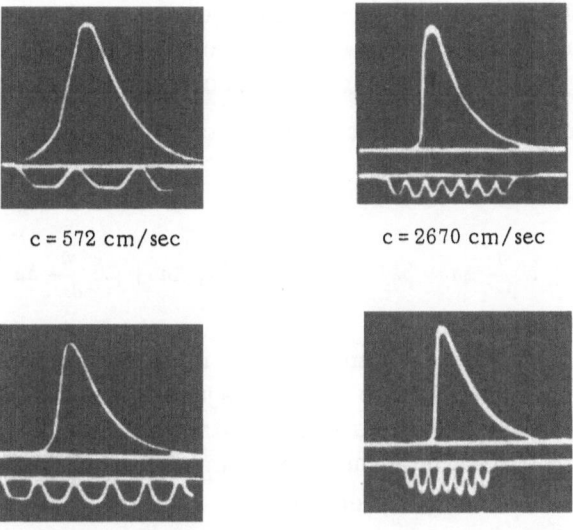

c = 572 cm/sec   c = 2670 cm/sec

c = 1335 cm/sec   c = 3820 cm/sec

Fig. 177. Dependence of the signal asymmetry on the velocity of motion of the injection signal (illuminated strip);
calibration signal is shown below each curve and its frequency is proportional to the velocity c [12].

We shall consider the shape of the signal quantitatively. [This will
give us, in particular, the formulas (61.1) and (61.2)]. We shall assume that a uniform electric field E is established in the sample. *

We shall restrict ourselves to the one-dimensional case, which differs from that in § 57 only by the fact that the injection region moves along
the x axis at a constant velocity, c. Then it is obvious that the process
considered is described by the same equations (57.1) and (57.2) (the dif-
ference is only that now, unlike in §57, we cannot assume that $\frac{\partial}{\partial t} \Delta n = 0$).

Substituting Eq. (57.2) into Eq. (57.1), we obtain

$$\frac{\partial}{\partial t} \Delta n = D \frac{\partial^2}{\partial x^2} \Delta n + \mu E \frac{\partial}{\partial x} \Delta n - \frac{\Delta n}{\tau} + \beta k I. \qquad (61.3)$$

Here I, in contrast to Eq. (57.3), depends on time.

We shall introduce a system of coordinates moving together with the
injection region at the velocity c. The formulas relating the old variables
x, t to the new ones x', t' are written in the form

$$x = ct + x' \quad (\text{or} \quad x = ct' + x'), \quad t = t'.$$

* We shall show later that the application of an electric field gives us the opportunity of using
the null method in the determination of the mobility.

Calculating the derivatives of Δn with respect to the new variables, we find that on the transformation of x, t into x', t' the first and second derivatives with respect to x become, respectively, the first and second derivatives with respect of x', but $\frac{\partial}{\partial t} \Delta n$ becomes $\frac{\partial}{\partial t'} \Delta n + c \frac{\partial}{\partial x'} \Delta n$. After suitable substitutions Eq. (61.3) becomes

$$\frac{\partial}{\partial t'} \Delta n + c \frac{\partial}{\partial x'} \Delta n = \beta kI - \frac{\Delta n}{\tau} + D \frac{\partial^2}{\partial x'^2} \Delta n + \mu E \frac{\partial}{\partial x'} \Delta n. \qquad (61.4)$$

Here I is independent of time.

From now on, we shall assume that the minority carrier packet has been found formed some time ago and consequently the density in the moving system of coordinates has reached its steady-state value, i.e., it is independent of t' (this happens if the sample is sufficiently long). Then $\frac{\partial}{\partial t'} \Delta n = 0$ and Eq. (61.4) becomes

$$D \frac{\partial^2}{\partial x'^2} \Delta n + (\mu E - c) \frac{\partial}{\partial x'} \Delta n - \frac{\Delta n}{\tau} = - \beta kI. \qquad (61.5)$$

The above equation is almost completely identical with Eq. (57.3): the only difference is that the coefficient of the first derivative has, instead of the drift velocity μE in Eq. (57.3), the difference of the drift and injection-region velocities. Consequently, in analogy with Eqs. (57.6), (57.7), (57.9), and (57.10), we can write for the dark regions I and II:

for region I

$$\Delta n = \text{const } e^{k_1 x'}, \qquad (61.6)$$

where

$$k_1 = \frac{\mu E - c}{2D} - \sqrt{\frac{(\mu E - c)^2}{4D^2} + \frac{1}{D\tau}} \qquad (61.7)$$

(of the two values of k we select the negative one because $\Delta n \to 0$ when $x' \to +\infty$);

for region II

$$\Delta n = \text{const } e^{k_2 x'}, \qquad (61.8)$$

where

$$k_2 = \frac{\mu E - c}{2D} + \sqrt{\frac{(\mu E - c)^2}{4D^2} + \frac{1}{D\tau}} \,. \qquad (61.9)$$

We shall return now to the old variables

$$x = x' + ct.$$

Then in region I we have

$$\Delta n = \text{const } e^{k_1 x} \cdot e^{-ck_1 t} \tag{61.10}$$

or, if we are interested in the density at a point fixed in a stationary system of coordinates, i.e., at the collector point,

$$\Delta n = A_1 e^{-ck_1 t} = A_1 e^{\tfrac{t}{\theta_1}}, \tag{61.11}$$

where

$$\theta_1 \equiv \frac{1}{-ck_1} = \frac{1}{-\dfrac{c(\mu E - c)}{2D} + c\sqrt{\dfrac{(\mu E - c)^2}{4D^2} + \dfrac{1}{D\tau}}} > 0. \tag{61.12}$$

For region II we find that

$$\Delta n = \text{const } e^{k_2 x} e^{-ck_2 t} = A_2 e^{-\tfrac{t}{\theta_2}}, \tag{61.13}$$

where

$$\theta_2 \equiv \frac{1}{ck_2} = \frac{1}{\dfrac{c(\mu E - c)}{2D} + c\sqrt{\dfrac{(\mu E - c)^2}{4D^2} + \dfrac{1}{D\tau}}} > 0. \tag{61.14}$$

Here A_1 and A_2 are constants.

Thus, the signal at the collector in the case of uniform motion of a narrow injection region in the presence of a field consists essentially of two exponential branches with different time constants, given by Eqs. (61.11) and (61.13), and a transition region for which we shall not give a formula.

In the absence of a field ($E = 0$), we have for region I

$$\theta_1 = \frac{1}{\dfrac{c^2}{2D} + c\sqrt{\dfrac{c^2}{4D^2} + \dfrac{1}{D\tau}}}, \tag{61.15}$$

and for region II

$$\theta_2 = \frac{1}{-\dfrac{c^2}{2D} + c\sqrt{\dfrac{c^2}{4D^2} + \dfrac{1}{D\tau}}}. \tag{61.16}$$

From these expressions, we can easily derive Eqs. (61.1) and (61.2) for τ and D.

The results of measuring the lifetime and mobility of minority carriers by the moving light-spot method agree satisfactorily with the results obtained by other methods described in § 58 and § 59. New possibilities arise from the use of an electric field.

A. Compensation (Null) Method for Measuring Mobility

The signal in the fixed collector circuit becomes asymmetric when the injection region moves (Fig. 176b).

On the other hand, it is known (§57) that in the presence of a uniform field the one-dimensional steady-state distribution of the minority carrier density is asymmetric even if the injection region is fixed.

In §57 we gave the appropriate curves (Fig. 168) whose branches in the dark regions are exponentials with different time constants.

Thus, the action of a field, like the effect of injection-region motion, gives rise to an asymmetry in the minority carrier distribution (Fig. 176c).

The question is now whether the asymmetry due to injection-region motion can be compensated by a suitably directed electric field.

The answer to this question follows from the initial equation (61.5), or from Eqs. (61.7) and (61.9). It is evident from Eq. (61.5) that if the sign and magnitude of the electric field are selected so that $\mu E = c$, then the second term disappears from Eq. (61.6) and the equation acquires the same form as in the case of diffusion only (naturally, the distribution in the case of diffusion is symmetrical). Substituting

$$\mu E = c,$$

into Eqs. (61.7) and (61.9) we find that

$$\theta_1 = \theta_2.$$

Consequently, by selecting the conditions so that the signal is symmetrical, we can determine the mobility from

$$\mu = \frac{c_0}{E_0}, \tag{61.17}$$

where c_0 and E_0 are the values of the injection-region velocity and field respectively at the moment when the signal becomes symmetrical.

We have considered so far the one-dimensional case but we can show that irrespective of the type of current spreading (cylindrical, linear, etc.), of the form of the light spot (provided it is symmetrical with respect to x'), and of the surface recombination and the sample cross section (provided they are independent of x), the signal is balanced or symmetrical when

$$\mu E = c.$$

Thus, the process of making the signal symmetrical can be used to measure the mobility [17].

Figure 178 shows the complete apparatus for the traveling-spot method, which is suitable also for measuring the mobility of the null method. These measurements are carried out as follows.

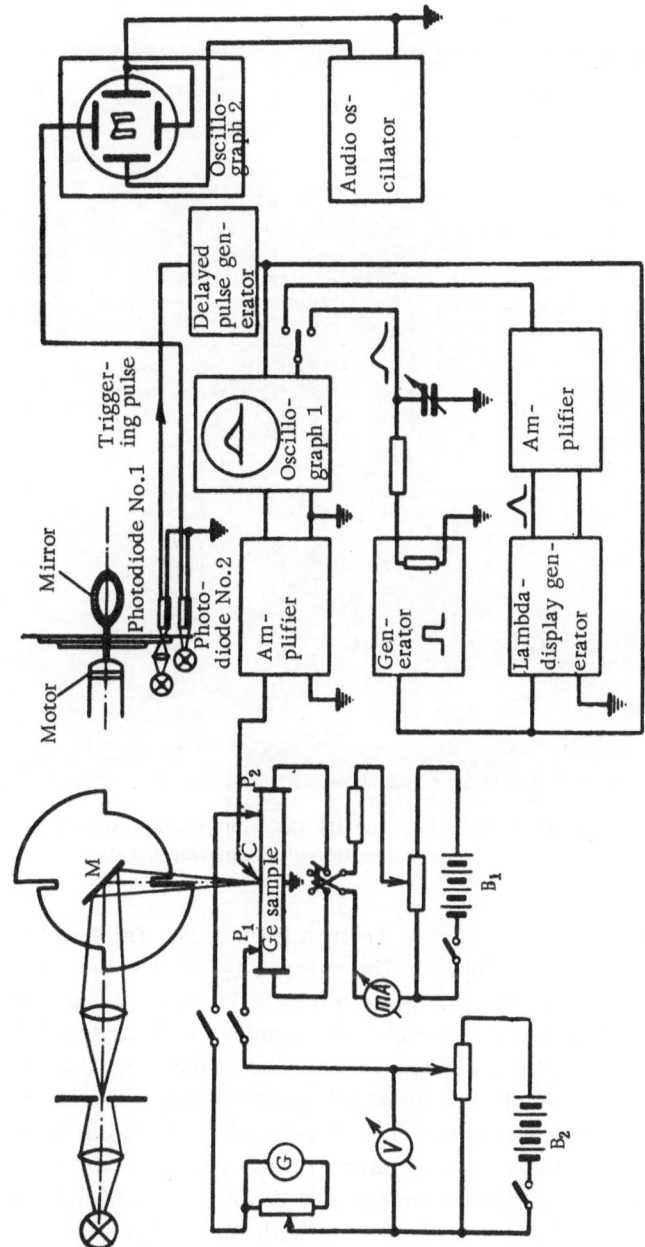

Fig. 178. Basic circuit for measuring μ and τ by the moving injection-region method.

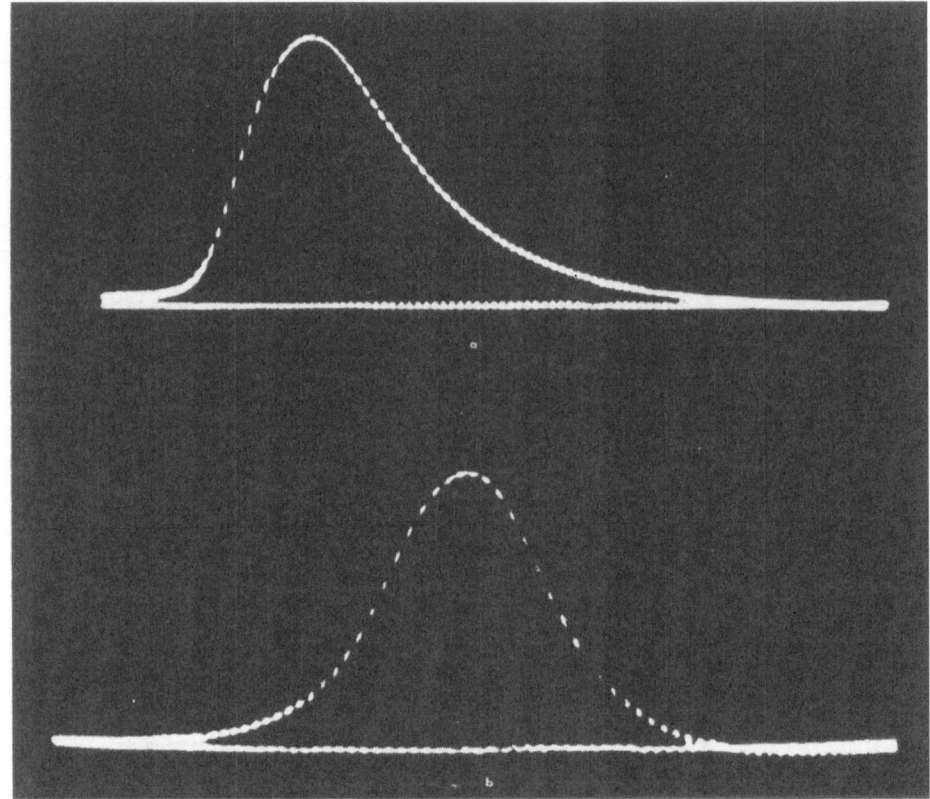

Fig. 179. Asymmetrical signal for a moving injection region in the absence of a field; b) the same on application of a compensating field.

A light beam, reflected from a mirror M, falls in the form of a narrow strip on a sample. The mirror is rotated by a motor and the light strip moves along the sample surface. The sample is provided with the usual end electrodes to which a constant voltage is applied from a battery B_1. In addition to these electrodes, three metal points are used: two probes P_1 and P_2 for measuring the voltage drop in the middle part of the sample (for accurate measurement of the field) and a collector C which records the minority carrier density. The amplified signal from the collector is observed on the screen of the oscillograph 1. The oscillograph display is synchronized with the signal by means of, for example, a germanium photodiode illuminated through a cutaway disk which is mounted on the same rotating axle as the mirror. The frequency of rotation (and, consequently, the linear velocity of motion of the illuminated strip across the sample) is found by comparing the signal from another photodiode with a calibrated signal from an audio oscillator (using the Lissajous figures on the screen of the oscillograph 2).

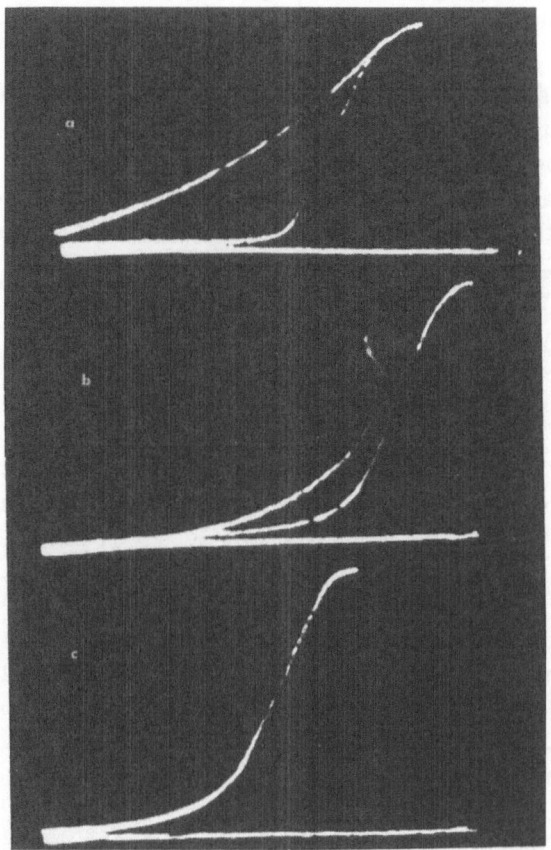

Fig. 180. Signal in the case of a lambda display. a) No compensating field; b) compensating field insufficient; c) complete compensation.

The oscillograms obtained (Fig. 179) show clearly the effect of making the "bell" symmetrical on application of a field. However, it is quite difficult to determine the actual moment of achieving symmetry by means of these curves. This difficulty is avoided by using a special "lambda" display which makes the signal into a triangular pulse rather like the letter lambda. Shifting the pulse along the time base, we can superimpose its vertex on the vertex of the bell-shaped signal. Then, obviously, the branches of the test signal are superimposed on one another and they coincide completely at the moment of achieving symmetry (Fig. 180).

Apart from enabling us to measure simultaneously the lifetime and mobility, the traveling light beam method has other notable features which distinguish it from the methods described in § 58 and § 59. We shall return to this point in § 62.

§62. QUANTITATIVE DISCUSSION OF DIFFUSION AND DRIFT FOR ANY RATIO OF THE ELECTRON AND HOLE DENSITIES

In homogeneous semiconductors (of not too low conductivity), which are homogeneous at equilibrium, the nonequilibrium processes occur in such a way as to obey the neutrality law.

This means that if excitation produces nonequilibrium electrons of density Δn then in the same region excess holes should appear, Δp in the valence band and $\sum_k \Delta p_k$ at the impurity levels, to neutralize the electron charge:

$$\Delta n = \Delta p + \sum_k \Delta p_k. \tag{62.1}$$

One frequently has the conditions such that the role of impurities is unimportant when considering neutrality* and therefore the neutrality is conserved if

$$\Delta n = \Delta p. \tag{62.2}$$

We shall discuss the problem of diffusion and drift in the case when the densities of nonequilibrium electrons and holes are equal but the total electron and hole densities n and p may be arbitrary [9]. For simplicity, we shall consider only the plane case (as in §§52, 57), i.e., we shall assume that the density gradient and the electric field are directed along the x axis.

We shall also assume that there is no excitation: $\beta kI = 0$ (the region is in darkness). In contrast to the cases discussed in §52 and §57, the electrons and holes are now "equivalent" carriers, i.e., there is no "asymmetry" in their mobilities or densities, so that the principal equations apply to electrons as well as to holes.

The continuity equations are:

$$\frac{\partial}{\partial t} \Delta p = -\frac{\Delta p}{\tau} - \frac{1}{e} \frac{\partial}{\partial x} j_p, \tag{62.3}$$

$$\frac{\partial}{\partial t} \Delta n = -\frac{\Delta n}{\tau} + \frac{1}{e} \frac{\partial}{\partial x} j_n. \tag{62.4}$$

Here j_p and j_n are the electron and hole components of the current density, τ is the lifetime of electrons and holes (or electron-hole pairs). †

---

* For example, at low impurity concentrations or in the case of impurity levels close to the edges of the allowed bands (large value of N_{cM}), when the carrier density at these levels is practically constant. These conditions are frequently obtained in important semiconducting materials such as germanium or silicon.

† We have shown in Chap. V that the neutrality condition $\Delta n = \Delta p$ implies equal lifetimes of electrons and holes.

For the current densities we have

$$j_p = \sigma_p E - e D_p \frac{\partial}{\partial x} \Delta p, \tag{62.5}$$

$$j_n = \sigma_n E + e D_n \frac{\partial}{\partial x} \Delta n. \tag{62.6}$$

Here

$$\left. \begin{array}{l} \sigma_p = e \mu_p p, \\ \sigma_n = e \mu_n n \end{array} \right\} \tag{62.7}$$

are, respectively, the hole and electron components of the conductivity. Assuming that the external field E is uniform and much stronger than the internal field (which is justified in view of the neutrality), * differentating Eq. (62.5) and (62.6) with respect of x, and substituting into Eqs. (62.3) and (62.4) the quantities $\frac{\partial}{\partial x} j_p$ and $\frac{\partial}{\partial x} j_n$,we obtain

$$\frac{\partial}{\partial t} \Delta p = D_p \frac{\partial^2}{\partial x^2} \Delta p - \mu_p E \frac{\partial}{\partial x} \Delta p - \frac{\Delta p}{\tau}, \tag{62.8}$$

$$\frac{\partial}{\partial t} \Delta n = D_n \frac{\partial^2}{\partial x^2} \Delta n + \mu_n E \frac{\partial}{\partial x} \Delta n - \frac{\Delta n}{\tau}. \tag{62.9}$$

Multiplying next Eq. (62.8) by σ_n, and Eq. (62.9) by σ_p, and adding both equations, we obtain the following equation, after replacing Δn with Δp ($\Delta n = \Delta p$) and restricting the treatment to the steady-state case $\left(\frac{\partial}{\partial t} \Delta n = \frac{\partial}{\partial t} \Delta p = 0 \right)$:

$$\frac{D_p \sigma_n + D_n \sigma_p}{\sigma_n + \sigma_p} \frac{\partial^2}{\partial x^2} \Delta p + \frac{\mu_n \sigma_p - \mu_p \sigma_n}{\sigma_n + \sigma_p} E \frac{\partial}{\partial x} \Delta p - \frac{\Delta p}{\tau} = 0. \tag{62.10}$$

It is easily seen that Eq. (62.10) is identical in form with the equation for the diffusion and drift of minority carriers discussed in § 57: Eq. (57.3). The only difference is that instead of the diffusion coefficient in front of the second derivative and the mobility in front of the first derivative, as in Eq. (57.3), Eq. (62.10) has complicated quantities which we shall call the ambipolar diffusion coefficient D and the ambipolar drift mobility μ_E, respectively. †

These quantities can be written in several ways:

$$D = \frac{D_p \sigma_n + D_n \sigma_p}{\sigma_n + \sigma_p} = \frac{n_0 + p_0}{\frac{n_0}{D_p} + \frac{p_0}{D_n}} = \frac{kT}{e} \frac{n_0 + p_0}{\frac{n_0}{\mu_p} + \frac{p_0}{\mu_n}}, \tag{62.11}$$

* This assumption is valid only at low excitation levels when $n + p \approx n_0 + p_0$. We shall therefore assume that the excitation level is low.

† The term "ambipolar" (or "bipolar") denotes the simultaneous diffusion or drift of electrons and holes.

$$\mu_E = \frac{\mu_n \sigma_p - \mu_p \sigma_n}{\sigma_n + \sigma_p} = \frac{p_0 - n_0}{\dfrac{n_0}{\mu_p} + \dfrac{p_0}{\mu_n}}. \tag{62.12}$$

The mobility μ_E may change its sign depending on the ratio of n_0 to p_0 and vanishes when $n_0 = p_0$. It is evident from Eq. (62.11) that if the ambipolar diffusion coefficient D is represented in accordance with the Einstein relationship in the form

$$D = \frac{kT}{e} \mu_D, \tag{62.13}$$

then the expression corresponding to the mobility is

$$\mu_D = \frac{n_0 + p_0}{\dfrac{n_0}{\mu_p} + \dfrac{p_0}{\mu_n}}, \tag{62.14}$$

which is not identical with the ambipolar drift mobility μ_E of Eq. (62.12).

In contrast to μ_E the quantity μ_D is called the ambipolar diffusion mobility. The significance of these terms will be explained below.

Comparison of Eq. (62.10) with Eq. (57.3), which has been solved earlier, gives us qualitative information on the nature of the diffusion and drift when $\Delta n = \Delta p$. The nature of the distribution is the same as that given by Eqs. (57.6), (57.7), and (57.8) and shown in Figs. 167 and 168.

We shall consider some special cases.

For an intrinsic semiconductor ($n_0 = p_0$), we have from Eqs. (62.11) and (62.12)

$$D = 2 \frac{D_p D_n}{D_p + D_n} = 2 \frac{kT}{e} \cdot \frac{\mu_p \mu_n}{\mu_p + \mu_n}, \tag{62.15}$$

$$\mu_D = 2 \frac{\mu_p \mu_n}{\mu_p + \mu_n}, \tag{62.16}$$

$$\mu_E = 0. \tag{62.17}$$

Equation (62.17) shows that under these conditions the drift in an electric field is absent, and consequently, the external electric field does not affect the density distribution. This is because the forces of attraction between electrons and holes are considerably stronger than the forces acting on charges due to the external field (this assumption is implicit in the condition $\Delta n = \Delta p$), * and therefore because of the symmetry ($n = p$) the effect of the field on the electrons is balanced by the effect on the holes to which the former are bound by attraction. †

* This is valid only in fields which are not too strong.
† It can be shown that in the special case of an intrinsic semiconductor, Eqs. (62.15)-(62.17) are valid for any excitation level.

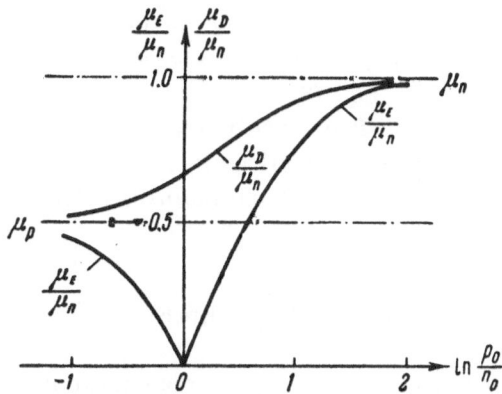

Fig. 181. Dependence of μ_D and μ_E on the ratio of n_0 to p_0 at low excitation level in germanium. Here it was assumed that $\mu_n = 3900$ cm$^2 \cdot$ V$^{-1} \cdot$ sec^{-1} and $\mu_p = 1900$ cm$^2 \cdot$ V$^{-1} \cdot$ sec^{-1}.

The diffusion is given by a coefficient which depends on the diffusion coefficients (mobilities) of both types of carrier.

If the mobilities (diffusion coefficients) are very different for electrons and holes, the diffusion is governed mainly by the less mobile carriers which retard the motion of those which are more mobile. For example, if $D_n \gg D_p$, then Eq. (62.15) gives $D = 2D_p$.

We shall consider next the very interesting case of a strongly extrinsic semiconductor when, for example, $n_0 \gg p_0$.

Then Eqs. (62.11), (62.12), and (62.14) give

$$\left. \begin{array}{l} D = D_p \\ \mu_D = |\mu_E| = \mu_p \end{array} \right\}, \qquad (62.18)$$

i.e., the diffusion and drift are completely determined by the motion of minority carriers (holes, in our case).

This conclusion has been obtained earlier, in § 57, from qualitative considerations.

Figure 181 shows that if p_0 and n_0 are very different, then μ_E and μ_D are identical and equal to the minority-carrier mobility. When $n_0 = p_0$ the quantity μ_E vanishes, i.e., the external electric field ceases to have any effect on the nature of the carrier distribution, which is then governed completely by the diffusion. The ambipolar drift mobility μ_E changes sign at $n_0 = p_0$ [cf. Eq. (62.12)] due to a change of the sign of the carriers which dominate the drift (i.e., the minority carriers). *

In § 61 we have considered the method for determining the lifetime and mobility by means of a traveling spot and we described two methods

* The change of sign cannot be seen in Fig. 181 because it gives the absolute value of μ_E.

for finding the mobility and calculated it by means of two different for-mulas: Eqs. (61.2) and (61.17). It can be shown that Eq. (61.2) gives the diffusion mobility, defined by Eq. (62.14), while Eq. (61.17) gives the drift mobility of Eq. (62.12). Thus, the traveling-spot method allows us to determine simultaneously μ_D as well as μ_E.

The direct method of determining the mobility, described in §59, gives μ_E.

§63. DEMBER EMF

If the excitation (illumination) of a semiconductor is nonuniform and therefore there is a gradient of the nonequilibrium carrier density, the resultant diffusion currents disturb the neutrality and give rise to an elec-tric field and an emf.

Such volume emf's which appear during the nonuniform excitation of a homogeneous semiconductor are known as Dember emf's (§52).

We shall calculate the steady-state Dember field and Dember emf for an isolated semiconductor with ambipolar conductivity.

We shall consider the one-dimensional case. As usual, the electron and hole components of the current densities are:

$$j_n = en\mu_n E + \mu_n kT \frac{dn}{dx}, \qquad (63.1)$$

$$j_p = ep\mu_p E - \mu_p kT \frac{dp}{dx}. \qquad (63.2)$$

Here, E is the internal field due to the departure from neutrality. We have neglected this field in considering the carrier distribution in §62.

The Dember field has a weak effect on the carrier distribution but it can be detected experimentally. The total current is

$$j = j_n + j_p = e(n\mu_n + p\mu_p)E + kT\left(\mu_n \frac{dn}{dx} - \mu_p \frac{dp}{dx}\right). \qquad (63.3)$$

If the external circuit is open (the isolated semiconductor case, $j = 0$), the Dember field is given by the following general expression:

$$E = \frac{kT}{e}\frac{\mu_n \dfrac{dn}{dx} - \mu_p \dfrac{dp}{dx}}{n\mu_n + p\mu_p}. \qquad (63.4)$$

In the special case of unipolar conductivity (for example, when $p = 0$)

$$E = \frac{kT}{e}\frac{1}{n}\frac{dn}{dx} \qquad (63.5)$$

and the Dember emf between the points x_1 and x_2 is

$$V_{1,2} = \int_{x=x_1}^{x=x_2} E\,dx = \frac{kT}{e} \int_{x=x_1}^{x=x_2} \frac{dn}{n} = \frac{kT}{e} \ln \frac{n_2}{n_1}, \qquad (63.6)$$

where n_2 and n_1 are the electron densities at x_2 and x_1. The expression (63.6) has been considered earlier (Chap. XII).

In the ambipolar case we shall assume that, in the first approximation, the neutrality is conserved and the charge at local levels is unimportant; therefore, the neutrality condition can be written in the form $\Delta n = \Delta p$.

Since

$$\frac{dn}{dx} = \frac{dp}{dx} = \frac{d}{dx}\Delta n, \qquad (63.7)$$

simple transformations of Eq. (63.4) give

$$E = \frac{kT}{e} \frac{\mu_n - \mu_p}{\mu_n + \mu_p} \frac{1}{\sigma} \cdot \frac{d}{dx}\Delta\sigma, \qquad (63.8)$$

where

$$\sigma = e\,(\mu_n n + \mu_p p), \qquad (63.9)$$

$$\Delta\sigma = e\,(\mu_n + \mu_p)\,\Delta n, \qquad (63.10)$$

and the Dember emf is

$$V_{1,2} = \int_{x_1}^{x_2} E\,dx = \frac{kT}{e} \frac{\mu_n - \mu_p}{\mu_n + \mu_p} \ln \frac{\sigma_2}{\sigma_1}. \qquad (63.11)$$

Thus, the greater the difference between the electron and hole mobilities, the higher the value of the Dember emf. When the mobilities are equal, the Dember emf vanishes.

The expression (63.11) simplifies in the case when σ_2 and σ_1 are not too different, i.e.,

$$\frac{\sigma_2 - \sigma_1}{\sigma_1} \ll 1. \qquad (63.12)$$

Then, using the notation $\delta\sigma = \sigma_2 - \sigma_1$, we expand in series the logarithm in Eq. (63.11) and, taking only the first term, we find

$$V_{1,2} = \frac{kT}{e} \frac{\mu_n - \mu_p}{\mu_n + \mu_p} \frac{\delta\sigma}{\sigma_1}, \qquad (63.13)$$

i.e., at low excitation levels the Dember emf between two points is directly proportional to the difference of the conductivities at these two points and decreases with increase of the average conductivity of the semiconductor.

This reduction of the Dember emf with increase of the conductivity of the semiconductor (which is characteristic also of other emf's in semiconductors, for example, the thermoelectric emf, the photomagnetic emf, etc.) may be explained by the following simple considerations.

The density gradient produced by nonuniform illumination and the diffusion current of charged particles j_D, which is proportional to this gradient, disturb the neutrality and give rise to an electric field E as well as to a conduction current $j_E = \sigma E$ proportional to the field. Under steady-state conditions, j_D is equal to j_E:

$$j_D = \sigma E. \tag{63.14}$$

Under the given conditions (j_D fixed), the higher the conductivity σ, the lower the Dember field E, which is necessary to satisfy Eq. (63.14).

There is an apparent contradiction in discussions of the origin of the Dember emf. The emf is due to the appearance of a space charge, i.e., due to a departure from the neutrality. However, in deriving Eq. (63.8) we used the assumption that $\Delta n = \Delta p$ and so the neutrality is conserved. In fact, the contradiction is only apparent. The equality of Δn and Δp is not used in Poisson's equation (where it would lead to the conclusion that the space charge is identically zero). In our case, the Dember field was calculated from the equations for the current, in which we can assume that $\Delta n \approx \Delta p$ because of the small difference between these two quantities.

Chapter XIV

SOME PHOTOMAGNETOELECTRIC
AND PHOTOMAGNETODENSITY EFFECTS

(i) Photomagnetoelectric Effects

§ 64. AMBIPOLAR PHOTOMAGNETOELECTRIC EFFECT
OF KIKOIN AND NOSKOV

When one of the surfaces of a semiconductor sample is illuminated with light which is absorbed sufficiently strongly, a carrier density gradient appears along the direction of propagation of the light and this is accompanied by the diffusion of nonequilibrium carriers from the front (illuminated) surface of the sample into the interior (Fig. 182).

If such a sample is placed in a magnetic field H perpendicular to the direction of propagation of the light, a Lorentz force acts on the diffusing electrons and holes, deflecting them at right angles to the direction of their motion, the deflection for electrons being opposite to that for holes (Fig. 183).

In this way, charge of opposite sign is accumulated at the two opposite sides of a sample (a and b) and, consequently, an electric field and an emf appear at right angles to the directions of the propagation of the light and the magnetic field.   The charge will continue to be accumulated and the emf will continue to rise until the conduction current (flowing under the action of this emf) from b to a balances the "magnetodiffusion" current.

This photomagnetoelectric effect (also known as the photomagnetic effect) was first detected by Kikoin and Noskov in cuprous oxide in 1935 [1] and explained by Ya. I. Frenkel' [2].   This effect is nowadays used to measure some semiconductor parameters.

We shall calculate the photomagnetoelectric emf and the short-circuit current for some conditions which are easily realized experimentally, and we shall show that it is possible to use the photomagnetoelectric effect to determine the lifetime τ and the surface recombination velocity s.

A.   Photomagnetoelectric EMF and Short-Circuit Current [7, 9, 21]

It is known that on the application of magnetic field at right-angles to the current the moving carriers are deflected from the initial direction by the Hall angle θ (Fig. 184), where

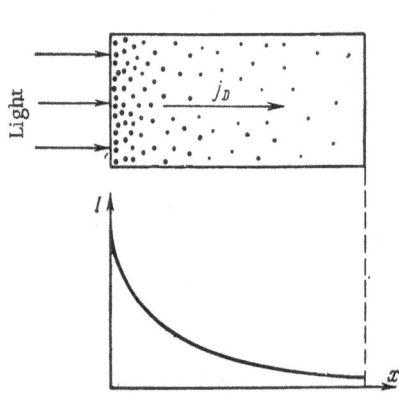

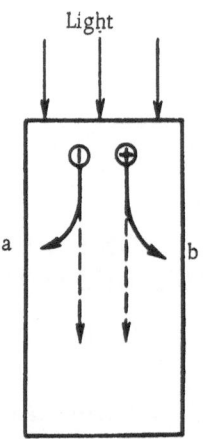

Fig. 182. Diffusion current during illumination of one of the surfaces of a semiconductor sample with strongly absorbed light; x shows the direction of propagation of the light.

Fig. 183. Mechanism of the appearance of the photomagnetoelectric emf. The magnetic field is perpendicular to the plane of the diagram.

$$\left.\begin{array}{l} \tan \theta_n = -\dfrac{\mu_n^* H}{c}, \\[2mm] \tan \theta_p = \dfrac{\mu_p^* H}{c}. \end{array}\right\} \tag{64.1}$$

Here, θ_n and θ_p are the Hall angles for electrons and holes, respectively: μ_h^* and μ_p^* are the Hall mobilities of electrons and holes, which are related to the drift mobilities by: $\mu_n^* = \kappa\mu_n$, $\mu_p^* = \kappa\mu_p$, where κ is a coefficient of the order of unity, which depends on the scattering mechanism.

Consequently, the currents of electrons (j_{nx}) and holes (j_{px}) diffusing from the illuminated surface along the x axis should be deflected by the angles θ_n and θ_p respectively. This gives rise to "magnetodiffusion" components of the currents along the y axis (j_{ny} and j_{py}) and consequently, to a photomagnetoelectric emf or a short-circuit current.

We shall assume that the magnetic field is weak ($\mu H/c \ll 1$).

Then the magnetodiffusion currents are:

$$j_{py} = j_{px}\,\tan \theta_p = \frac{\mu_p^* H}{c}\,j_{px}, \tag{64.2}$$

$$j_{ny} = j_{nx}\tan \theta_n = -\frac{H}{c}\,\mu_n^* j_{nx}. \tag{64.3}$$

The total magnetodiffusion current at any point is

$$j_y = j_{py} + j_{ny} = \frac{\kappa H}{c}\,(\mu_p j_{px} - \mu_n j_{nx}). \tag{64.4}$$

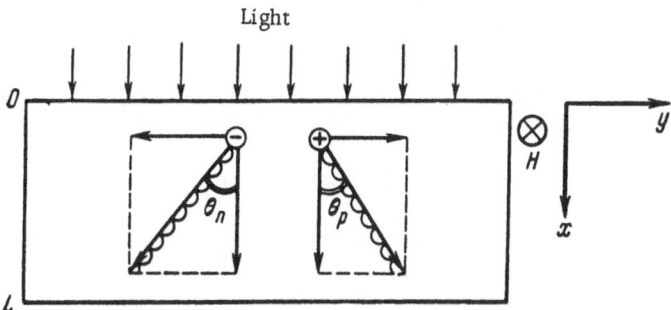

Fig. 184. Effect of a magnetic field on the diffusion currents of electrons and holes, which give rise to the "magnetodiffusion" components of the currents along the y axis.

It is evident from Eqs. (64.2), (64.3), and (64.4) that the required currents along the y axis are completely determined by the diffusion currents along the x axis. Consequently, we must first find these diffusion currents.

It is easily seen that under the conditions of approximate neutrality, when $\Delta n = \Delta p$, we have

$$j_{px} = - j_{nx} = - eD \frac{d}{dx} \Delta p = - eD \frac{d}{dx} \Delta n, \qquad (64.5)$$

where $D = \dfrac{n+p}{\dfrac{n}{D_p} + \dfrac{p}{D_n}}$ is the ambipolar diffusion coefficient (§ 62). There-

fore,

$$j_y = - eD \frac{H}{c} \left(\mu_p^* + \mu_n^* \right) \frac{d}{dx} \Delta p. \qquad (64.6)$$

We shall assume now that D is a constant, i.e., that we are dealing with the case described in § 62.

To find the magnetodiffusion current j_y which gives the photomagnetoelectric effect, we must find the distribution of the density Δn or Δp (and consequently, $\frac{d}{dx} \Delta n = \frac{d}{dx} \Delta p$) along the x axis (i.e., along the direction of diffusion).

We shall find $\Delta n(x) = \Delta p(x)$ using Eq. (57.3) or (62.10)* without the second term because there is no external electric field in the present case.

We shall also bear in mind that the rate of generation of nonequilibrium pairs decreases along the x axis due to decreasing absorption of light (Fig. 182) according to the law

---

* The distribution is sought in the absence of a magnetic field on the assumption that it is not greatly affected by weak magnetic fields.

$$\beta kI = \beta kI_0 e^{-kx}. \tag{64.7}$$

Then we have for Δp (and consequently for Δn, because $\Delta n = \Delta p$)

$$D \frac{d^2}{dx^2} \Delta p - \frac{\Delta p}{\tau} + \beta kI_0 \exp(-kx) = 0. \tag{64.8}$$

The solution of the above equation is

$$\Delta p = A \exp\left(-\frac{x}{l_D}\right) + B \exp\left(\frac{x}{l_D}\right) + C \exp(-kx). \tag{64.9}$$

The constants of integration are found from the boundary conditions, according to which the currents at the front and rear faces of the sample should be governed by the surface recombination current, i.e.,

$$D \left(\frac{d}{dx} \Delta p\right)_{x=0} = s_0 p_{(0)}, \tag{64.10}$$

$$-D \left(\frac{d}{dx} \Delta p\right)_{x=L} = s_L p_{(L)}. \tag{64.11}$$

Here s_0 and s_L are the surface recombination velocities at the front (illuminated) and rear faces of the sample.

General but very cumbersome expressions are obtained for A, B, and C from Eqs. (64.9)-(64.11), but these expressions simplify in real special cases. The general expressions are

$$A = -\frac{\beta kI_0}{D\left(k^2 - \frac{1}{l_D^2}\right)}$$

$$\times \frac{(s_0 + kD)(s_L + D/l_D)\exp\frac{L}{l_D} - (s_0 - D/l_D)(s_L - kD)\exp(-kL)}{(s_0 + D/l_D)(s_L + D/l_D)\exp(L/l_D) - (s_0 - D/l_D)(s_L - D/l_D)\exp(-L/l_D)}. \tag{64.12}$$

$$B = \frac{\beta kI_0}{D(k^2 - 1/l_D^2)}$$

$$\times \frac{(s_0 + kD)(s_L - D/l_D)\exp(-L/l_D) - (s_0 + D/l_D)(s_L - kD)\exp(-kL)}{(s_0 + D/l_D)(s_L + D/l_D)\exp(L/l_D) - (s_0 - D/l_D)(s_L - D/l_D)\exp(-L/l_D)}. \tag{64.13}$$

$$C = \frac{\beta kI_0}{D(k^2 - 1/l_D^2)}. \tag{64.14}$$

Using Eqs. (64.9), (64.12)-(64.14), we can find Δp and $\frac{d}{dx} \Delta p$, and then, using Eq. (64.6), we can determine the magnetodiffusion current j_y.

The final expression for Δp for the very frequent case of very strong absorption of light $(kL \gg 1)$ is

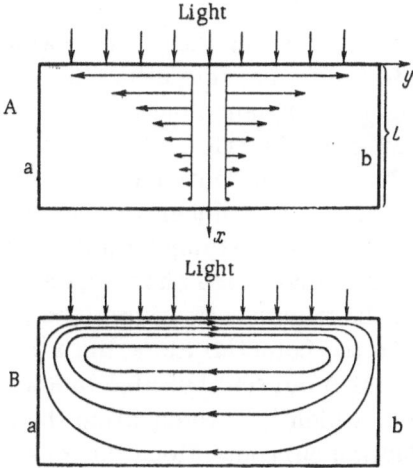

Fig. 185. A) Variation of the magnetodiffusion current density along the x-axis; B) flow of the current in the sample.

$$\Delta p = \frac{\beta I_0}{s_0} \left[\frac{\dfrac{D}{l_D s_L} \cosh \dfrac{L-x}{l_D} + \sinh \dfrac{L-x}{l_D}}{\left(1 + \dfrac{D^2}{l_D^2 s_0 s_L}\right) \sinh \dfrac{L}{l_D} + \dfrac{D(s_0 + s_L)}{l_D s_0 s_L} \cosh \dfrac{L}{l_D}} \right]. \qquad (64.15)$$

We shall find now the relationship between the magnetodiffusion current j_y and the experimentally measured photomagnetoelectric emf (with the circuit open) and the short-circuit current.

The density gradient and the current j_x decrease with distance from the front illuminated surface and therefore the current j_y, which is proportional to j_x, also decreases (Fig. 185A). Clearly, when the sides a and b are connected up, the short-circuit current in the external circuit should be equal to the total magnetodiffusion current through the side surface (the cross section of the sample in the xz plane), i.e., using Eq. (64.6), we obtain

$$J_{sc} = g \int_0^L j_y \, dx = geD \frac{\varkappa H}{c} (\mu_p + \mu_n)[\Delta p(0) - \Delta p(L)]. \qquad (64.16)$$

Here, g is the dimension of the sample along the z axis.

Thus, the photomagnetoelectric current under short-circuit conditions is governed only by the difference between the nonequilibrium carrier densities at the front (illuminated) and rear surfaces of the sample.

We shall now consider how the photomagnetoelectric emf appears in an isolated sample. In this case, the magnetodiffusion current gives rise to the accumulation of charges of opposite signs at the faces a and b, i.e., it establishes an electric field along the y axis, which produces a conduction current opposite to the magnetodiffusion current. Under steady-state conditions, the currents balance out and this corresponds to a certain value of the photomagnetoelectric emf.

The magnetodiffusion current decreases with distance from the illuminated surface (Fig. 185A). Consequently, the photomagnetoelectric emf should also depend on x, decreasing with distance from the illuminated surface. This means that there should be also a field and a potential difference along the x axis. However, the photomagnetoelectric emf along the y axis and the potential along the x axis are governed not only by the magnetodiffusion current but also by the conduction currents along the y and x axes, which tend to equalize the potential along these directions. The conduction currents along the x and y axes are governed by the resistance and, therefore, by the length of the sample along these directions.

Thus, in samples which are thin in the direction of the x axis and long in the direction of the y axis the potential along the x axis is equalized more effectively than that along the y axis. Such samples are frequently used in experiments. We can assume that in these samples the field along the y axis and the potential difference between the side surfaces are approximately independent of x, i.e., we can assume that the whole side surface a is approximately at one potential and the side surface b at a different potential. Then the density of the conduction current caused by the photomagnetoelectric emf is almost independent of x. At the same time, the magnetodiffusion current decreases strongly with increase of x (Fig. 185A). Consequently, the currents are not balanced at every point: at low values of x, the magnetodiffusion current is greater than the conduction current, and at high values of x, the opposite obtains. This means that a current flows through the sample as shown in Fig. 185B [17]. Under steady-state conditions, the total current through the whole cross section of the sample in the xz plane is zero, i.e., the total magnetodiffusion current through the same cross section is equal to the short-circuit current of Eq. (64.16). The total magneto-diffusion current is also equal to the total conduction current, which can be written in the form:

$$J_{cond} = V_{pm} \sigma_{ab} = - J_{sc}. \tag{64.17}$$

where σ_{ab} is the total conductivity of the sample between the planes a and b, and V_{pm} is the photomagnetoelectric emf.*

_____

* The expression (64.17) is approximately valid for isolated samples, "extended" along the x axis. If the ends of the sample a and b are provided with metal electrodes which completely equalize the potential at the ends, the expression (64.17) becomes exact for samples of arbitrary dimensions along the x and y axes.

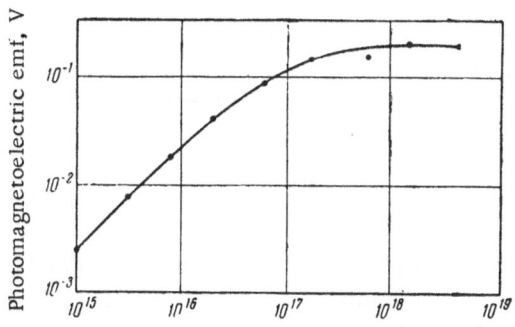

Fig. 186. Experimental dependence of the photomagnetoelectric emf on the intensity of illumination of germanium, illustrating the linear rise of
V_{pm} at low intensities and the saturation at high
intensities.

Since

$$\sigma_{ab} = \frac{gL}{h}\,\sigma_0 + \frac{g}{h}\int_0^L \Delta\sigma\,dx = \frac{gL}{h}\,\sigma_0 + \frac{g}{h}\,e\,(\mu_n + \mu_p)\int_0^L \Delta p\,dx, \qquad (64.18)$$

where σ_0 is the dark conductivity, we obtain using Eq. (64.16), the photomagnetoelectric emf

$$V_{pm} = h\,\frac{eD\,\dfrac{\kappa H}{c}\,(\mu_n + \mu_p)\,[\Delta p\,(0) - \Delta p\,(L)]}{L\sigma_0 + e\,(\mu_n + \mu_p)\displaystyle\int_0^L \Delta p\,dx}. \qquad (64.19)$$

Thus, V_{pm} is proportional to the magnetic field (in weak fields, which
we are considering here) and is governed by the nonequilibrium carrier
distribution Δp.

Analysis of Eq. (64.19) shows that with increase of the excitation intensity (and, consequently, with increase of Δp) the photomagnetoelectric
emf tends to saturation. The numerator of Eq. (64.19) is proportional
to Δp and the denominator has two terms, one of which ($L\sigma_0$) is constant
and the other is proportional to Δp. At low excitation levels, the second
term in the denominator is small and the photomagnetoelectric emf is
proportional to Δp, i.e., to the excitation intensity. At high excitation
levels, the first term in the denominator may be neglected and after can-
celing Δp in the numerator and denominator the photomagnetoelectric
emf is found to be independent of Δp.

Figure 186 shows the transition from the linear rise of V_{pm} at low
illumination intensities to saturation at high intensities.

The expression for Δp has been given previously [cf. Eq. (64.9), as well as Eqs. (64.12)-(64.14)]. Substituting the value of Δp into Eq. (64.19), we obtain quite a cumbersome general expression for V_{pm}.

We shall also consider some special cases for which Eq. (64.19) simplifies considerably so that the photomagnetoelectric effect can be used to measure the diffusion length and the surface recombination velocity.

We shall assume that the depth of penetration of light $1/k$ is small compared with l_D (this condition is easily realized in substances such as germanium and silicon when excitation is by wavelengths in the fundamental region) and the sample thickness L is much greater than l_D, i.e.,

$$\frac{1}{k} \ll l_D \ll L. \tag{64.20}$$

Under these conditions, the generation occurs near the illuminated surface, the situation at the rear surface (x = L) is unimportant and the carrier density decreases along the x axis because of diffusion in accordance with

$$\Delta p = \Delta p\,(0)\,e^{-\frac{x}{l_D}}. \tag{64.21}$$

We shall consider two extreme cases.

a) High Excitation Level $[\Delta p(x) \gg n_0 + p_0]$. Neglecting the first term in the denominator of Eq. (64.19), we have

$$V_{pm} = hD\,\frac{\kappa H}{c}\,\frac{\Delta p\,(0) - \Delta p\,(L)}{\int_0^L \Delta p\,dx}. \tag{64.22}$$

Since $L \gg l_D$, we find, using Eq. (64.21),

$$\Delta p\,(L) \approx 0; \quad \int_0^L \Delta p\,dx \approx \Delta p\,(0)\,l_D$$

and

$$V_{pm} = hD\,\frac{\kappa H}{c}\,\frac{1}{l_D}. \tag{64.23}$$

Thus, the saturation value of the photomagnetoelectric emf, reached at high excitation levels, can be used to determine directly the diffusion length

$$l_D = \frac{1}{V_{pm}}\cdot hD\,\frac{\kappa H}{c}. \tag{64.24}$$

This quantity has definite meaning only if we satisfy the very rigorous condition

$$\Delta p(x) \gg n_0 + p_0, \qquad (64.25)$$

i.e., only when at any x the nonequilibrium carrier density is greater than the equilibrium density. Then l_D is the diffusion length and D, given by Eqs. (64.22)-(64.24) and equal to $D = 2D_n D_p/(D_n + D_p)$ [cf. Eq. (62.15)], is the diffusion coefficient for high excitation levels.

If near the illuminated surface $\Delta p \gg n_0 + p_0$ and far from it $\Delta p < n_0 + p_0$, the density distribution along the x axis is no longer exponential, l_D becomes a function of x and the experiments give the value of this length averaged over the distance x and, consequently, over the range of excitation levels.

b) Low Excitation Level [$\Delta p(x) \ll n_0 + p_0$]. Neglecting the second term in the denominator of Eq. (64.19) and the quantity $\Delta p(L)$ in the numerator, we find

$$V_{pm} = hD \frac{\kappa H}{c} \frac{e \, \Delta p(0)(\mu_n + \mu_p)}{L\sigma_0}. \qquad (64.26)$$

We shall now find $\Delta p(0)$ by assuming that the surface recombination velocity s_0 is negligibly small.*

In this case, the total number of nonequilibrium carriers in the sample, taken per unit area of the illuminated surface, is

$$\int_0^L \Delta p \, dx = \Delta p(0) \int_0^L e^{-\frac{x}{l_D}} dx = \Delta p(0) l_D, \qquad (64.27)$$

which can be written in the form $\beta I_0 \tau$. Consequently,

$$\Delta p(0) = \frac{\beta I_0 \tau}{l_D} = \beta I_0 \frac{l_D}{D}. \qquad (64.28)$$

Substituting Eq. (64.28) into Eq. (64.26), we obtain

$$V_{pm} = \frac{h}{L} \frac{\kappa H}{c} \frac{e(\mu_n + \mu_p)\beta I_0}{\sigma_0} l_D. \qquad (64.29)\dagger$$

The expression (64.29) allows us to determine l_D at low excitation levels:

* It can be shown that the criterion of smallness of s_0 is $s_0 \ll D/l_D$.
† The short-circuit current is then

$$J_{st} = g \frac{\kappa H}{c} e(\mu_n + \mu_p)\beta I_0 l_D. \qquad (64.29')$$

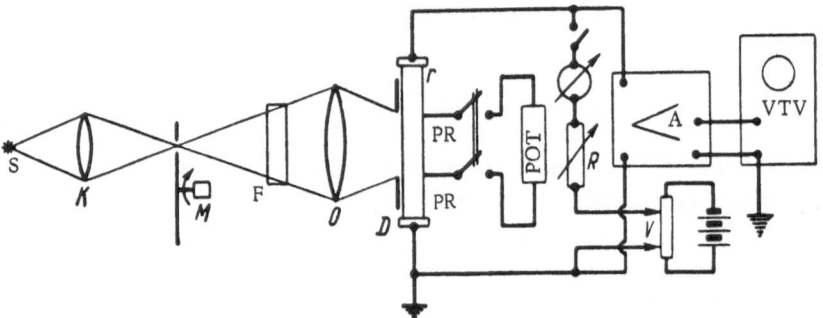

Fig. 187. Circuit for measuring the diffusion length using the compensation of
the photomagnetoelectric emf by varying the photoconductive voltage drop across
the sample. S is the source of light; K, O are the objectives; M is the light modulator; F is the filter; D is the diaphragm; r is the sample; POT is the dc potentiometer; R is the limiting variable resistance; A is the resonance amplifier; VTV
is the vacuum-tube voltmeter; PR are the probes.

$$l_D = V_{\text{pm}} \frac{c}{\kappa H} \frac{L}{h} \frac{\sigma_0}{e(\mu_n + \mu_p)\beta I_0}. \tag{64.30}$$

In conclusion, we shall give the general expression for the photomagnetoelectric emf for a sample with metal electrodes and arbitrary
dimensions along the x axis and arbitrary surface recombination velocities, with the sole restriction that the absorption is strong:

$$V_{\text{pm}} = \frac{A\beta I_0}{B + \beta I_0}, \tag{64.31}$$

where A and B are given by

$$A = h \frac{\kappa H}{c} \frac{D}{l_D} \frac{\dfrac{D}{l_D s_L}\left(\cosh\dfrac{L}{l_D} - 1\right) + \sinh\dfrac{L}{l_D}}{\dfrac{D}{l_D s_L}\sinh\dfrac{L}{l_D} + \cosh\dfrac{L}{l_D} - 1}, \tag{64.32}$$

$$B = \frac{L}{l_D} \frac{\sigma_0}{e(\mu_n + \mu_p)} \frac{s_0\left(1 + \dfrac{D^2}{L^2 s_0 s_L}\right)\sinh\dfrac{L}{l_D} + \dfrac{D(s_0 + s_L)}{Ls_0 s_L}\cosh\dfrac{L}{l_D}}{\dfrac{D}{l_D s_L}\sinh\dfrac{L}{l_D} + \cosh\dfrac{L}{l_D} - 1}. \tag{64.33}$$

It is easily seen that Eq. (64.31) reduces to Eq. (64.29) at low excitation levels, and to Eq. (64.23) at high excitation levels.

B.   Compensation Method for Measuring l_D Under Weak Excitation

As just shown, it is possible to measure l_D at low excitation levels
[cf. Eq. (64.30)] in sufficiently thick samples if the generation rate βI_0
is known and the surface recombination velocity s_0 is small.

There is, however, a "compensator" method in which we do not need to know βI_0 to determine l_D and the value of s_0 need not be small. This method uses compensation of the photomagnetoelectric emf by varying the photoconductive voltage drop across the sample [11].

The measuring circuit is shown schematically in Fig. 187. The sample is illuminated by light. A constant voltage V can be applied to the circuit containing a large resistance R so that the same current J always flows through the sample.

During illumination, the voltage drop across the sample varies with the frequency of modulation because of the photoconductivity. In the presence of a magnetic field perpendicular to the plane of the diagram, a modulated photomagnetoelectric emf appears between the end electrodes of the sample. Obviously, we can select the direction and magnitude of the steady current through the sample so that the photoconductive voltage drop at its end electrodes compensates the photomagnetoelectric emf; therefore, the amplifier records no modulated signal. At the moment of compensation there is a simple relationship between the steady current through the sample and the quantity l_D; we shall prove this below.

The amplitude of the alternating signal between the electrodes, related to the photoconductivity, is given by

$$v = J \Delta r \cong J r^2 \Delta\sigma^*; \tag{64.34}$$

here r is the resistance of the sample (which is approximately the same in darkness and during illumination), and $\Delta\sigma^*$ is the total change of the sample's conductivity during illumination.

It is easily seen that

$$\Delta\sigma^* = \frac{g}{h} \int_0^L \Delta\sigma \, dx = \frac{g}{h} e (\mu_n + \mu_p) \int_0^L \Delta p \, dx. \tag{64.35}$$

At a low excitation level, in the case of strongly absorbed light and sufficiently thick (along the x axis) samples, we find, using Eq. (64.21), that

$$\int_0^L \Delta p \, dx = \Delta p(0) \, l_D. \tag{64.36}$$

and, consequently,

$$v = J r^2 \frac{g}{h} e (\mu_n + \mu_p) \Delta p (0) \, l_D. \tag{64.37}$$

On the other hand, the photomagnetoelectric emf under these conditions is given by Eq. (64.26). Equating the right-hand parts of Eqs. (64.26) and (64.37), we then find that at compensation

$$l_D = h\,\frac{\kappa H}{c}\,\frac{D}{Jr} = h\,\frac{\kappa H}{c}\,\frac{D}{W}\,, \qquad\qquad (64.38)$$

where W is the constant voltage drop across the sample due to the passage of the current J.

The process of measuring l_D is thus reduced to the compensation of the signal by varying V (the moment of compensation is found to correspond to the minimum reading of the vacuum-tube voltmeter, denoted by VTV in Fig. 187) and the measurement of the voltage drop across the sample W. To avoid the influence of the contacts,* they are frequently placed outside the illuminated region so that W represents the voltage drop across the illuminated part of the sample. To determine W accurately, the probes PR are sometimes used and the voltage between them is measured with a potentiometer POT (Fig. 187).

Since the theory of the method applies to steady-state conditions and the photoconductivity kinetics may differ from the kinetics of the photomagnetoelectric emf, the compensation should be obtained under steady-state conditions. This restricts the frequency of light modulation: the durations of the light pulses and of the intervals between them (or the period in the case of sinusoidal modulation) should be considerably longer than the photoconductivity and photomagnetoelectric emf rise times.

Using the compensation method at a low excitation level and measuring V_{pm} at a high excitation level [cf. Eq. (64.24)] we can determine relatively simply the value of l_D at high and low excitation levels eliminating the need for measuring the intensity of light and the influence of the surface recombination, which is a very important advantage.

When light is strongly absorbed, the recombination of the illuminated surface produces an apparent reduction of the intensity of illumination and, consequently, of the magnitudes of the observed effects, without affecting, however, the nature of the distribution of nonequilibrium carriers in the interior of the sample.

It follows that when the intensity of illumination does not occur in an expression [cf, for example, Eqs. (64.24) and (64.38)], the surface recombination velocity is also absent from this expression.

C. Use of the Photomagnetoelectric Effect for Determining the Surface Recombination Velocity

In all the cases considered above, we have assumed that the density gradient along the x axis is related to the volume recombination of the nonequilibrium carriers diffusing from the illuminated surface and, be-

---

*The contacts distort the assumed theoretical distribution of the minority carriers of Eqs. (64.9), (64.12), (64.13), and (64.14) and give rise to "parasitic" photo-emf's.

cause the effective diffusion depth is l_D, one can determine this quantity l_D by measuring the photomagnetoelectric emf.

It is obvious that if a sample is thin along the x axis, i.e., $L \ll l_D$, then the recombination in the interior of the sample is unimportant and the nonequilibrium carrier distribution, carrier density gradient, diffusion current along the x axis and, consequently, the magnetodiffusion current along the y axis are all governed by the surface recombination velocity (s_L) on the side of the sample opposite to the illuminated surface. This makes it possible to determine s_L directly from measurements of the photomagnetoelectric emf under the conditions of strong excitation [9].

If I_0 is sufficiently large, we then have from Eqs. (64.31) and (64.32)

$$V_{pm} = A = h \frac{\kappa H}{c} \frac{D}{l_D} \frac{\dfrac{D}{l_D s_L}\left(\cosh\dfrac{L}{l_D} - 1\right) + \sinh\dfrac{L}{l_D}}{\dfrac{D}{l_D s_L}\sinh\dfrac{L}{l_D} + \cosh\dfrac{L}{l_D} - 1}. \qquad (64.39)$$

Substituting $L \ll l_D$ into the above equation, we find that

$$V_{pm} = L \frac{\kappa H}{c} s_L. \qquad (64.40)$$

D. Photomagnetoelectric Effect for an Arbitrary Relationship between the Nonequilibrium Electron and Hole Densities

When $\Delta n \neq \Delta p$, i.e., when the trap concentration is high, so that nonequilibrium electrons and holes accumulate at traps in quantities comparable with Δn and Δp, the lifetimes of the nonequilibrium electrons and holes are, in general, unequal (§ 23B). Then the lifetime of the nonequilibrium conductivity state τ_{neq}, determined from the photoconductivity relaxation, is given by Eq. (49.1) or, in the steady-state case, by Eq. (49.3). *

In the foregoing subsections, A and B, we have considered the possibility of using the photomagnetoelectric method to determine τ for the case $\Delta n = \Delta p$, i.e., when the electron and hole lifetimes are equal and τ represents the pair lifetime. The question is, what do the values of τ represent if they are obtained for $\Delta n \neq \Delta p$, i.e., when $\tau_n \neq \tau_p$. Obviously, one would expect that in an extrinsic semiconductor the short-circuit photomagnetoelectric current should be governed mainly by the minority carriers which control the diffusion processes. Consequently,

---

* It is easily seen that if the electron and hole lifetimes differ very strongly, then τ_{neq} is equal to the greater of these lifetimes.

in such semiconductors the lifetime, which enters into the expression (64.30) for l_D, should represent the minority carrier lifetime.

We can show that, in general, τ represents in these expressions the quantity

$$\tau_{pm} = \frac{p_0 \tau_n + n_0 \tau_p}{p_0 + n_0}. \tag{64.41}$$

It is evident from the above equation that, for example, when $p_0 \gg n_0$, $\tau_{pm} \approx \tau_n$, i.e., it is equal to the minority carrier lifetime.

Comparison of Eqs. (49.3) and (64.41) shows that the "lifetimes" found from the photoconductivity are, in general, not equal to the values found from the photomagnetoelectric effect; they are equal only if $\Delta n = \Delta p$ [25, 30].

The quantity τ determined by the compensation of the photomagneto-electric emf with the photoconductivity (subsection B above) also has complex meaning when $\Delta n \neq \Delta p$.

An analysis shows that the quantity obtained by this compensation method, which we shall denote by τ_c, is

$$\tau_c = \frac{(\tau_{neq})^2}{\tau_{pm}}. \tag{64.42}$$

Comparison of the experimentally determined values of τ_{neq}, τ_{pm}, and τ_c, may be used to obtain information on the relationship between Δn and Δp (i.e., between τ_n and τ_p) and, consequently, can be used to establish a model which describes the properties of the investigated material.

§65. UNIPOLAR TRANSIENT PHOTOMAGNETOELECTRIC EFFECT

In the case of unipolar photoconductivity (for example, in the case of impurity excitation or fundamental excitation, when the electron and hole lifetimes are very different), there should be no photomagnetoelectric effect under steady-state conditions.

In the photomagnetoelectric effect, the magnetic field deflects the charges moving along the x axis. In the ambipolar (bipolar) case under steady-state conditions, there are two equal and mutually compensating components of the current (the hole and electron components), i.e., there are real electron and hole currents along the x axis although the total current along the x axis is zero. The deflection of these currents by a magnetic field is responsible for the appearance of the photomagnetoelectric effect. In the unipolar case, the zero value of the current along the x axis is due to the compensation of the diffusion current by the conduc-

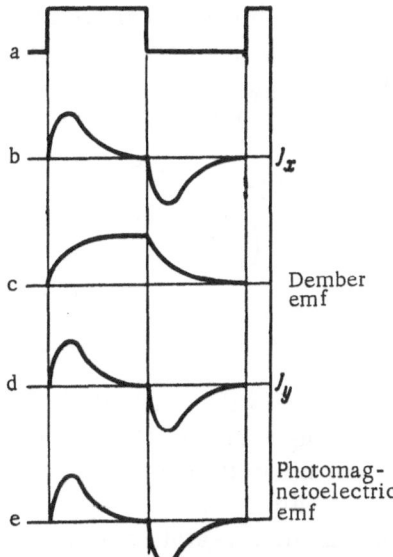

Fig. 188. Variation with time of:
b) the diffusion current, c) Dember
emf, d) magnetodiffusion current,
and e) photomagnetoelectric emf,
during illumination of a semiconductor with a square pulse of
light (a).

tion current. The directed motion of carriers is then absent and, consequently, the
magnetic field does not deflect charges. *

However, even in the unipolar case we
can, in principle, observe the photomagnetoelectric effect but under transient conditions. When illumination of a sample
begins (Fig. 188a), a diffusion current J_X
appears along the x axis because of a density gradient. This current gives rise to
a space charge and a field opposing the
current so that finally the diffusion current (under steady-state conditions) decays
to zero (Fig. 188b). However, before it
decays to zero, the carriers are deflected
along the y axis by the magnetic field and,
consequently, there is a photomagnetoelectric effect (Figs. 188d and 188e). When
the illumination stops, a current representing the dispersal of the space charge appears along the x axis; this current is opposite in direction to the diffusion current
which had been observed during illumination.

Consequently, a photomagnetoelectric effect of opposite sign is observed. An experimental study of the unipolar photomagnetoelectric effect is difficult because in the case of excitation in the impurity region
(which ensures the unipolar conduction) the absorption of light and, consequently, the density gradient are usually weak. Because of this, the
diffusion and magnetodiffusion (along the y axis) currents are small. To
avoid these difficulties, only a part of the sample is illuminated uniformly
throughout its volume, while the other part remains in darkness (Fig. 189).
At the boundary between these two regions, one would expect a considerable density gradient which would be governed not by the depth of penetration of light but by the screening length.

A calculation for a low excitation level shows that for this geometry

and the conditions $L_x \gg l_e$ and $\frac{L_1}{L_2} \gg l_e$ the dependence of the photomag-

* The photomagnetoelectric effect is in many ways similar to the thermomagnetic Nernst-
Ettingshausen effect. However, the latter is observed also in the unipolar case because of the
difference of carrier energies in the opposing diffusion and conduction currents.

Since, during the observation of the photomagnetoelectric effect we can assume that the
nonequilibrium carrier temperature is always the same and equal to the lattice temperature, it
is obvious that the steady-state effect should be absent in the unipolar case.

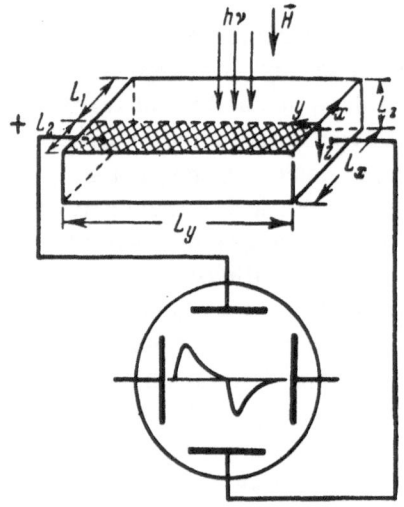

Fig. 189. Circuit for observing the unipolar transient photomagnetoelectric effect.

netoelectric field on time (if illumination begins at t = 0) is given by the expression [41]:

$$E_y = -\frac{\mu H}{c}$$

$$\times \frac{4\pi e D}{\varepsilon L_x} \Delta n_{st} \tau \frac{\left[\left(\frac{1}{\tau} - \frac{1}{\tau_\sigma}\right)t - 1\right]e^{-\frac{t}{\tau_\sigma}} + e^{-\frac{t}{\tau}}}{\left(1 - \frac{\tau}{\tau_\sigma}\right)^2},$$

$$(65.1)$$

where $\tau_\sigma = \varepsilon/4\pi\sigma$ is the Maxwell time; $\tau = [\gamma_n(N_cM + n_0 + M - m_0)]^{-1}$ is the lifetime in the impurity photoconductivity case [cf. Eq. (42.8)]; $\Delta n_{st} = m_0 q I_0 \tau$ is the steady-state density in the interior of the illuminated part of the sample.

Figure 190 shows the dependence of E_y on time, plotted using Eq. (65.1) for two values of the ratio τ/τ_σ. It is evident that the photomagnetoelectric effect increases strongly on reduction of τ/τ_σ.

It is convenient to estimate the value of E_y from its maximum value E_y^{max}, which is given by

$$E_y^{max} = -\frac{\mu H}{c}\frac{4\pi e D}{\varepsilon L_x} \Delta n_{st} t_{max} e^{-\frac{t_{max}}{\tau_\sigma}}, \qquad (65.2)$$

where t_{max} is defined by

$$\exp\left[\left(\frac{1}{\tau_\sigma} - \frac{1}{\tau}\right)t_{max}\right] = \left(1 - \frac{t_{max}}{\tau_\sigma} + \frac{\tau}{\tau_\sigma} \cdot \frac{t_{max}}{\tau_\sigma}\right). \qquad (65.3)$$

If the Maxwell relaxation time τ_σ is considerably smaller than the lifetime τ, it follows from Eq. (65.3) that

$$\exp\frac{t_{max}}{\tau_\sigma} = \frac{\tau\, t_{max}}{\tau_\sigma^2} + 1, \qquad (65.4)$$

so that in place of Eq. (65.2), we obtain

$$E_y^{max} \approx -\frac{\mu H}{c}\frac{4\pi e D}{\varepsilon L_x} \Delta n_{st}\frac{\tau_\sigma^2}{\tau}. \qquad (65.5)$$

To estimate the magnitude of the unipolar effect, it is of interest to compare E_y^{max} with the usual ambipolar photomagnetoelectric effect E_{pm},

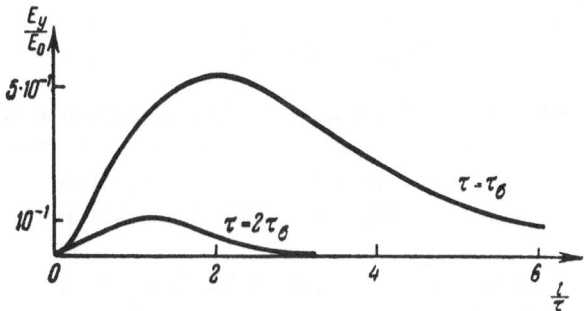

Fig. 190. Relaxation curves of the unipolar photomag-

netoelectric effect; $E_0 = \dfrac{\mu H}{c} \dfrac{4\pi e D}{\varepsilon L x} \Delta n_{\text{st}} \tau$.

assuming that the steady-state densities of nonequilibrium carriers are equal in both cases.

It is found that

$$\frac{E_y^{\max}}{E_{\text{pm}}} = \frac{\tau_\sigma}{\tau}. \tag{65.6}$$

Thus in the case of semiconductors with high conductivity, for which τ_σ is small, the unipolar photomagnetoelectric effect is small compared with the ambipolar effect.

It is evident from Eqs. (65.2) and (65.5) that if the condition $L_x \gg l_e$ is satisfied then $E_y^{\max}$ decreases with increase of L_x. This is because the diffusion occurs in a small region, defined by $\sim l_D$, and the remaining part of the semiconductor of thickness $L_x - l_D \approx L_x$ acts as a ballast which shunts the photomagnetoelectric emf. The expression for the short-circuit current is independent of L_x and has the form

$$J_y = \frac{\mu H}{c} e D L_z \frac{e^{-\frac{t}{\tau_\sigma}} - e^{-\frac{t}{\tau}}}{(1 - \tau/\tau_\sigma)} \Delta n_{\text{st}}. \tag{65.7}$$

The short-circuit current, like the photomagnetoelectric emf, varies with time in the way indicated by the curve in Fig. 188d. The maximum value of the current is

$$J_y^{\max} = \frac{\mu H}{c} e D L_z \Delta n_{\text{st}} \left[\frac{\tau}{\tau_\sigma} \right]^{\frac{\tau}{(\tau_\sigma - \tau)}}. \tag{65.8}$$

The ratio of $J_y^{\max}$ to the short-circuit current in the case of the normal photomagnetoelectric effect is

$$\frac{J_y^{max}}{J_{pm}} = \left[\frac{\tau}{\tau_\sigma}\right]^{\frac{\tau}{(\tau_\sigma - \tau)}}. \qquad (65.9)$$

When $\tau/\tau_\sigma \gg 1$, we have $J_y^{max}/J_{pm} \approx \tau_\sigma/\tau$, and when $\tau/\tau_\sigma \sim 1$, this ratio is also ~1. Thus we can expect that the transient unipolar photomagnetoelectric effect is large in poorly conducting materials (for example, in compensated germanium and silicon at sufficiently low temperatures).

In conclusion, we note that the saturation of E_y with increase of the intensity of illumination found in the case of the normal photomagnetoelectric effect is also observed in the unipolar photomagnetoelectric effect. Our formulas do not reflect this property because we have considered the solution for a low excitation level. The unipolar photomagnetoelectric effect has not yet been studied experimentally.

§66. NEGATIVE PHOTOCONDUCTIVITY IN A MAGNETIC FIELD

In some cases, a semiconductor which exhibits positive photoconductivity under normal conditions, shows negative photoconductivity (a reduction of the conductivity on illumination) in a magnetic field directed at right angles to the flow of the current [9]. A circuit for observing this effect is shown in Fig. 191. The sample is connected up in the usual circuit for measuring photoconductivity. For convenience, it is illuminated with short light pulses separated by long periods of darkness. The signal, taken from a load resistance R, is analyzed with an oscillograph. Typical oscillograms obtained for germanium are shown in Figs. 192a and 192b. Oscillogram a was obtained in the absence of a magnetic field and represents the normal "positive" photoconductivity (positive conduction pulses). When a magnetic field is applied (oscillogram b), the photoconductivity changes sign (the conduction pulses are negative).

A qualitative explanation of this effect is as follows. When carriers drift in a transverse magnetic field, their trajectories, over a mean path,

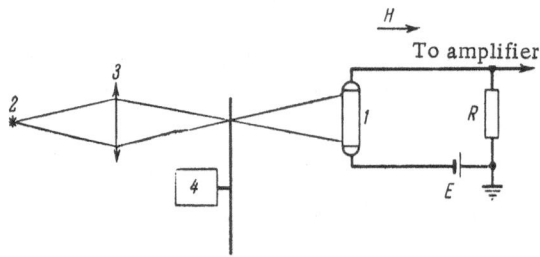

Fig. 191. Circuit for observing the negative photoconductivity in a magnetic field. 1) Sample; 2) source of light; 3) objective; 4) light modulator.

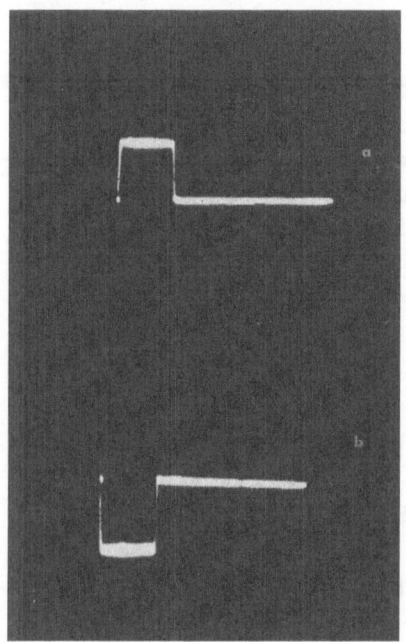

Fig. 192. Photoconductivity oscillograms.
a) Positive photoconductivity pulses in the
absence of a magnetic field; b) negative
photoconductivity pulses in a magnetic field.
The results were obtained for p-type germanium with ρ = 12 Ω · cm at T = 77°K;
H = 20,000 Oe; excitation level $\Delta n / p_0$
was ∼ 0.005.

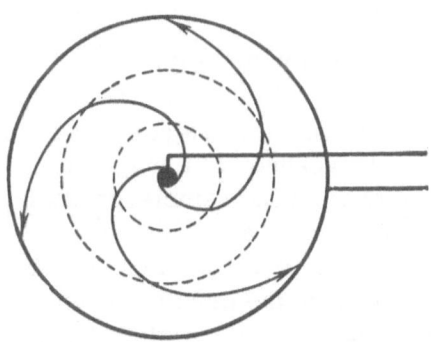

Fig. 193. Sample in the form of a Corbino
disk.

are bent.   This reduces the mobility and increases the resistance in
a magnetic field.   However, the Hall
field partly "rectifies" the carrier
trajectories and thus limits the resistance rise in a magnetic field.

If the Hall field is reduced in
some way, the bending of the trajectories becomes greater and the
resistance of the semiconductor
rises.

The Hall field may be reduced
in principle by:

a) "short-circuiting" the Hall
emf with a low external resistance;

b) using samples in the form
of Corbino disks in which the Hall
field does not appear because of the
sample geometry (Fig. 193).

However, there is another method which is suitable in the impurity
conduction case, when the density
of carriers of one sign (for example,
holes, p_0) is considerably greater
than the density of the other carriers
(electrons, n_0), and, consequently,
the Hall coefficient at equilibrium
is governed by the quantity p_0.   If
such a semiconductor is illuminated
with light from the fundamental absorption band, electron-hole pairs
are generated.   The related increase
of the electron density reduces the
Hall field and this results in a resistance rise, i.e., we have the
negative photoconductivity.

Thus, on the one hand, illumination raises the conductivity by the
generation of nonequilibrium carriers, and on the other, the conductivity falls because of the reduction of the Hall field and the con-
sequent reduction of the equilibrium carrier mobility.   If the second effect predominates, we have the negative photoconductivity.

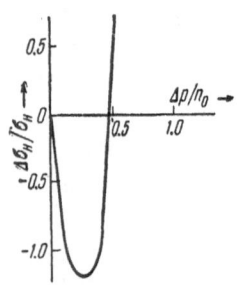

Fig. 194. Dependence of $\Delta\sigma_H/\sigma_H$ on the excitation level $\Delta p/n_0$ for n-type germanium at T = 77°K; H = 10,000 Oe. The negative photoconductivity is observed at low excitation levels (up to $\Delta p/n_0 \approx 0.5$).

Calculations show that the negative photoconductivity should occur in the two cases of unipolar equilibrium conductivity and of ambipolar nonequilibrium conductivity at low excitation levels.

Figure 194 gives the theoretical calculated curve of the relative change of the photoconductivity of n-type germanium. The curve shows that $\Delta\sigma_H$ remains negative until the injection level ($\Delta p/n_0$) reaches 0.46.

The negative photoconductivity observed experimentally (Fig. 192) does indeed appear under these conditions.

(ii) Magnetodensity Effects

Diffusion in a magnetic field (§§ 64 and 65) gives rise to space charge and emf's due to redistribution of charged carriers in space. However, apart from observing these emf's, one can observe directly and study changes of the density distribution of the diffusing nonequilibrium carriers in a magnetic field, i.e., the "magnetodensity effects." One of them is the Suhl effect which represents the change in the density distribution of minority carriers drifting in an electric field on application of a transverse magnetic field; this effect is used mainly to determine the surface recombination characteristics (cf. § 68).

Another effect, the description of which is given in § 67, is the "focusing" of the diffusing carriers by a longitudinal magnetic field. There are also other magnetodensity effects.

§ 67.   LONGITUDINAL MAGNETODENSITY EFFECT

If a point light probe generates nonequilibrium carriers at some point on the surface of a semi-infinite sample, in the absence of a magnetic field these carriers will diffuse into the interior of the sample, producing a spherically symmetrical distribution with its center at the injection point, as shown in Fig. 195 (it is assumed that the recombination velocity on the illuminated surface is zero). We shall place the origin of coordinates at the injection point and direct the z axis at right angles to the sample surface.

On the application of a magnetic field parallel to the z axis, the diffusion-current components perpendicular to this field, i.e., the components in any plane at right angles to the z axis, will be subject to a Lorentz force which deflects them by an angle $\theta = \mu H/c$. Thus, if we look from above (Fig. 195), the macrotrajectories of the nonequilibrium

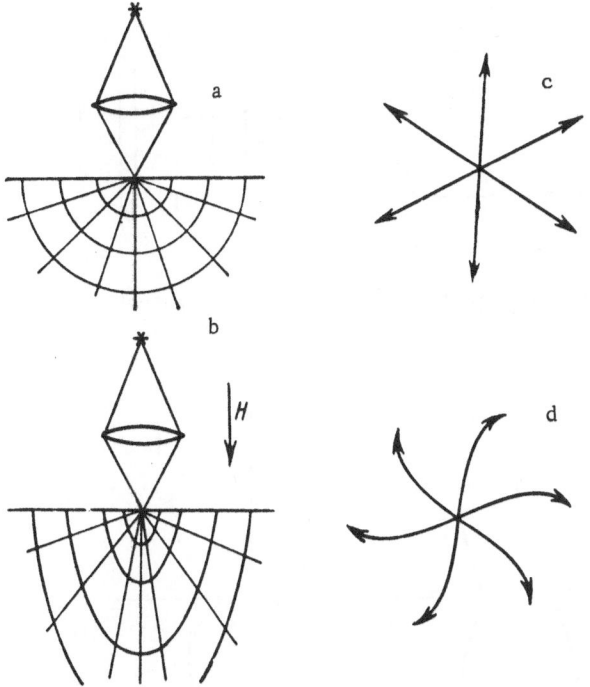

Fig. 195. Schematic representation of the diffusion of nonequilibrium carriers generated by a point light probe on the surface
of a semiinfinite sample in the absence (a, c) and presence (b, d)
of a magnetic field; a and b give the view from the side of the
sample; c and d give the view from above.

carriers are bent, as shown in Fig. 195d. Consequently, in the direction at right angles to H, the carriers diffuse during their lifetime to distances from the injection point which are shorter than in the case when
H = 0.   This is equivalent to the reduction of the diffusion length in the
planes which are parallel to the xy plane (if the lifetime is constant) and
should lead to a rise of the carrier density along the z axis.

The change of the carrier density distribution in a magnetic field is
related in a definite way to the carrier mobility. A calculation – which
will not be reproduced here – shows that in a strongly extrinsic semiconductor (for example, a p-type one) this distribution, for $\Delta n = \Delta p$ and
a low excitation level, has the form [31, 37]

$$\Delta n_H = \frac{\Delta N}{2\pi l_D^2}\left[1+\left(\frac{\mu_n H}{c}\right)^2\right]\times \frac{\exp\left(-\frac{1}{l_D}\sqrt{(x^2+y^2)\left[1+\left(\frac{\mu_n H}{c}\right)^2\right]+z^2}\right)}{\sqrt{(x^2+y^2)\left[1+\left(\frac{\mu_n H}{c}\right)^2\right]+z^2}},\quad (67.1)$$

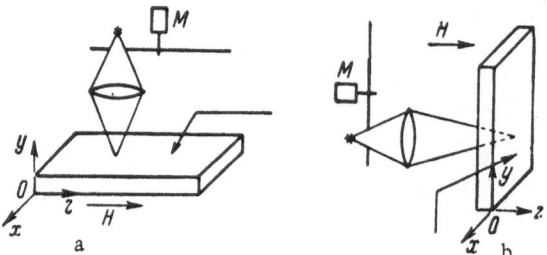

Fig. 196. Layouts for investigating the magnetodensity
effect. M is a light-beam modulator.

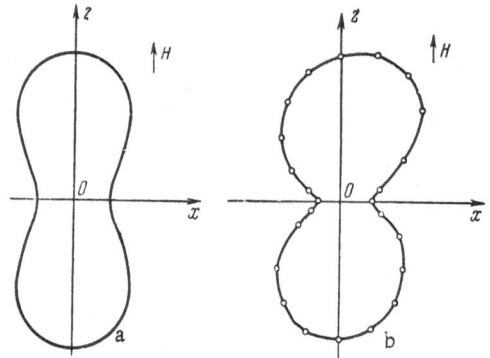

Fig. 197. a) Calculated dependence of the collector
current on the angle of rotation of the sample about
the y axis; b) results of measurements for one sample.

where $\Delta n = \beta I_0 \tau$ is the total density of nonequilibrium pairs in the semiconductor, which is obviously independent of the magnetic field; l_D and
μ_n are, respectively, the diffusion length for $H = 0$ and the Hall mobility
of minority carriers.*

In the absence of a magnetic field, Eq. (67.1) transforms into

$$\Delta n_0 = \frac{\Delta N}{2\pi l_D^2} \frac{\exp\left(-\dfrac{r}{l_D}\right)}{r},\tag{67.2}$$

where $r = \sqrt{x^2 + y^2 + z^2}$ is the distance from the injection point. From
Eqs. (67.2) and (67.1), it follows that the equal-density surfaces are
spheres when $H = 0$ but in the presence of a magnetic field they become
ellipsoids of revolution with their major axes directed along the magnetic
field (Fig. 195b). The equation for these ellipsoidal surfaces is

---

* A more rigorous treatment with allowance for the velocity distribution of carriers shows that
the quantity μ_n is more complex (this is about to be discussed).

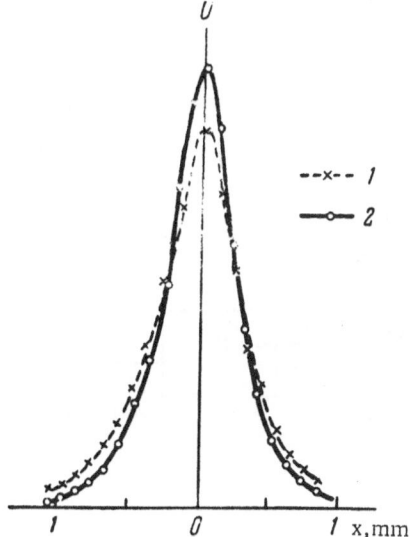

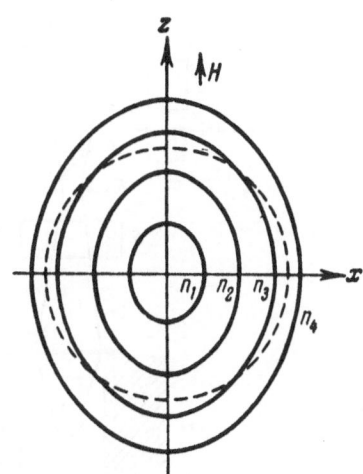

Fig. 198. Dependence of the collector signal on the distance between the collector and the injection point in a plane perpendicular to the direction of the magnetic field. 1) H = 0; 2) H = 15,000 Oe.

Fig. 199. Section of an equal-density surface by the x0z plane. The dashed circle represents the position of the collector when the sample is rotated about the y axis.

$$(x^2 + y^2)\left[1 + \left(\tfrac{\mu_n H}{c}\right)^2\right] + z^2 = \text{const.} \qquad (67.3)$$

From Eq. (67.3), it is evident that the major axis is $\sqrt{1 + \left(\tfrac{\mu_n H}{c}\right)^2}$ times greater than the minor axis. The effect of the magnetic field thus reduces to "focusing" of the diffusing carriers along the magnetic field direction. An experimental determination of the ratio of the ellipsoid axes gives us the quantity μ_n.

A change in the density distribution in a magnetic field may be detected experimentally in the following way. A sample in the form of a plane-parallel slab is illuminated with a point light probe (Figs. 196a and 196b). The minority-carrier density is found from the signal at a point collector probe. The magnetodensity effect is represented by a relative change of the density on application of a magnetic field

$$\xi = \frac{v_H - v_0}{v_0}, \qquad (67.4)$$

where v_H and v_0 are the signals at the collector with and without a magnetic field.

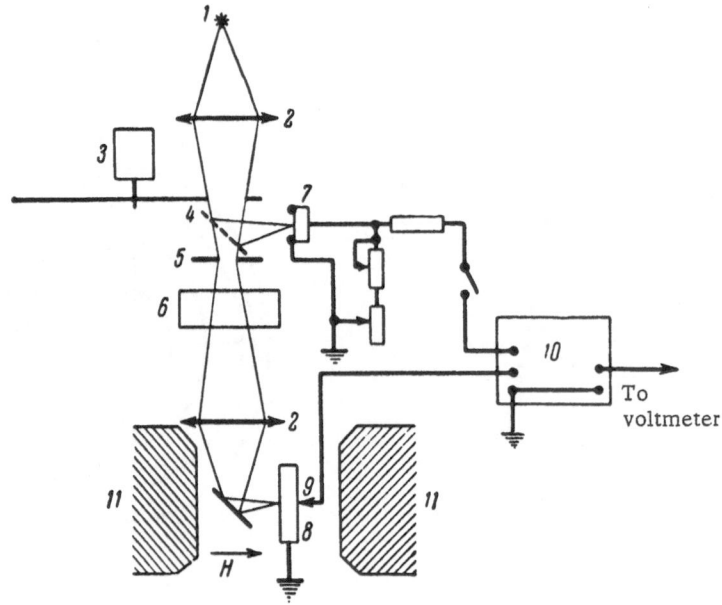

Fig. 200. Measurement of the longitudinal magnetodensity effect.
1) Source of light; 2) objective; 3) modulator; 4) semitransparent
mirror; 5) diaphragm; 6) water filter; 7) photodiode; 8) sample; 9)
collector; 10) amplifier; 11) electromagnet poles.

Depending on the sign of ξ, the magnetodensity effect may be either
positive (increase of the carrier density with H) or negative. Figures
197b and 198 show curves representing the change in the minority carrier
distribution in a magnetic field for germanium slabs placed parallel to
the xOz and xOy planes.

In the xOz plane, the distribution should be ellipsoidal. Figure 199
shows the equal density surfaces for this plane at a certain value of H.
If the sample is rotated about an axis perpendicular to the magnetic field
and passing through the injection point (i.e., an axis coinciding with the
light beam, Fig. 196a), this is equivalent to moving the collector along
a circle — shown dashed in Fig. 199 — in the xOz plane. The collector
then intersects lines of various densities and the magnitude of the signal
depends on the angle through which the sample is rotated.

Figure 197a shows a calculated curve, plotted in polar coordinates,
and Fig. 197b shows a typical experimental dependence obtained for germanium and showing a qualitative agreement with the calculated curve.

When the sample is placed parallel to the xOy plane, at right angles
to the magnetic field (Fig. 196b), the rotation of the sample about the
light beam should not alter the magnitude of the signal. Therefore, it is
sufficient to carry out measurements along any direction which passes
through the injection point.

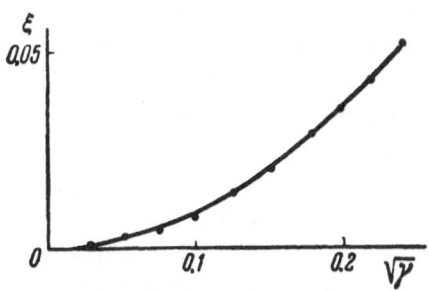

Fig. 201. Dependence of the magnitude of the longitudinal magnetodensity effect in p-type germanium on the magnetic field; $\gamma = 9\pi\mu_0 H/16c$. The theoretical curve is continuous and the points represent the experimental results.

Figure 198 illustrates as expected, that on the application of a magnetic field the carrier density increases near the injection point but decreases far from it (the "focusing" effect).

All these results from measuring the carrier distribution in the xOz and xOy planes confirm qualitatively the ellipsoidal symmetry of the injected carrier distribution and the conclusion that the action of a magnetic field on diffusing carriers reduces to a bending of their macrotrajectories. However, a quantative comparison of these results with the theory is difficult. In order to find the ratio of the ellipsoid semiaxes, one could compare the theoretical and experimental dependences of the collector current on the angle of rotation shown in Fig. 197. However, one must remember that a study of ellipsoid cross sections, using samples in the form of plates, can give only a qualitative picture because the conditions for carrier spreading are different from those in an infinite semiconductor. In these experiments there is the unavoidable influence of the plane boundaries of the sample on the measured currents. In the case of strong surface recombination, the carriers deflected to the surface of the plate recombine faster and their density decreases with distance from the injection point faster than in an infinite sample. The influence of the surface is not eliminated even when the surface recombination velocity is zero because carriers will still be reflected from the sample boundaries. Measurements of the longitudinal magnetodensity effect, i.e., measurements of the increase of the density along the z axis, are free of these limiations. It follows from Eq. (67.1) that along the z axis, i.e., for x = y = 0,

$$\Delta n_z = \frac{\Delta N}{2\pi l_D^2} \frac{e^{-\frac{z}{l_D}}}{z} \left[1 + \left(\frac{\mu_n H}{c}\right)^2\right]. \qquad (67.5)$$

Then the quantity ξ is given by

$$\xi = \frac{\Delta n_H - \Delta n_0}{\Delta n_0} = \left(\frac{\mu_n H}{c}\right)^2. \qquad (67.6)$$

Figure 200 is a schematic representation of the setup measuring this effect. Light from a source 1, modulated by a disk chopper 3, falls

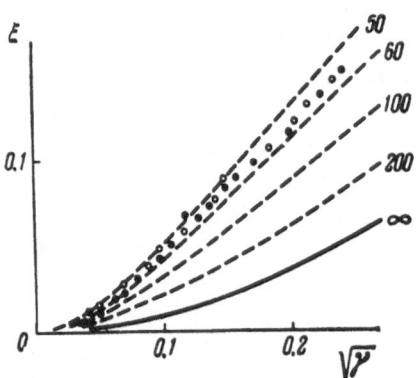

Fig. 202. Dependence of the magnitude
of the longitudinal magnetodensity effect
in n-type germanium on the magnetic
field; $\gamma = 9\pi\mu_0 H/16c$.

on a diphragm with a very small aperture, * and its image is projected by a mirror onto the surface of a semiconductor slab 8 placed between the poles of an electromagnet 11. A water filter 6 absorbs the long-wavelength radiation, which could generate carriers in the interior of the sample. The signal from a point collector 9, placed on the dark side of the sample exactly opposite the illuminated spot on the other side, is fed to a narrow-band amplifier 10, tuned to the frequency of modulation of the light beam. To increase the accuracy of measuring the difference between the collector signals with and without a magnetic field, we can use a compensation circuit: light is directed by a semitransparent mirror 4 to a photocell (for example, a photodiode) 7 and the photocell signal is also fed to the amplifier in antiphase with the collector signal. If these signals compensate each other at H = 0, we can determine directly the signal due to the action of a field H.

A theoretical analysis shows that for the selected geometry the surface recombination and the sample thickness along the z axis should not affect the quantity ξ.

It follows from Eq. (67.6) that we can deduce the value of μ_n from the value of ξ. The exact theory of the magnetodensity effect allowing for the carrier velocity distribution shows that μ_n is not the Hall mobility (as would follow from the phenomenological theory) but is a complex function of the true microscopic mobility μ_n^0, magnetic field H and the power exponent ν in the dependence of the relaxation time τ on the energy ε: $\tau \sim \varepsilon^{\nu}$.

The nature of the dependence of μ_n on H is thus governed by the scattering mechanism represented by the value of ν. If this mechanism is known, we can calculate and plot the theoretical dependence μ (H) [or ξ(H)] from which the microscopic mobility μ_n^0 can be determined by comparison with experiment.

Figure 201 shows the theoretical dependence ξ (H) for p-type germanium (the minority carriers are electrons and $\nu = -1/2$). The points shown on this curve were obtained experimentally in a study of the longi-

---

* The relationships (67.1), (67.5), etc., obtained previously are valid only if the dimensions of the region where carriers are generated is small compared with l_D.

tudinal magnetodensity effect. The agreement between the experimental
points and the theoretical curve indicates that

$$\mu_n^0 = 3650 \ \text{cm}^2 \cdot \text{V}^{-1} \cdot \text{sec}^{-1}.$$

The magnetodensity effect is very sensitive to the presence of carriers with an effective mass different from the mass of the majority
carriers. It is known that in germanium there are usually two types of
hole ("heavy" holes and a small number of "light ' ones) due to the degeneracy of the valence band. The effective masses of these holes differ
by a factor of about 8. This gives rise to a considerable difference between the observed magnetodensity effect in n-type germanium (where the
minority carriers are holes) and the same effect in the same material
calculated on the assumption that only the heavy holes are present.
    The lowest (continuous) curve in Fig. 202 is a theoretical one cal-
culated on the assumption that only the heavy holes are present. * It is
clear from Fig. 202 that the experimental points lie far from this curve.
The dashed curves represent the ξ (H) dependences calculated allowing
for the presence of light holes. The numbers by the curves represent the
ratio of the density of the heavy holes to that of the light holes. The experimental points are seen to lie between the ratios of 50 and 60 (more
exactly, 57).
    The presence of a small number of the light holes ($\sim 2\%$) markedly
alters the magnitude of the magnetodensity effect (and the stronger H, the
greater the effect) and this makes it possible to use the magnetodensity
effect for the experimental determination of the ratio of the densities of
carriers having different effective masses.

§68. SUHL MAGNETODENSITY EFFECT

The Suhl effect is due to the change in direction of the drift of nonequilibrium minority carriers in a transverse field [10].
    To demonstrate clearly the characteristics of this effect, we shall
consider first the problem of the influence of a transverse magnetic field
on the drift of majority equilibrium carriers, i.e., essentially, we shall
return to the Hall effect in a unipolar semiconductor.
    When a magnetic field is applied at right angles to the plane of the
diagram in Fig. 203A, the majority carriers drifting in an electric field

---

* More exactly, this is a universal curve for electrons or holes scattered on the lattice vibrations when the carriers of other effective masses are absent. In particular, this curve is identical with that given in Fig. 201.

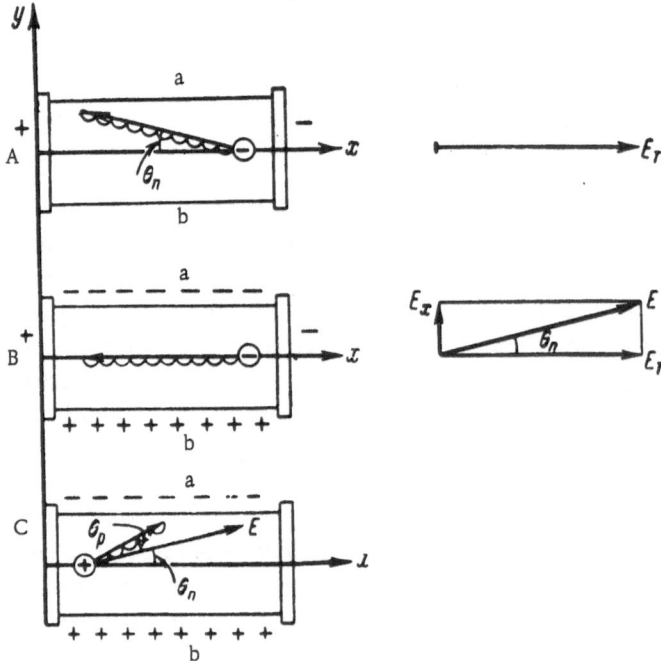

Fig. 203. Suhl effect mechanism. A) Hall effect for majority carriers (electrons); B) motion of majority carriers under steady-state conditions; C) motion of minority carriers.

(electrons in Fig. 203) are deflected in the y direction, i.e., they move at a certain angle θ_n with respect to the direction of motion before the application of the magnetic field, i.e., with respect to the direction of the external electric field. As a result, a charge appears at the surfaces a and b and, therefore, a Hall field is established. The deflection from the initial direction of motion will continue to increase until the Hall field is able to compensate the deflecting action of the magnetic field. After the establishment of a steady state, the direction of motion of the majority carriers remains the same as initially (Fig. 203B).

Thus, in the first approximation, the motion of the majority carriers in the presence of a transverse magnetic field proceeds in the same direction as in the absence of such a field (Fig. 203B). The total field, consisting of the external and Hall fields, makes an angle θ_n with the direction of motion of the carriers (right-hand part of Fig. 203B).

We shall now find what happens to the minority carriers (holes) which enter the sample (Fig. 203C). The conditions for the minority carriers are different from those for the majority carriers. The minority carriers are injected even in the presence of a Hall field in the sample. Moving from left to right in Fig. 203, they are deflected upward by the magnetic field (to side a). The Hall electric field (which opposes such deflection in the case of the majority carriers) then further deflects the minority carriers in the same direction.

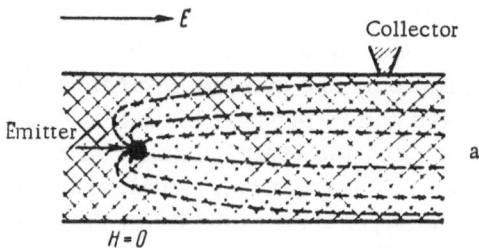

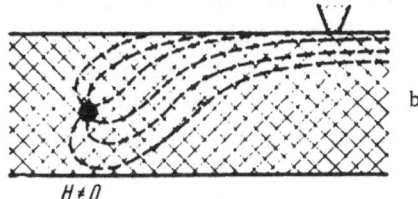

Fig. 204. Trajectories of minority carriers: a) In the absence of a magnetic field; b) on the application of a magnetic field.

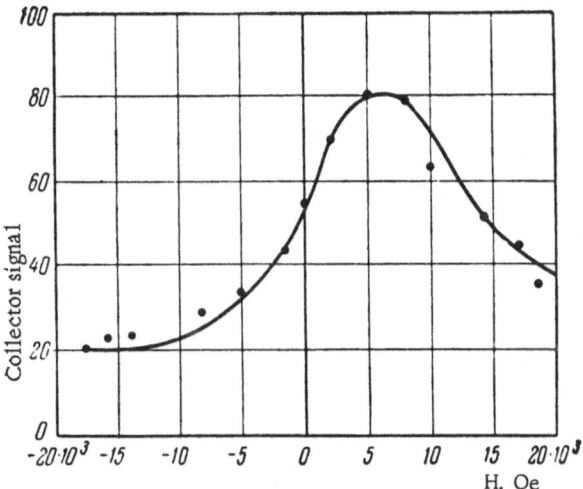

Fig. 205. Dependence of the minority-carrier density at the collector on the magnetic-field intensity (Suhl effect).

Consequently, the minority carriers, drifting along the sample, tend to veer toward the side a and move mainly near that side: this is known as the Suhl effect. The accumulation of the minority carriers on one side of the sample does not give rise to a space charge and a field, which would oppose such an accumulation (such opposition does arise in the case of the majority carriers).

The accumulation of the minority charge is obviously limited by the recombination rate and by the diffusion from a to b.

To observe the Suhl effect, the sample is furnished with two rectifying contacts (usually sharp points pressed against the sample), one of which is used for injection and the other, located on a side surface, for the detection of the minority carriers (Fig. 204). On the application of a magnetic field, the minority-carrier density at the side a and the signal of the collector electrode both increase (Fig. 205). However, if the rate of recombination at the surface a exceeds the rate of recombination in the interior, then the stronger the effect of the magnetic field which drives the minority carriers toward the side a, the faster the annihilation of the minority carriers in their motion to the right.

It follows that in sufficiently strong fields the minority-carrier density at the collector point may even decrease, as indicated by the descending part of the curve in Fig. 205.

Chapter XV

PHOTOELECTROMOTIVE FORCES IN INHOMOGENEOUS SEMICONDUCTORS

Under some conditions, a potential difference, known as the photo-emf, appears between separate points of an illuminated semiconductor sample (§§ 63 and 64). The appearance of an emf in a semiconductor is obviously related only to the appearance of a space charge, i.e., to a separation in space of positive and negative charges (electrons and holes).

However, the absorption of light and photoionization alter (increase) directly only the energy of electrons and holes but do not separate them in space. * When an electron is transferred by light from a local impurity level to the conduction band, an electron-hole pair, which remains at the impurity level, becomes separated in the energy space but remains together in the geometrical space.

A photo-emf appears only when there are additional factors causing the separation of the nonequilibrium charges of different sign. Among such factors are, for example:

a) the difference of the mobilities of carriers of different sign so that a nonuniform illumination separates the charges due to their different diffusion velocities; this factor is responsible for the appearance of the Dember photo-emf, discussed in § 63;

b) the presence of a magnetic field so that diffusing charges are separated by deflection in opposite directions; this is the basis of the photo-magnetoelectric Kikoin-Noskov effect, discussed in § 64.

These two factors causing the separation of carriers of different sign are typical of homogeneous (before illumination) semiconductors. However, a very effective separation of nonequilibrium carriers occurs in inhomogeneous semiconductors or semiconducting systems, in particular, in p-n junctions and other types of blocking layers.

We shall consider in the next section the photoelectric phenomena in inhomogeneous semiconductors using as an example a semiconductor with

---

* We shall neglect the very small effects due to the pressure of light.

329

a p-n junction. * We shall analyze the characteristics of the most highly perfected of the contemporary semiconducting devices: signal-converters in the form of photodiodes and phototransistors; energy-converters in the form of p-n photocells.

§69. PRINCIPLE OF OPERATION OF p-n JUNCTION PHOTOCELLS

A.   Energy Model of a p-n Junction

Figure 206 illustrates the energy-band structure of p- and n-type semiconductors before they are brought into contact. It is assumed that the semiconductors differ only in the nature of their impurities. The edges of the potential wells shown at the same level represent the potential energy of an electron outside the semiconductor. The energy gaps between the Fermi levels and the potential-well edges, indicated by arrows in Fig. 206, define the work function. Clearly, the work function of the n-type semiconductor is smaller than that of the p-type. Consequently, when the two semiconductors are brought into contact, electrons move from the n- to the p-type semiconductor. A double charged

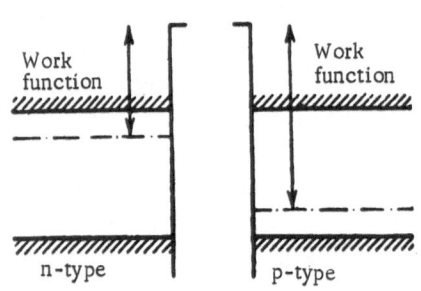

Fig. 206. Energy-band structure of n- and p-type semiconductors before they are brought into contact.

layer appears at the contact: its positive component is in the n-type half and its negative component in the p-type half; consequently, a potential difference, known as the contact potential difference, is established. As more electrons move from the n- to the p-type side, this potential difference increases and the levels in the p-type semiconductor are shifted upward until the Fermi levels in both semiconductors become equal.

As a result of this, the energy-band structure becomes that shown in Fig. 207a. In the space-charge region near the contact, the energy separation of the Fermi level from the edges of the allowed bands is greater than in isolated n- and p-type semiconductors and, consequently, the carrier density and conductivity are smaller than in this region. The region is known also as the blocking layer.

Without giving a detailed description of the electrical properties of a p-n junction established in this way at the contact between n- and p-type semiconductors (such junctions are widely used as rectifiers), we shall return directly to the problem of the use of p-n junctions in photocells.

* The discussion of the general case is given in Lashkarev's work [5].

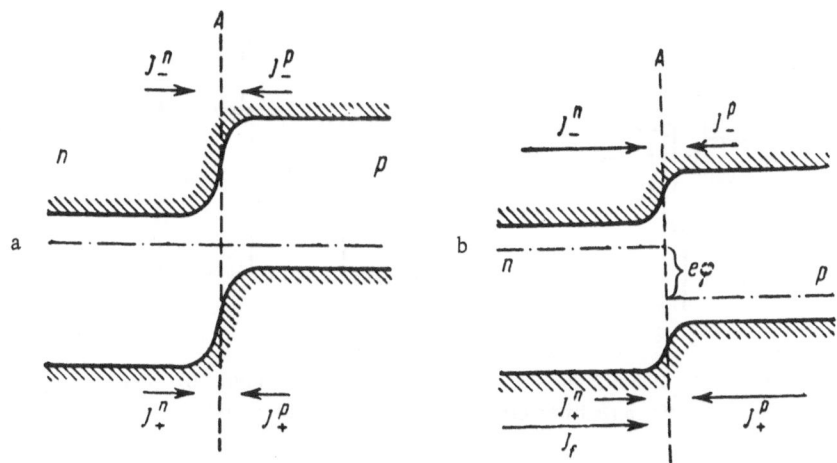

Fig. 207. Electron and hole currents through a p-n junction: a) At equilibrium;
b) during illumination.

A typical structure of a p-n photocell (photodiode) is shown in cross section in Fig. 208a. The mechanism of the photodiode action is very simple. Light incident on the photocell surface (in our case, on the n-type semiconductor, as shown in Fig. 208a) generates electron-hole pairs. The "minority" carriers (holes in the n-type part) diffuse across the n-type region toward the junction. On reaching the junction, the holes are driven across it by the junction field into the p-type region. This process continues until the positive charge in the p-type region reaches a value which stops further motion of holes across the junction. At this moment, a certain potential difference is established between the n- and p-type regions, which is known as the barrier photo-emf.

Photodiodes which use p-n junctions are simply a variant of the well-known barrier photocell. However, there is an essential difference between p-n junction photodiodes and other types of barrier photocell: photodiodes can be used at considerable voltages applied in reverse. This cannot be done in the case of barrier photocells. The operation under high reverse voltages is known as the "photodiode" regime, in contrast to the condition without an external voltage, which is known as the "barrier" regime (Fig. 208b and c). The photodiode regime is so important in practice, that photodiodes can be regarded as an essentially new type of photoelectric detector.

We shall give below the phenomenological theory of photodiodes working in the barrier and photodiode regimes, and describe their principal properties.

B.   Fundamental Equation of a Photodiode

We shall consider quantitatively the mechanism for the appearance of a barrier photo-emf at a p-n junction. We shall assume that the p-n

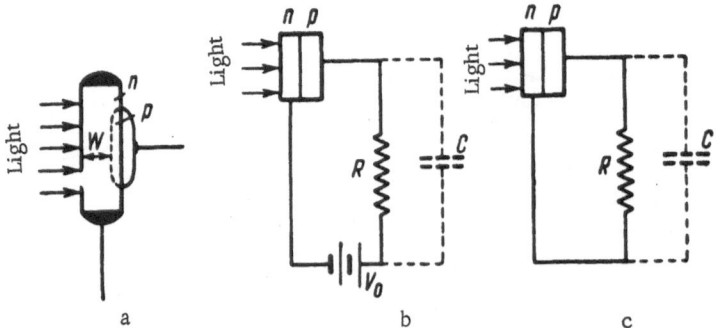

Fig. 208. Photodiode circuits: a) Photodiode structure; b) "photodiode regime" circuit; c) "barrier regime" circuit.

junction is at equilibrium in darkness (Fig. 207a). We shall consider the
magnitude of the current across the junction. For simplicity, we shall
use the "diode" theory. The current through the junction is then the algebraic sum of the thermal currents of charge carriers, passing per unit
time through the junction, multiplied by the electronic charge. At equilibrium, the sum of the thermal currents is zero. Moreover, the principle of detailed balancing indicates that the electron and hole currents
across the junction are each equal to zero.

Consequently, denoting the absolute values of the electron and hole
currents reaching the junction A from the n-type region by J_-^n and J_+^n,
respectively (the currents from the p-type region will be denoted by J_-^p
and J_+^p), we find that at equilibrium (Fig. 207a)

$$-J_-^n + J_+^n + J_-^p - J_+^p = 0, \tag{69.1}$$

$$-J_-^n + J_-^p = 0, \tag{69.2}$$

$$J_+^n - J_+^p = 0. \tag{69.3}$$

We shall introduce the following notation for the equilibrium values of
these currents:

$$J_-^n = J_-^p = J_{ns}, \tag{69.4}$$

$$J_+^n = J_+^p = J_{ps}. \tag{69.5}$$

We shall also assume that one of the semiconductors, for example,
the n-type, is illuminated. This generates electron-hole pairs. In the
first approximation, we can neglect the change in the electron density due
to illumination because the nonequilibrium electrons rapidly assume the
distribution corresponding to the sample temperature and only a negligible
fraction of these electrons is capable of reaching the p-type region after

overcoming the potential barrier of the junction. Thus, the effect of illumination appears mainly as an increase in the density of minority carriers – in our case, holes – which do not have to overcome a potential barrier. Consequently, the hole current reaching the junction A increases. We shall denote this hole current due to illumination by J_f. *

The current J_f disturbs the equilibrium by charging the p-type region positively with respect to the n-type region and, consequently, reducing the electron energy levels in the p-type region (Fig. 207b). The Fermi levels, which are related directly to the densities of carriers generated by thermal motion, are no longer at the same energy positions in the two regions of the photodiode. The gap between the Fermi levels is $e\varphi$, where φ is the potential difference due to illumination. The displacement of the energy levels during illumination raises the equilibrium electron current flowing from the n- to the p-type region and the equilibrium hole current in the opposite direction. Figure 207b shows that such a displacement increases the number of electrons in the n-type region (and the number of holes in the p-type region) whose energies are sufficient to overcome the potential barrier at the junction.† Thus, with increase in the energy gap between the Fermi levels, the "charging" effect of the photocurrent is compensated more and more by a corresponding increase of the reverse majority-carrier "thermal" current.

Under steady-state conditions, the total current through a p-n junction is zero. To determine the corresponding barrier photo-emf, we can write

$$J_f - J_-^n + J_+^n + J_-^p - J_+^p = 0, \tag{69.6}$$

where J_-^n, J_+^n, and J_-^p, J_+^p are the equilibrium carrier currents during illumination. It is easily seen from Fig. 207b that J_+^n and J_-^p are unaltered by the illumination:

$$J_+^n = J_{ps},$$

$$J_-^p = J_{ns}. \tag{69.7}$$

On the other hand, the currents J_-^n and J_+^p are altered by the illumination, due to the displacement of the energy levels and change in the potential barriers, becoming:

$$J_-^n = J_{ns} e^{\frac{e\varphi}{kT}}, \tag{69.8}$$

---

* Here and later, we shall use the generally accepted notation of photodiode characteristics.
† The smaller the energy gap between the Fermi level and the top of the energy barrier, the greater this number.

$$J_+^p = J_{ps} e^{\frac{e\varphi}{kT}}.$$ (69.9)

Substituting Eqs. (69.7)-(69.9) into Eq. (69.6), we obtain

$$J_f - J_{ns}\left(e^{\frac{e\varphi}{kT}} - 1\right) - J_{ps}\left(e^{\frac{e\varphi}{kT}} - 1\right) = 0,$$ (69.10)

or using the notation

$$J_{ns} + J_{ps} = J_s,$$ (69.11)

we find that

$$J_f - J_s\left(e^{\frac{e\varphi}{kT}} - 1\right) = 0.$$ (69.12)

Hence, the barrier photo-emf becomes [12, 14]

$$\varphi = \frac{kT}{e} \ln\left(\frac{J_f}{J_s} + 1\right).$$ (69.13)

In general, if the n- and p-type regions of the photodiode are closed by an external circuit, we should write J in place of 0 in the right-hand part of Eq. (69.12); J represents the branch current in accordance with Kirchhoff's law. We then obtain the general photodiode equation applicable under any conditions:

$$J_f - J_s\left(e^{\frac{e\varphi}{kT}} - 1\right) = J.$$ (69.14)

This equation may be written also in the form

$$\varphi = \frac{kT}{e} \ln\left(1 + \frac{J_f - J}{J_s}\right).$$ (69.15)

Here, as before, φ denotes the potential difference at the p-n junction [which, in general, is not equal to the photo-emf in an open circuit, given by Eq. (69.13)]. If the external circuit of the photocell comprises simply a resistance R, then J = φ/R and Eq. (69.14) may be rewritten in the form

$$J_f - J_s\left(e^{\frac{e\varphi}{kT}} - 1\right) = \frac{\varphi}{R}.$$ (69.16)

The expression (69.16) is the general equation for a photodiode working under barrier conditions. It may be analyzed in the way usually employed in discussing the operation of barrier photocells (see below).

The photocurrent J_f is equal to the sum of the current through the external resistance, J = φ/R, and the "leakage" current through the barrier layer, $J_y = J_s\left(e^{\frac{e\varphi}{kT}} - 1\right)$:

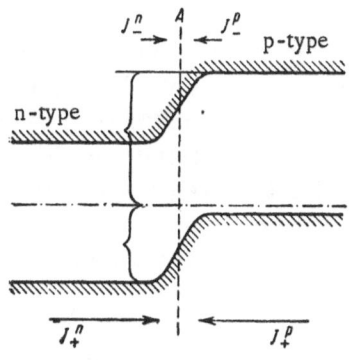

Fig. 209. "Asymmetric" p-n junction.

$$J_f = J + J_y.$$

If the external resistance is very low (short-circuit conditions), $J \gg J_y$ and, consequently, $J \approx J_f$. If R is high (photo-emf regime), $J_y \approx J_f$.

In contrast to the older barrier photocells, photodiodes may be used successfully under high reverse voltages (photodiode regime). The expressions (69.14) and (69.15) are obviously applicable to this case. Thus, if, apart from the resistance R, the external circuit contains also a series-connected voltage source (a typical photodiode regime circuit is shown in Fig. 208b), the current in the external circuit, which should be substituted into the right-hand part of Eq. (69.14), is given by the expression

$$J = \frac{\varphi - V}{R}, \tag{69.17}$$

where V is the voltage from the source in the external circuit.

The fundamental equation for a photodiode working under photodiode conditions is written in the form

$$J_f - J_s \left(e^{\frac{e\varphi}{kT}} - 1 \right) = \frac{\varphi - V}{R}. \tag{69.18}$$

If V = 0, the above equation reduces to Eq. (69.16) and if R = ∞, it reduces to Eq. (69.12). Finally, if $J_f = 0$, Eq. (69.18) reduces to the well-known diode equation.

In conclusion, we note that the usual photodiodes have a strongly "asymmetric" p-n junction, i.e., the conductivities of the n- and p-type regions are very different.

It is easily seen that under these conditions the processes occurring at the junction will be governed by currents of either holes only (if $\sigma_p \ll \sigma_n$) or electrons only (if $\sigma_p \gg \sigma_n$). Figure 209 illustrates the contact at equilibrium in the case $\sigma_p \gg \sigma_n$; it shows that the separation of the Fermi level from the top of the barrier for holes is less than the corresponding separation for electrons. Consequently, the hole density and current through the junction A are greater than the electron density and current. Thus, the total current J_s is governed by the hole component.

The relationships obtained in the present section are obviously applicable to any (including asymmetric) p-n junctions. The relationships deduced from diode theory do not differ in their form from those obtained from the diffusion theory, which better represents the conditions in a p-n junction. However, the diode theory is easier to follow.

In the diffusion theory, one calculates the diffusion current of minority carriers flowing to the p-n junction. The density gradient which governs this current is found by solving Eq. (57.5) for the n- and p-type regions with the field E assumed to be zero. The boundary conditions for the densities at the boundaries between the n- and p-type regions and the space-charge region are found from the Boltzmann relationship.

The diffusion theory leads to the following expression for the current J_s:

$$J_s = J_{ps} + J_{ns} = \left(q p_n \frac{l_p}{\tau_p} + q n_p \frac{l_n}{\tau_n} \right) S. \qquad (69.19)$$

Here, p_n and n_p are, respectively, the densities of the minority carriers; l_n and l_p are the diffusion lengths in the n- and p-type regions, respectively; and S is the p-n junction area. The ratios l_p/τ_p and l_n/τ_n represent the velocities of motion (diffusion) of the minority carriers.

*   *   *

We should distinguish two principal uses of semiconductor photocells:

1) conversion of light into electrical signals;

2) conversion of light (for example, solar) energy into electrical energy.

The former use is very important in automation, measurement techniques, computers, etc.

The latter use has become important in connection with space flight and seems promising for terrestrial applications.

The following sections will deal with the main characteristics of photodiodes used as signal converters (analyzing particularly their kinetics), and briefly with the efficiency of photocells used as energy converters.

§70. PRINCIPAL CHARACTERISTICS OF p-n JUNCTION PHOTOCELLS [13]

The experimental results about to be given were obtained mainly for germanium photodiodes, now mass-produced.

Figures 210 and 211, which display some photodiode characteristics under the barrier regime, show that with increase in the resistance R the potential difference φ across the photocell also increases, tending, as R → ∞, towards a value given by Eq. (69.13), while the external-circuit current J decreases from J_f when R = 0 and $\varphi = 0$ [cf. Eq. (69.14)] to zero.

Figure 212 shows that for a given illumination intensity the low currents correspond to the maximum value of the photo-emf.

We shall now consider the operation of photocells under the photodiode regime.

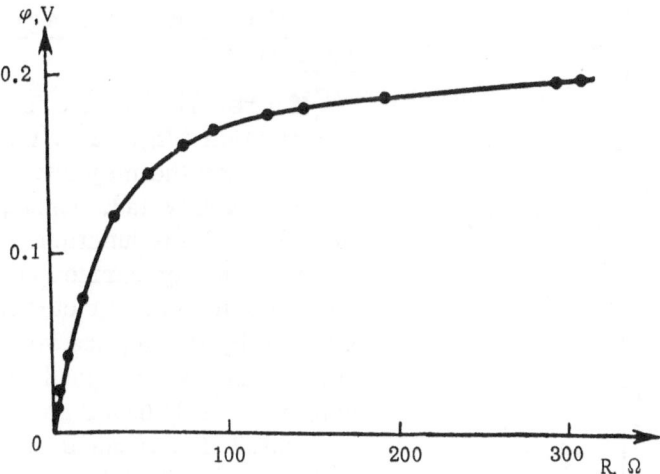

Fig. 210. Load characteristic of a p-n junction photocell under barrier conditions: $\varphi = f(R)$.

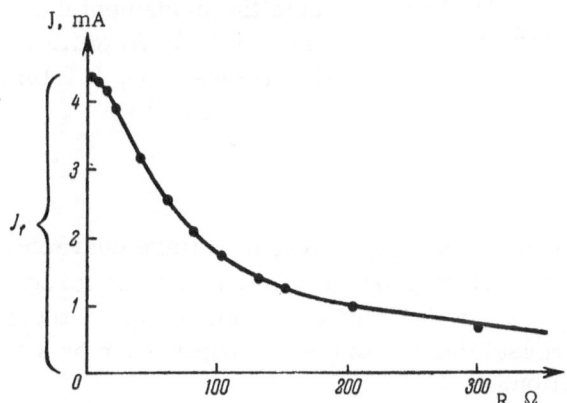

Fig. 211. Barrier conditions: $J = f(R)$.

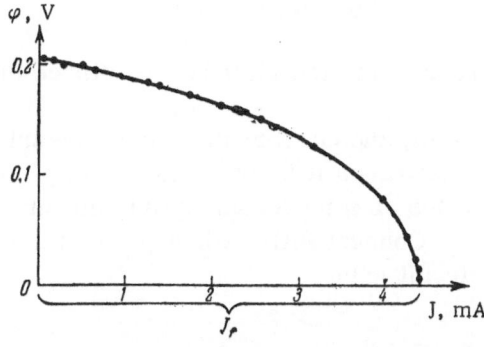

Fig. 212. Current-voltage characteristic of a p-n junction photocell under barrier conditions.

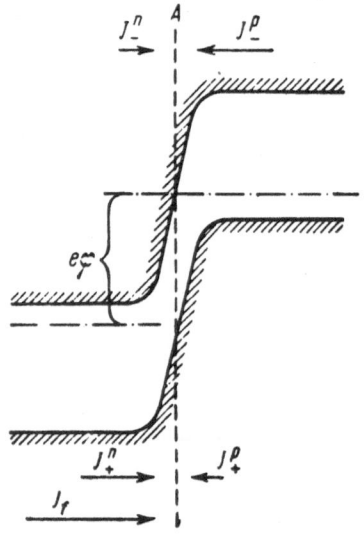

Fig. 213. p-n junction photocell working in the reverse direction (photodiode regime).

A. Current-Voltage Char-acteristic

If a reverse voltage is applied to a p-n junction (Fig. 213), the potential barriers for the majority carriers increase sharply and, consequently, the flow through the junction is governed by the minority-carrier currents. Since the minority-carrier currents are not affected by the applied voltage, * the current is also independent of this voltage, i.e., it behaves as a "saturation current." In darkness, the current is J_S but during illumination it increases to J_f because of the appearance of additional minority carriers. This follows from the fundamental photodiode equation (69.14). At sufficiently high negative values of φ, it follows from Eq. (69.14) that

$$J = J_s + J_f. \tag{70.1}$$

The experimentally obtained current-voltage characteristics of germanium photodiodes are in good agreement with the foregoing conclusions (Fig. 214). However, in the region of high voltages, there is a deviation (current increase) due to carrier multiplication by strong electric fields in the junction.

At low voltages ($\varphi \ll kT/e$), we find, after expanding in a series the expression in the parentheses in Eq. (69.14),

$$J = J_f + J_s \frac{e\varphi}{kT} = J_f + \frac{\varphi}{R_0}, \tag{70.2}$$

where $R_0 = kT/eJ_S$ represents the internal resistance of the junction at low values of φ.

In darkness ($J_f = 0$), the current increases linearly with φ until $\varphi \approx kT/e$ is reached, and then it tends to saturate. During illumination, the current for any value of φ increases by an amount J_f compared with its value in darkness. Consequently, when $\varphi = 0$, the current is J_f, which is the short-circuit value.

---

* There is no potential barrier for the minority carriers and the current is proportional to their density, which is governed by the rates of thermal and optical generation.

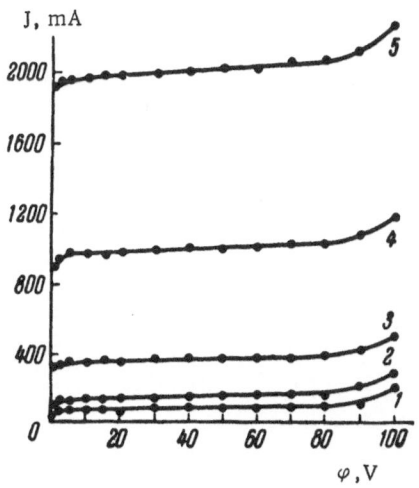

Fig. 214. Current-voltage characteristics in the photodiode regime: 1) In darkness; 2-5) at increasing illumination intensities.

B. Lux-Ampere Characteristic

The short-circuit current under barrier conditions and the additional current under photodiode conditions are proportional to the density of the minority nonequilibrium carriers generated by illumination.

Under steady-state conditions, this density is usually proportional to the rate of generation βkI multiplied by the average lifetime of the minority carriers in the base.

This lifetime is limited by two annihilation processes: recombination and leakage through the p-n junction. The influence of recombination may be neglected* for well-constructed photodiodes with a sufficiently thin base because all the minority carriers are capable of leaking away through the junction in a time much shorter than τ. It is easily seen that under these conditions the current J_f is equal to the product of the electronic charge and the total number of carriers generated by light per unit time in the total volume V:

$$J_f = e\beta k \int_V I dV. \tag{70.3}$$

Thus the value of J_f is directly proportional to I for any excitation level. This conclusion is in good agreement with experiment (Fig. 215).

Under the photodiode regime, the relationship between J_f and I is strictly linear up to very high illumination intensities. This is an important advantage of photodiodes. Under the barrier regime and short-circuit conditions, the initial part of the lux-ampere characteristic is also linear but at considerable illumination intensities (this limit will be different for different samples) there is a very pronounced departure from linearity. This is because the finite value of the resistance of germanium, its contacts and external circuit begins to play an important part at high illuminations, gives rise to a departure from the short-circuit conditions.

It follows from Eq. (69.13) that in an open circuit, the photo-emf depends logarithmically on J_f and the illumination intensity.

_____

*Otherwise, one could observe, in principle, a weak dependence of the photocurrent on the illumination intensity due to the changes in τ and l_D on transition from low to high excitation levels.

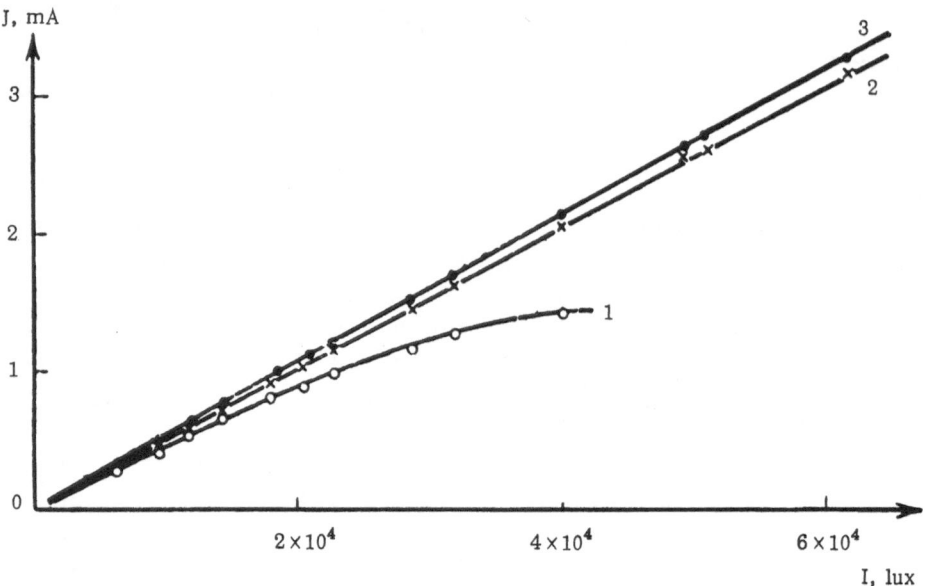

Fig. 215.   Typical lux-ampere characteristics of germanium photodiodes: 1) Barrier regime;
2, 3) photodiode regime for different voltages.

C.   Photosensitivity Spectrum

The spectral response curves of two germanium photodiodes are
given in Fig. 216. The long-wavelength edge of the response corresponds
to the fundamental absorption edge. The spectral responses of different
samples differ a little, particularly at wavelengths shorter than 1 μ. However, we can assume approximately that the response extrapolated toward
short wavelengths passes through the origin in Fig. 216. The curves of
Fig. 216 were reduced to unit incident energy and therefore we see that
the number of carriers generated in germanium is not proportional to
the photon energy but to the number of photons. To carry out rough calculations, we can represent the response spectrum in the form of a tri-
angle with its vertex at 1.5 μ and a base extending from 0 to 2 μ (dashed
lines in Fig. 216). *

D.   Temperature Dependence of the Dark Current,
Photocurrent, and Photo-emf

The photocurrent J_f in a well-constructed photodiode is governed by
the total number of absorbed quanta and, if this number is constant, the
photocurrent should be independent of temperature.

---

* In fact, as pointed out in Chap. IV, for quanta of ~2.3 eV and at shorter wavelengths the
reduction in sensitivity stops due to the effects of impact ionization.

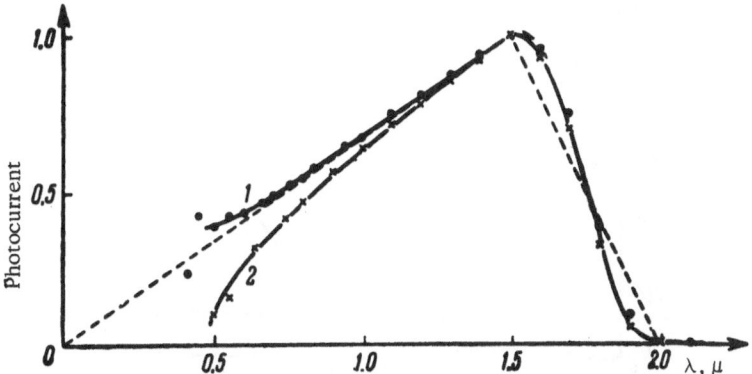

Fig. 216. Photocurrent spectrum for two photodiodes.

The experimental data presented in Fig. 217 confirm this in principle. The weak rise of J_f with temperature is obviously governed by the reduction of the forbidden bandwidth, which means that if white light is used, the number of quanta in the spectrum capable of generating minority carriers is increased. The weak temperature dependence of the photocurrent is an important advantage of photodiodes. A disadvantage is the very strong temperature dependence of their dark current J_S (Figs. 217a and 217b). This is due to the strong temperature dependence of the minority-carrier density. The majority-carrier density (for example, electron density) is constant because in the usual working range of temperatures the impurities are completely ionized, and, therefore, the expression

$$n_0 p_0 = n_i^2 = N_c P_v e^{-\frac{\Delta \mathscr{E}}{kT}} \tag{70.4}$$

gives for the minority-carrier density

$$p_0 = \frac{1}{n_0} N_c P_v e^{-\frac{\Delta \mathscr{E}}{kT}} \sim e^{-\frac{\Delta \mathscr{E}}{kT}}. \tag{70.5}$$

Thus p_0 and, consequently, J_S rise rapidly (approximately exponentially) with temperature and the total forbidden bandwidth occurs in the power index of the exponential function.

From this discussion, it is clear that to reduce the dark current we must either lower the temperature or use a material with a wide forbidden band.

Let us examine approximately the temperature dependence of the open-circuit emf at a given illumination intensity (J_f).

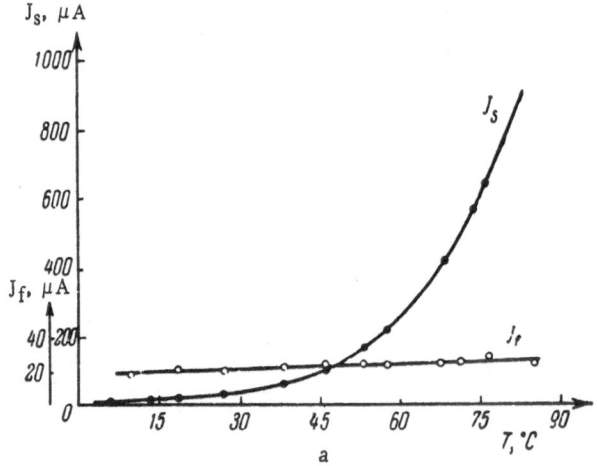

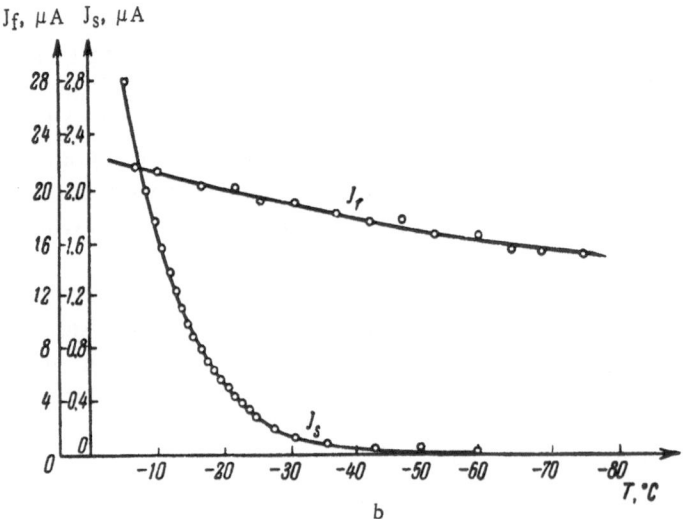

Fig. 217. Temperature dependence of the saturation current J_s and photocurrent J_f of a germanium photodiode: a) Above room temperature; b) below room temperature.

Substituting into Eq. (69.13) $J_s = Ae^{-\frac{\Delta\mathscr{E}}{kT}}$ [cf. Eq. (70.5)], we find that, at high temperatures when $J_f / J_s \ll 1$,

$$\varphi = \frac{kT}{e} \frac{J_f}{A} e^{\frac{\Delta\mathscr{E}}{kT}} . \tag{70.6}$$

It follows that φ increases approximately exponentially on cooling. This is because J_s decreases on cooling and to retain the compensation of the photocurrent J_f, the potential barrier must decrease considerably, i.e., φ must increase.

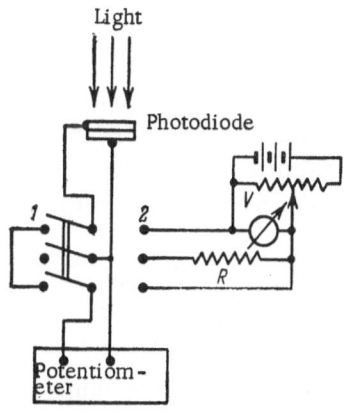

Fig. 218. Circuit for checking the principal relationships for a photo-diode.

At sufficiently low temperatures, when $J_f / J_S \gg 1$, the continuing rise in φ is greatly retarded.

E. Experimental Verification of the Relationship for the Barrier Photo-emf

The foregoing comparison of the experimental photodiode characteristics with those predicted theoretically has shown that they are in qualitative agreement.

To carry out a quantitative check, one usually employs the relationship (69.13). If φ, J_f, and J_S are measured at different illumination intensities and the results are plotted so that one axis gives φ and the other $\ln[(J_f/J_S)+1]$, then, according to Eq. (69.13), the points should lie on a straight line which passes through the origin of the coordinates and whose slope is kT/e, i.e., ~0.026 V at room temperature. Therefore the measurement of this slope gives the required check of Eq. (69.13) and of the general assumptions of the photodiode theory.

A typical circuit for measuring φ, J_f, and J_S is given in Fig. 218. With the switch in position 1, the barrier photo-emf is measured with a potentiometer; in position 2, the photodiode is connected in series with a variable voltage source V and resistance R.

The voltage applied in the reverse direction is selected so that it is sufficient to ensure saturation in the photodiode circuit, i.e., so that the term $\exp(e\varphi/kT)$ in Eq. (69.14) becomes negligibly small compared with unity. The current in the external circuit in the absence of illumination is then J_S and during illumination it is $J_S + J_f$. Consequently, if the voltage drop across a known resistance R is measured with a potentiometer, we can find both J_S and J_f. In practice, J_S and J_f are found by recording separately the current-voltage characteristics at low voltages in darkness and during illumination, the values of J_S and $J_S + J_f$ being found by extrapolation of the saturation region to V = 0.

Figure 219 shows the dependence of φ on $\ln[(J_f/J_S) + 1]$ at room temperature for two germanium photodiodes and Fig. 220 presents the same information for the temperature of liquid nitrogen. The experimental points fit straight lines satisfactorily. The numbers by these lines indicate their slopes, which are quite close to the value kT/e predicted theoretically.

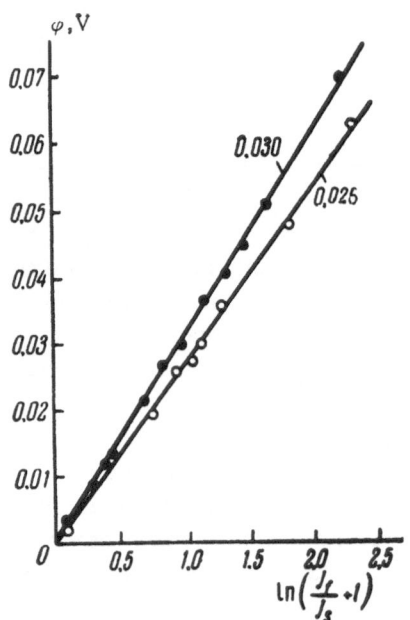

Fig. 219. Results of checking Eq. (69.13).
Room temperature.

It should be pointed out, however, that the experimental data for photodiodes made of other materials (silicon, gallium arsenide) differ considerably from the predictions of the foregoing theory. *

§71. PHOTODIODE KINETICS

One of the principal advantages of photodiodes is their relatively fast response. However, the "time constant" and the nature of the relaxation processes in a photodiode depend considerably on the operational regime. This is illustrated by the oscillograms in Fig. 221, which were obtained during the illumination of a germanium photodiode with square light pulses. Oscillogram a represents the photodiode regime, i.e., the operation of a photodiode under a voltage applied in the reverse direction (Fig. 208b); oscillogram b corresponds to the barrier regime with a large resistance in the external circuit (Fig. 208c) (i.e., the photo-emf regime). These oscillograms show that the effective time constant in the latter case is considerably greater than in the former, i.e., the barrier photo-emf regime shows more inertia than the photodiode regime. Moreover, in the barrier-regime case, the photo-emf decay is considerably slower than the photo-emf rise.

Sometimes, the relaxation curves in the photodiode-regime case are more complex (Fig. 221c).

Curves with kinks similar to that shown in Fig. 221c are sometimes found also in the barrier-regime case. We shall give below a qualitative treatment which accounts for the difference between the relaxation curves in the photodiode and barrier cases and which establishes the cause of, and the conditions for, the appearance of complex relaxation curves.

A.   Photodiode Regime

The usual structure of germanium photodiodes, which will be used to discuss the kinetics, is shown in Fig. 208a. The nonequilibrium minor-

---

* This is due to the neglecting of recombination in the space-charge region.

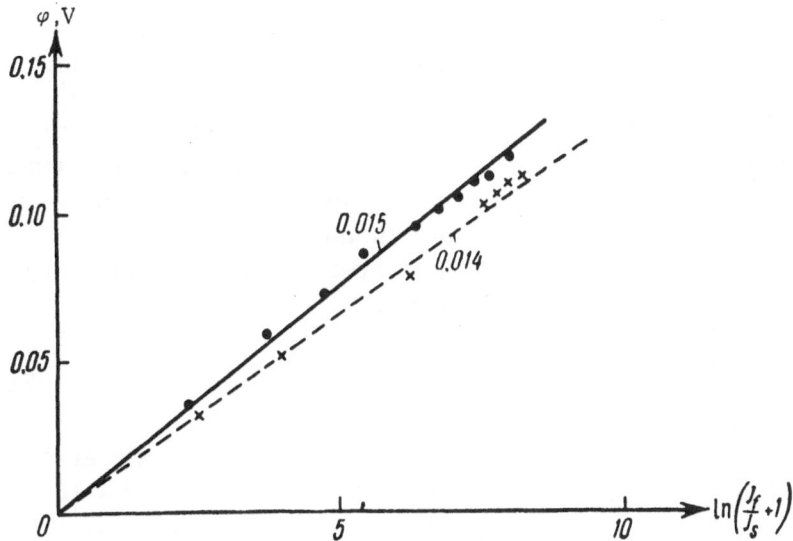

Fig. 220. Results of checking Eq. (69.13). Liquid-nitrogen temperature.

ity carriers, generated by light at the illuminated surface of n-type germanium, diffuse to the p-n junction, covering a distance w, equal to the n-type region thickness. The carriers which reach the junction are driven by the junction field into the p-type region. In the photodiode-regime case, the kinetics should be governed by the "transit" time t_0 of the nonequilibrium minority carriers from the illuminated surface to the junction. Under these conditions, i.e., with a large voltage φ applied to the junction in the reverse direction, the measured current in the external circuit is equal to the sum of the saturation current in darkness J_s and the short-circuit photocurrent J_f

$$J = J_s + J_f. \tag{71.1}$$

The relaxation of the current in the external circuit is governed by the relaxation of the photocurrent J_f, the value of which is, at every moment of time, proportional to the density of the nonequilibrium minority carriers at the p-n junction. The annihilation of these carriers (for example, when the illumination stops) in well-constructed photodiodes, for which the diffusion length is $l_D \gg w$, is governed mainly not by the recombination but by the carrier leakage across the junction. Consequently, the time constant which determines the relaxation process is not the lifetime τ but the "collection" or "transit" time t_0 of the minority carriers moving toward the p-n junction.* This time can be found approximately

---

* In general, when t_0 and τ are comparable, the time constant is a complex function of both these quantities.

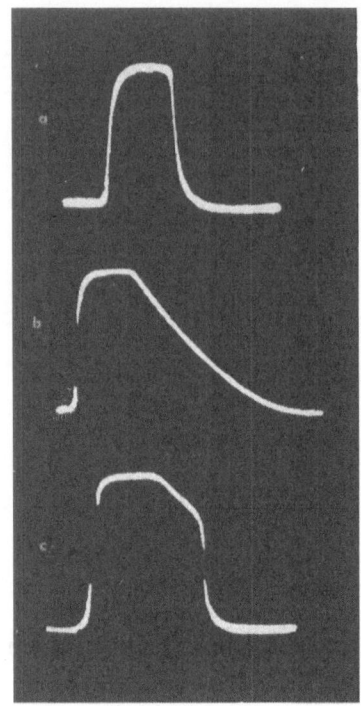

Fig. 221. Relaxation curves of a germanium photodiode [18]. a) Photodiode regime; b) barrier photo-emf regime; c) "hybrid" regime.

from the relationship which gives the path l travelled by the diffusing carriers during a time t: $l = \sqrt{Dt}$.

In the fundamental absorption region, the carriers are generated mainly near the illuminated surface, i.e., at a distance w from the p-n junction. To travel a distance w, they require a time

$$t_0 \approx \frac{w^2}{D}. \qquad (71.2)$$

For base thicknesses used in mass-produced photodiodes $(\sim 2 \times 10^{-2}$ cm), we find, assuming that $D = 49$ cm^2/sec for holes, that $t_0 \sim 10^{-5}$ sec. The same values of the relaxation time constant for the photodiode regime have been found experimentally.

It is obvious that in the barrier regime, under short-circuit conditions, when the measured current in the external circuit is $J = J_f$ and the relaxation is determined by the relaxation of J_f, the time constant has the same value t_0.

The photodiode regime circuit (Fig. 208b) usually includes a load resistance R as well as the capacitance C of the junction and the leads. Consequently, the relaxation may be governed also by the quantity RC.

To determine the nature of the relaxation curve in a real circuit, we can, in principle, use the general photodiode equation

$$J_f - J_s\left(e^{\frac{e\varphi}{kT}} - 1\right) = J. \qquad (71.3)$$

In the transient case, the current J consists of a current through the resistance R: $J_R = (\varphi - V_0)/R$ (where V_0 is the voltage from an external source), and a current which charges the capacitance C, $J_C = C(d\varphi/dt)$,

$$J_f - J_s\left(e^{\frac{e\varphi}{kT}} - 1\right) = \frac{\varphi - V_0}{R} + C\frac{d\varphi}{dt}. \qquad (71.4)$$

In the photodiode case (high negative φ), we can neglect the term $\exp(e\varphi/kT)$ compared with unity and, therefore,

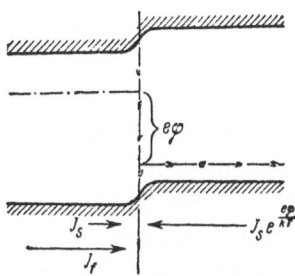

Fig. 222. Band structure of an "asymmetric" p-n junction during illumination of the n-type region.

$$\frac{d\varphi}{dt} + \frac{\varphi}{RC} = \frac{V_0 + J_s R}{RC} + \frac{1}{C} J_f. \qquad (71.5)$$

To solve Eq. (71.5), we must know the dependence of J_f on time. For simplicity, we shall assume that its dependence is exponential (the exact dependence is calculated in [15]) and that the time constant is t_0. In that case, the rise of J_f from the moment of starting illumination obeys the law $J_f = J_f^{st} [1 - \exp(-t/t_0)]$, where J_f^{st} is the steady-state value of J_f at a given illumination intensity, and the decay after illumination is given by the law $J_f = J_f^{st} \exp(-t/t_0)$.

The solution of Eq. (71.5) for the decay curve may be given by

$$\Delta\varphi = \Delta\varphi_{st} \left(1 - \frac{t_0}{t_0 - RC}\right) e^{-\frac{t}{RC}} + \Delta\varphi_{st} \frac{t_0}{t_0 - RC} e^{-\frac{t}{t_0}}, \qquad (71.6)$$

where $\Delta\varphi = \varphi - V_0 - J_s R$ is the voltage drop across the p-n junction which is related to the illumination, i.e., the signal recorded during the modulated illumination of the photodiode. It follows from Eq. (71.6) that the relaxation is governed by the quantities t_0 and RC. * If $RC \ll t_0$, Eq. (71.6) transforms into

$$\Delta\varphi = \Delta\varphi_{st} e^{-\frac{t}{t_0}} \qquad (71.7)$$

and the relaxation time constant is given by the transit time t_0.

B. Barrier Photo-emf Regime

If the p-n junction is strongly asymmetric (for example, if the n-type germanium conductivity is less than that of p-type germanium, as in mass-produced germanium photodiodes), the band structure of n-type germanium during the illumination of an open-circuit photodiode (barrier photo-emf regime) is that shown in Fig. 222. [The electron current is small and we can neglect it (cf. § 69).]

During illumination in the steady-state case, the hole current from the n-type region to the p-type region, equal to the sum of the saturation current J_s and the steady-state short-circuit photocurrent J_f^{st}, is balanced by the reverse current from the p-type region $J_s \exp(e\varphi_{st}/kT)$:

_____

*In the case of a "linear" component of the reverse current, which is equivalent in the first approximation to the presence of a resistance R' shunting the p-n junction, R is understood to be the equivalent resistance of R' and of the load, connected in parallel.

$$J_f^{st} + J_s = J_s e^{\frac{e\varphi_{st}}{kT}}. \tag{71.8}$$

We shall consider the problem of relaxation in the barrier-regime case [18]. In contrast to the photodiode regime, where we have investigated the relaxation of the current in the external circuit [i.e., essentially the relaxation of J_f, cf. Eq. (71.1)], in the barrier photo-emf case we measure the emf value φ and, therefore, should analyze the relaxation of the photo-emf. We shall now determine the nature and sequence of the processes which govern the kinetics of φ in the barrier-regime case. Let us assume that at a time t = 0 the illumination stops and that the band structure during illumination is that given by Fig. 222. After illumination, the nonequilibrium hole density in the n-type region (which is proportional to the current J_f) begins to decrease because of the recombination.

This reduces the nonequilibrium minority carrier current J_f and this, in turn, disturbs the balance of the hole currents through the p-n junction. Due to the reduction in J_f, the hole current from the p-type region, $J_s \exp(e\varphi/kT)$, begins to exceed the hole current from the n-type region, $J_f + J_s$. Consequently, a positive charge moves from the p- to the n-type region* and the potential drop φ across the p-n junction decreases. This process (recombination hole current from the p- to n-type region and reduction in φ) continues until the photo-emf φ decays to zero.

It follows that the nonequilibrium hole density in the n-type region, which governs the current J_f, varies as a result of two processes: a) the recombination, and b) the hole motion from the n- to p-type region.

It can be shown (see below) that in many real cases under open-circuit conditions:

(1) we can neglect the influence of the hole motion from the n- to p-type region, therefore the nonequilibrium hole density (and the hole current J_f) decreases after illumination only because of recombination, i.e., we have a time constant τ in

$$J_f = J_f^{st} e^{-\frac{t}{\tau}}, \tag{71.9}$$

(2) the decay of J_f for which the time constant is τ, causes an almost simultaneous change in φ. Thus the time constant τ representing the recombination process is longer than the time constant for the establishment of the diffusion-drift equilibrium or the time constant for the discharge of the p-n junction capacitance.†

---

* Consequently, the p-n junction capacitance C is discharged.
† This constant, as will be shown below, is governed by the quantity $R_0 C$ where R_0 is the resistance of the p-n junction at low applied voltages.

The second condition means that the relaxation of φ occurs under conditions conserving the "quasi-steady state" when at any moment the value of the "slowly" varying current J_f corresponds to a "rapidly" established value of φ, which is related to J_f by the equation (69.13), as in the steady-state case.

On the other hand, it follows from the first condition mentioned above that J_f changes only because of recombination, so that the rise and decay curves of J_f are, respectively,

$$J_f = J_f^{st}\left(1 - e^{-\frac{t}{\tau}}\right),\qquad(71.10)$$

and

$$J_f = J_f^{st} e^{-\frac{t}{\tau}}.\qquad(71.11)$$

Substituting the above expressions for J_f in Eq. (69.13), we obtain the relaxation law for the barrier photo-emf:

during illumination

$$\varphi = \frac{kT}{e}\ln\left[\frac{J_f^{st}}{J_s}\left(1 - e^{-\frac{t}{\tau}}\right) + 1\right],\qquad(71.12)$$

after illumination

$$\varphi = \frac{kT}{e}\ln\left(\frac{J_f^{st}}{J_s}e^{-\frac{t}{\tau}} + 1\right).\qquad(71.13)$$

Figure 223 gives several theoretical curves plotted using Eqs. (71.12) and (71.13) for various ratios J_f^{st}/J_f. This figure shows that with increase in this ratio, i.e., with increase in the illumination intensity, the rise and decay curves differ more and more, the rise process becoming faster and the decay process slower. The difference between the rise and decay curves is not due to any essential change in the process itself (both rise and decay are governed by recombination with the same time constant τ). However, the experimentally measured value of φ is not related in the same way to τ in the rise and decay cases [cf. Eqs. (71.12) and (71.13)]. This is the reason for the considerable difference between the rise and decay curves.

Figure 224 shows oscillograms obtained for a germanium photodiode illuminated with square light pulses. These oscillograms were obtained for different illumination levels (represented by the ratio J_f^{st}/J_f) and agree well with the theoretical curves of Fig. 223. It is clearly seen that with increase in J_f^{st}/J_f the difference between the rise and decay rates of the photo-emf increases.

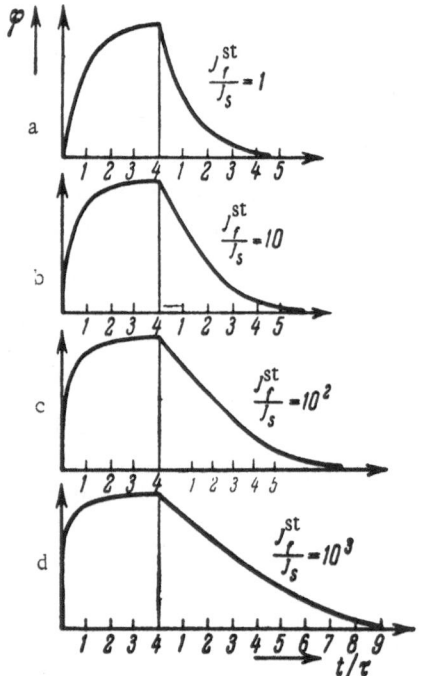

Fig. 223. Theoretical relaxation curves of the barrier photo-emf at various illumination levels.

It follows from Eqs. (71.12) and (71.13) that at low values of J_f ($J_f \ll J_s$), when the logarithm can be replaced by the first term in its expansion, φ is directly proportional to J_f and the relaxation curves are exponential with a time constant τ. Consequently, the experimental value of the lifetime may be found by investigating the relaxation of φ at low values of J_f / J_s. However, we can also use a more convenient method of determining τ from the photo-emf relaxation curves. It is evident from Eq. (71.13) that when $J_f / J_s \gg 1$, i.e., if the illumination is sufficiently intense,

$$\varphi = \frac{kT}{e} \ln \frac{J_f^{\mathrm{st}}}{J_s} - \frac{kT}{e\tau} t. \qquad (71.14)$$

The first term in Eq. (71.14) is a constant and, consequently, the time dependence of φ (in units of kT/e) is given by a straight line with a slope $1/\tau$. The rectilinear regions from which τ can be determined are clearly visible in the decay curves of Fig. 223 (when $J_f^{\mathrm{st}}/J_s \gg 1$) and Fig. 224.

Thus the "rectification" of the decay curves is automatic (without the use of a "taumeter") because the relationship between φ and J_f is logarithmic. The method described here is used to determine τ of photodiodes.

<u>Analysis of Assumed Conditions.</u> The foregoing treatment of the photo-emf kinetics is valid if the following three conditions are satisfied:

I. The p-n junction is "asymmetric," i.e., the conductivities of n- and p-type regions differ strongly so that we need consider only the hole or only the electron currents through the junction;

II. The relaxation of the nonequilibrium-carrier current J_f is of the recombination type, i.e., the density of the nonequilibrium minority carriers and the current J_f in an open-circuit photocell vary mainly because of the recombination;

III. "Quasi-steady-state" conditions obtain for the dependence of the photo-emf φ on J_f during relaxation, i.e., the time constant of the recombination process τ is much greater than the time constant for the discharge of the p-n junction capacitance.

The first of these conditions is usually ensured by the present methods of p-n junction manufacture.

To satisfy the second condition we must ensure that the change (during relaxation) in the number of nonequilibrium minority carriers (holes in n-type regions) is not affected by the motion of holes from the p-type region. It can be shown that this condition is equivalent to: $\tau \gg R_0 C$, where $R_0 = kT/eJ_s$ is the resistance of the p-n junction at low values of φ.

During illumination in the steady-state case, the total number of nonequilibrium holes in the n-type region is $J_f^{st}\tau/e$ and their charge is $J_f^{st}\tau$. After illumination, all these holes disappear. The change in the hole density in the n-type region after illumination may be due to two processes: the recombination and the motion across the p-n junction.

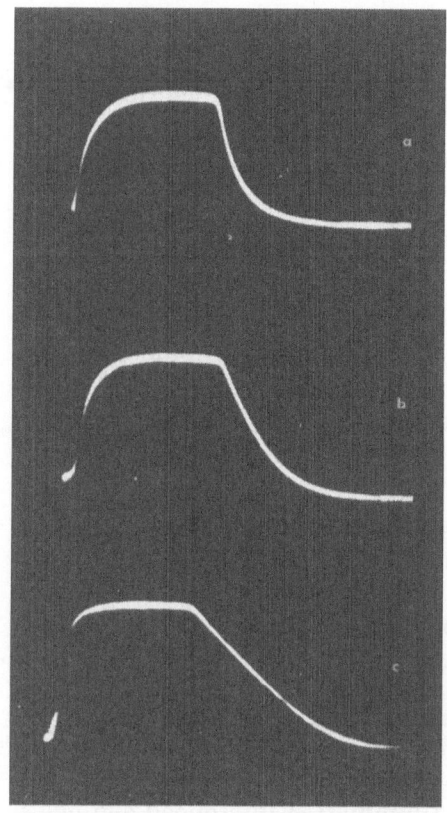

Fig. 224. Relaxation oscillograms of the barrier photo-emf at various illumination levels [18]: a) $J_f^{st}/J_s = 1$; b) $J_f^{st}/J_s = 10$; c) $J_f^{st}/J_s = 100$.

The total hole charge passing through the junction from the p- to n-type region due to the discharge of the junction capacitance C is $C\varphi_{st}$. If this charge is small compared with the total steady-state hole charge $J_f^{st}\tau$, it is clear that the change in the nonequilibrium hole density is mainly due to recombination. Consequently, condition II may be written in the following approximate form

$$J_f^{st}\tau \gg C\varphi_{st}. \tag{71.15}$$

Substituting into Eq. (71.15) the expression for φ_{st} from Eq. (69.13), we obtain

$$J_f^{st}\tau \gg C\,\frac{kT}{e}\ln\left(\frac{J_f^{st}}{J_s}+1\right). \tag{71.16}$$

To obtain a clearer criterion, we shall consider the case of small values of J_f^{st}, i.e., $J_f^{st} / J_S \ll 1$ (it is easily seen that the criterion for satisfying condition II then becomes even more rigorous). After expanding the logarithm as a series and using only the first term, we find that

$$\tau \gg C \frac{kT}{eJ_s},\qquad\qquad(71.17)$$

or

$$\tau \gg R_0 C.\qquad\qquad(71.18)$$

The above criterion is well satisfied by germanium photodiodes. For example, when $J_S \simeq 10^{-5}$ A, $C \simeq 10^{-10}$ F at room temperature ($kT/e \simeq 0.025$ V), we find that $R_0 = 2500$ Ω and $R_0 C = 2.5 \times 10^{-7}$ sec. Since $\tau \sim 10^{-5}$ sec for mass-produced devices, the criterion of Eq. (71.18) is well satisfied.

The criterion for condition III to be satisfied follows from the photodiode equation (71.3). In the barrier-emf case, the current J is the discharge (or charging) current of the p-n junction capacitance: $J = C(d\varphi/dt)$. We shall restrict ourselves to the case of weak emf's ($\varphi \ll kT/e$), when Eq. (71.3) can be written in the form

$$\frac{d\varphi}{dt} + \frac{\varphi}{R_0 C} = \frac{J_f}{C}.\qquad\qquad(71.19)$$

We shall consider the case when the illumination is stopped and assume that at t = 0, $\varphi = \varphi_{st}$, and $J_f = J_f^{st}$. In accordance with the condition II, J_f can then be written in the form

$$J_f = J_f^{st} e^{-\frac{t}{\tau}};\qquad\qquad(71.20)$$

substituting Eq. (71.20) into (71.19) and solving the latter for the initial conditions stated above, we obtain

$$\varphi = \varphi_{st}\left(1 - \frac{\tau}{\tau - R_0 C}\right) e^{-\frac{t}{R_0 C}} + \varphi_{st} \frac{\tau}{\tau - R_0 C} e^{-\frac{t}{\tau}}.\qquad\qquad(71.21)$$

It is clear that the quasi-steady-state condition, i.e., the condition for φ to follow changes in J_f without delay, is $R_0 C \ll \tau$. Then

$$\varphi = \varphi_{st} e^{-\frac{t}{\tau}}.\qquad\qquad(71.22)$$

It follows that the criteria for the fulfilment of conditions II and III are identical. We note that the condition $R_0 C \ll \tau$ is well satisfied by ger-

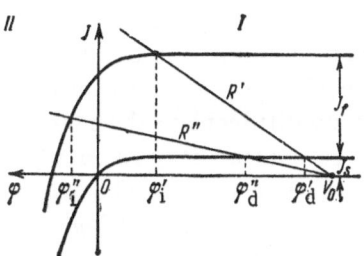

Fig. 225. Explanation of the "hybrid" relaxation.

manium photodiodes at room temperature but may not apply at low temperature because R_0 rises and τ decreases. It may not be satisfied so well by photocells made of other materials, for example, silicon. The kinetics in the barrier photo-emf case may be distorted strongly by a large "linear" component of the reverse current. *

The nature of the photocell kinetics in the case when the conditions just discussed are not satisfied is dealt with in § 72.

C. "Hybrid" Regime

In the case of a complex relaxation curve such as that shown in Fig. 221c, there are two quite distinct regions in the decay curve. When illumination is stopped, the signal decreases slowly at first and then it rapidly decays to zero. The "slow" initial region resembles the photo-emf decay in the barrier-regime case (Fig. 221b) and the subsequent "fast" region resembles the decay of the signal in the photodiode regime (Fig. 221a). As previously mentioned, this complex form of relaxation occurs under photodiode conditions, i.e., when a small reverse voltage V_0 is applied and the load resistance R is large (Fig. 208b).

In explaining the complex relaxation curve, we must bear in mind the strong difference between the relaxation rates under photodiode and barrier conditions (Fig. 221).

Figure 225 shows two current-voltage characteristics of a photodiode. One represents the conditions during illumination and the other the conditions in darkness. The portions of the curves in the quadrant I (the voltage is applied in the reverse direction) correspond to the photodiode regime, while those in the quadrant II correspond to the barrier regime. When the photodiode is connected in the usual way (Fig. 208b), the true voltage across it, φ, may be found from $\varphi = V_0 - JR$. The value of φ may be determined graphically by plotting the load curve, i.e., a straight line having a slope governed by the load resistance R and intersecting the horizontal axis at a point determined by the value of the externally applied voltage V_0. Figure 225 shows two such straight lines for two different load resistances R' and R" (R' < R"). The intersections of these lines with the current-voltage characteristics yield the steady-state voltages across the diode in darkness (φ_d) and during illumination (φ_i).

---

* In this case, the operational conditions apparently approach the "photodiode" regime.

Obviously, if under pulse illumination the voltage across the photodiode, which varies with time in accordance with a certain relaxation curve, changes from φ_d to φ_i during illumination and back to φ_d after illumination, the nature of the relaxation curves should be different in the rise and decay cases. For a small value of R', the relaxation of φ from φ_d' to φ_i' occurs under photodiode conditions and has, therefore, a short time constant, while for a large resistance R", the relaxation of φ''_d to φ''_i occurs partly under photodiode conditions (for the variation in φ from φ_d'' to 0) and partly under conditions of slow barrier-regime relaxation (when φ varies from φ_i'' to 0). Because of this, the relaxation curves become complex "hybrid," as shown in Fig. 221c and it follows that the inflection on the decay curve, where the "barrier" relaxation is replaced by "photodiode" relaxation, represents zero voltage across the photodiode.

The smaller the resistance R, the smaller should be the voltage V_0 to obtain the "hybrid" relaxation. The condition for the "hybrid" relaxation may be written in the form $V_0 - JR < 0$. However, the "hybrid" relaxation is very clearly observed at relatively low values of V_0 when the "width" of the photodiode region of the variation in φ does not exceed too much the "width" of the barrier region.

The "hybrid" relaxation should appear also as a kink on the rise curve. However, this kink is not so clear because the rise under the barrier regime is usually quite fast and does not differ greatly from that under the photodiode regime.

§72. BARRIER PHOTO-EMF KINETICS IN THE PRESENCE OF A LOAD FOR AN ARBITRARY RELATIONSHIP BETWEEN τ AND R_0C

In §71, we found the main features of the kinetics for the photodiode, barrier and "hybrid" regimes of photocell operation. However, an analysis of the barrier-regime case was carried out: 1) for an open circuit and 2) assuming that the capacitance current can be neglected. The first condition is very frequently violated in real photocell circuits. The second condition is equivalent to the inequality

$$\tau \gg R_0C \tag{72.1}$$

and is also often violated.

Because of this, there is great interest in the kinetics in the more general case of an arbitrary relationship between τ and R_0C and any value of the load R [23].

The emf relaxation process can be analyzed clearly in two limiting cases: (a) R = ∞ and (b) R finite but the capacitance current J_C much less than the resistance current J_R. We shall consider only the photo-emf de-

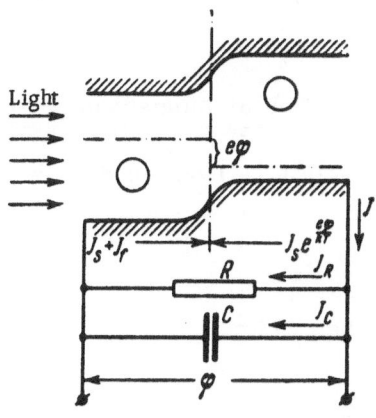

Fig. 226. Distribution of current in the photodiode circuit. The chain lines denote the Fermi levels which determine the equilibrium density.

cay curves in these two extreme cases. In case (a) when R = ∞, the relaxation is governed by the following processes. Immediately after cessation of illumination, the hole charge P begins to decay due to recombination. Consequently, the current J_f decreases, i.e., the equilibrium between the current $J_f + J_S$ (Fig. 226) and the reverse current $J_S \exp(e\varphi/kT)$ is disturbed. The difference between these currents is equal to the capacitance discharge current J_C. Thus the capacitance discharges and φ decreases. However, the kinetics will be governed not only by recombination and lifetime but also by the current because the number of nonequilibrium holes changes due to charging of the capacitance. The latter state of affairs brings holes into the n-type region and this disturbs the exponential decay law for holes, $P = P_0 \exp(-t/\tau)$.

Moreover, the decay time of φ is affected by the lag governed by the time constant of the discharge circuit $R_{in}C$; here, R_{in} is the internal differential resistance of the p-n junction, which may be calculated from the general photodiode equation (69.15):

$$\varphi = \frac{kT}{e} \ln\left(\frac{J_f - J}{J_s} + 1\right). \tag{72.2}$$

Since, in our case $J = J_C$, we find from Eq. (72.2) that

$$R_{in} = -\frac{d\varphi}{dJ} = \frac{R_0}{\dfrac{J_f - J}{J_s} + 1}. \tag{72.3}$$

However, we have shown in §71 that for τ greater than the maximum value of $R_{in}C$, which is equal to R_0C, we can neglect the capacitance current. Therefore, the law of decay for holes is exponential, φ follows the changes in P without delay and the relaxation is governed by the quantity τ.

In the opposite case, when $\tau \ll R_{in}C$, we must bear in mind that the charge P is replenished by holes entering the n-type region during the discharge of the capacitance. The process is then governed not by τ but by $R_{in}C$ because the latter is larger. The recombination of holes is "instantaneous" ($\tau \ll R_{in}C$) when they enter the n-type region and, consequently, J_f may be neglected and the capacitance discharge current may be written (Fig. 226)

$$C\frac{d\varphi}{dt} = -J_s\left(e^{\frac{e\varphi}{kT}} - 1\right).$$

(72.4)

In this case, the rate of relaxation at any particular moment is governed by the value of $R_{in}C$.

Substituting in Eq. (72.3) the value of $[(J_f - J)/J_s] + 1$ obtained from the general photodiode equation (72.2), we find

$$R_{in} = R_0 e^{-\Phi},$$

(72.5)

where $\Phi = e\varphi/kT$. Using Eq. (72.4), we can then write

$$\frac{d\Phi}{dt} = -\frac{e^{\Phi} - 1}{R_0 C} = \frac{1}{R_0 C} - \frac{1}{R_{in}C}.$$

(72.6)

Thus, depending on the relationship between τ and $R_{in}C$, we can clearly distinguish two limiting cases, in one of which ($\tau \gg R_0 C$) the decay of φ is governed only by τ, and in the other ($R_{in}C \gg \tau$) only by $R_{in}C$.

It is evident from Eq. (72.5) that the differential resistance R_{in} increases with decrease in φ. Hence, we can have the case when $\tau > R_{in}C$ in the initial stage, and $\tau < R_{in}C$ in the final stage of relaxation. We would then expect the relaxation law to change more or less suddenly at a time corresponding to $\tau \approx R_{in}C$ (cf. the oscillograms in Fig. 230).

In the other extreme case, referred to earlier in this section as b, we have $J_R \gg J_C$, i.e., the load resistance is low. Here again, we can predict the form of the decay curve.

When the illumination is stopped, the hole charge P decreases not only due to the recombination but also due to the loss of holes to the external circuit. Depending on the value of P, its variation may be governed either by the recombination or by the loss to the external circuit. Let us consider the decay process. When P decreases, the rate of recombination decreases more rapidly than the rate of loss of holes to the external circuit. This is because the recombination rate is proportional to P, but the loss of carriers to the external circuit, representing the current $J_R = \varphi/R$, varies as $\ln P$. Therefore, at high values of P and, consequently, of φ (the initial portion of the decay curve at a sufficient illumination intensity), the decay of P is governed by the recombination (as in the case $R = \infty$, $\tau \gg R_0 C$). At lower values of P and φ, the loss of carriers to the external circuit becomes dominant and this makes φ decay more rapidly than under open-circuit conditions. Consequently, a kink appears in the relaxation curve (cf. the corresponding oscillogram in Fig. 229). Obviously, the kink will appear earlier if the load resistance R is low.

An approximate quantitative solution of the problem confirms the qualitative conclusions (Figs. 227, 228).

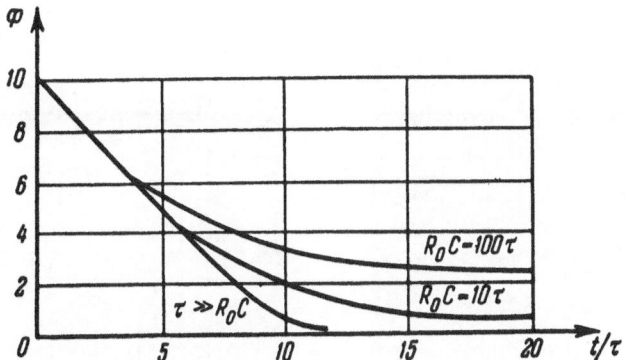

Fig. 227. Photo-emf relaxation curves in the open-circuit
case (calculations).

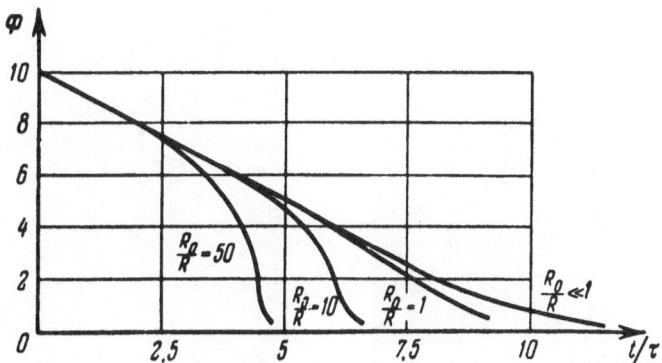

Fig. 228. Photo-emf relaxation curves in the presence of a
load (calculations).

The curves of Fig. 227 show clearly two relaxation regions: a linear region followed by slower decay. The linear decay $\Phi = \Phi_0 - t/\tau$ occurs for values of Φ for which $e^\Phi \gg 1$ and $\tau \gg e^{-\Phi}R_0C$ (i.e., $\tau \gg R_{in}C$), in full agreement with the qualitative treatment given earlier. In that region where $\tau \ll e^{-\Phi}R_0C$ the decay is governed by capacitance processes.

Thus, if the intensity of illumination is sufficiently strong, even if the inequality $\tau \gg R_0C$ is not obeyed, the relaxation curve still has a region which can be represented by τ. The moment of transition to slower decay corresponds to $e^\Phi \simeq R_0C/\tau$.

The curves in Fig. 228, obtained for different loads (but with the condition $\tau \gg R_0C$), have a kink which appears earlier and is more pronounced if R_0/R is large. It follows from the same figure that the open-circuit conditions are realized in practice if $R \gtrsim R_0$. It is also clear that if the value of Φ is sufficiently large, we can always find an initial linear relaxation region which can be used to determine the lifetime. This may be of practical importance in those cases when it is desirable to deter-

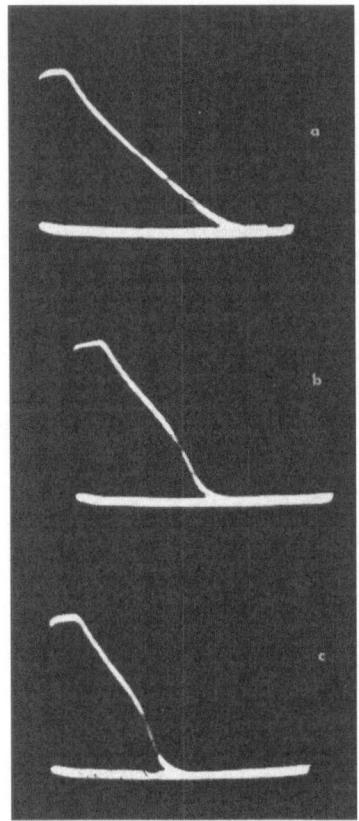

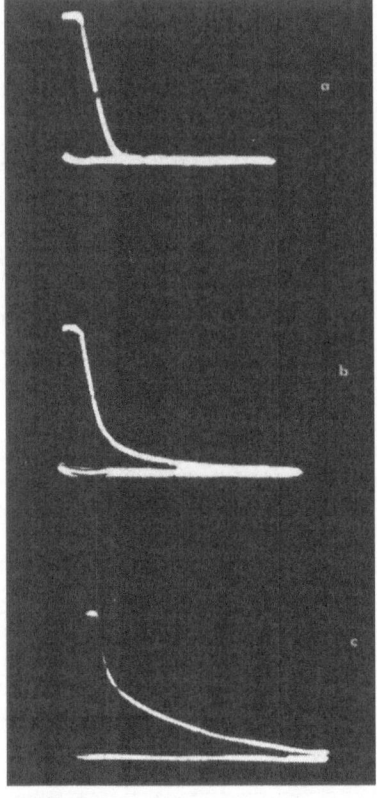

Fig. 229. Relaxation curves for various
loads [23]: a) $R_0/R \ll 1$; b) $R_0/R = 10$;
c) $R_0/R = 20$. Signal amplitudes was
$\Phi = \varphi/kT = 9$.

Fig. 230. Relaxation curves at various
temperatures [23]: a) $T = 300°K$,
$R_0C \ll \tau$; b) $T = 250°K$, $R_0C \approx 10\tau$;
c) $T = 220°K$, $R_0C \approx 10^3 \tau$.

mine the lifetime in devices containing a p-n junction with a considerable
linear component of the current.

The experimental results confirm satisfactorily the calculated results. The series of oscillograms in Fig. 229 illustrates the variation
of the relaxation curve with the load resistance (cf. Fig. 229). It is clear
that with a reduction in R the kink in the relaxation curves appears earlier and is more pronounced, and that beyond the kink the relaxation is
determined mainly by the loss of carriers through the resistance R.

Figure 230 presents a series of oscillograms for a single photodiode
at several temperatures. On cooling, the value of R_0 and, consequently,
the ratio R_0C/τ rise strongly, which slows down the decay in the final
stages. The slopes of the initial stages are also affected (in the oscillograms of Fig. 230, where the time scale is much compressed, this change
cannot be seen clearly).

In conclusion, we shall quote a convenient approximate formula which describes the relaxation process at low values of φ ($\varphi \ll kT/e$) and any relationship between τ and R_0C, as well as between R and R_0:

$$\varphi = \varphi_0 e^{-\frac{t}{\tau}\frac{1+R_0/R}{1+R_0C/\tau}}. \tag{72.7}$$

The expressions for the limiting cases can be deduced from Eq. (72.7):

1) $R = \infty$

a) $\tau \gg R_0C$, $\varphi = \varphi_0 e^{-\frac{t}{\tau}}$ is an exponential expression with time constant τ;

b) $\tau \ll R_0C$, $\varphi = \varphi_0 e^{-\frac{t}{R_0C}}$ – the decay is governed by the parameter R_0C.

2) $R \neq \infty$

a) $\tau \gg R_0C$, $\varphi = \varphi_0 e^{-\frac{t}{\tau}(1+R_0/R)}$ i.e., the decay is faster by a factor $\exp(-tR_0/\tau R)$ compared with the $R = \infty$ case;

b) $\tau \ll R_0C$, $\varphi = \varphi_0 e^{-t\left(\frac{1}{R_0C}+\frac{1}{RC}\right)}$ – the decay is governed by the quantity R_0C, where $1/R_e = [(1/R_0) + (1/R)]$.

The experimental data for germanium photodiodes are in good agreement with Eq. (72.7).

§ 73. INFLUENCE OF TRAPPING ON THE RELAXATION OF PHOTOCURRENT IN A PHOTODIODE

The current J_f is governed by the minority carrier density. Consequently, the trapping of the minority carriers should give rise to the same features in the kinetics of J_f as in the case of the photoconductivity kinetics, i.e., for example, the long "tails" might appear in the relaxation curve. However, experiments on germanium and silicon showed that the influence of trapping on the photodiode current is much smaller than its influence on the photoconductivity [39, 40]. Figure 231 illustrates the circuit for the simultaneous observation of the relaxation of the photoconductivity and of the photodiode current. The oscillograms of Fig. 232 show that the trapping, which strongly affects the photoconductivity relaxation, has practically no effect on the relaxation of J_f. This result may be explained as follows.

We shall consider the nature of the relaxation after the illumination is stopped. We shall assume that β-type trapping of the minority carriers (holes) occurs. When the generation of the free nonequilibrium holes stops, their density in the base decays because holes cross the p-n junction rapidly and move into the p-type region; the time constant for this process is small ($\sim t_0$). The photodiode current is proportional to

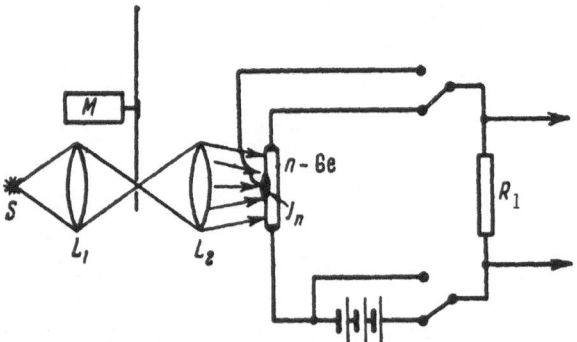

Fig. 231. Basic circuit for the simultaneous observation of the
relaxation of the photoconductivity and of the photodiode
current. S is the source of light; L_1, L_2 are lenses; M is the
modulator. A germanium sample has two ohmic contacts at
the ends and a fused indium contact forming the p-n junction.

the free hole density and therefore its value should also decay rapidly
with a time constant t_0. Simultaneously, the holes are replenished by
liberation from the trapping levels. Consequently, after a time $\sim t_0$, the
photodiode current is determined only by the rate of emptying these levels.
Because the time necessary to liberate holes from the β centers is long, the
current due to the liberated holes is small and practically imperceptible
although it continues for a long time.

In contrast to the photocurrent, the photoconductivity is governed not
only by the minority carrier (hole) density but also by the majority carrier density. After a short time $\sim t_0$, the density of holes in the band becomes very small. However, this density is high at the trapping levels
and decreases slowly due to thermal transitions to the band. In virtue
of the neutrality condition, the majority carrier (electron) density is equal
to the density of trapped holes and therefore the former is high and may
also decrease slowly with time (Fig. 233).

Thus, the photoconductivity for $t > t_0$ is mainly due to the majority
carriers, the relaxation of which displays long "tails."

When the trapping is of the α-type, we can distinguish two cases:
$t_0 \gg \theta$ and $t_0 \ll \theta$.* In the former, the illumination is stopped, the
minority carriers (both the band and the trapped carriers) cross the p-n
junction in a time $t = t_0 (1 + \alpha)$ and, consequently, the trapping affects the
time constant of the photocurrent J_f, changing it by a factor of $(1 + \alpha)$. †

---

*Moreover, it follows from the α-trapping condition that $\tau \gg \theta$ always and, in the case of well-
constructed photodiodes, $\tau \gg t_0$.

† This case is equivalent to α-trapping in the photoconductivity case discussed in § 27, with
the difference that now instead of the lifetime τ we have the transit time t_0 because the recombination is replaced by the process of carrier leakage across the p-n junction.

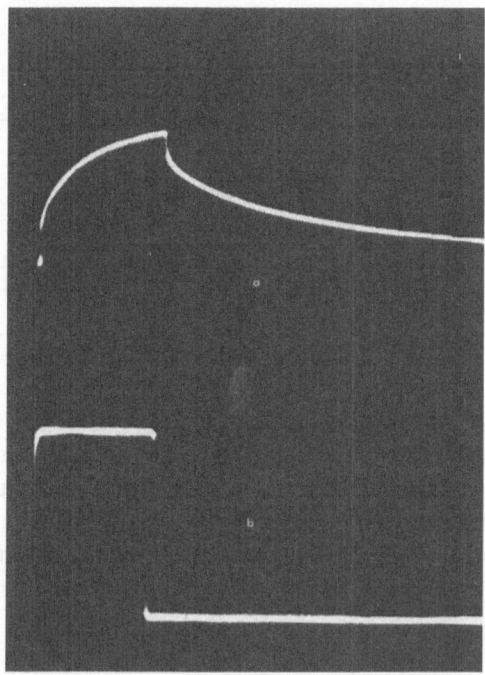

Fig. 232. Relaxation oscillograms: a) Photoconductivity; b) short-circuit current of a photodiode.
Pulse duration ∼150 μsec, T = 77°K.

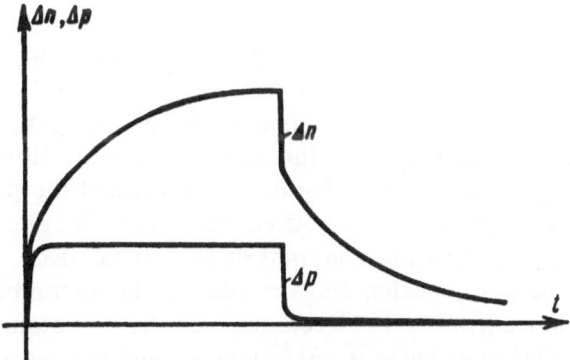

Fig. 233. Relaxation of the nonequilibrium density of the
minority (Δp) and majority (Δn) carriers in an n-type semiconductor containing hole-trapping levels.

However, the time constant of J_f is always smaller than the time constant of the photoconductivity, which is given by $\tau' = \tau(1 + \alpha)$, in the ratio t_0/τ, i.e., the relaxation curves of J_f should not exhibit slow "tails" extending as far as in the case of photoconductivity.

In the $t_0 \ll \theta$ case, after the illumination is stopped, the minority carriers leave the band "instantaneously" ($t_0 \ll \theta \ll \tau$) and cross the p-n junction; later their density in the band is governed only by the thermal transitions from the trapping levels, whose rate is relatively low because the liberation of holes from the trapping levels and the completion of the relaxation processes are governed by a large time constant

$$\frac{1}{\gamma P_{vM}} \gg \theta = \frac{1}{\gamma P_{vM} + \gamma M}.$$ Thus, the relaxation of J_f is completed mostly in

the time t_0, and then we have for a considerable period ($1/\gamma P_{vM}$) a small current which is negligible compared with the initial current J_f.

It is easily seen that this case is completely equivalent to the case of the effect of β trapping on the current J_f. Indeed, we have a condition, characteristic of β trapping, according to which the time for the establishment of equilibrium with the traps, $1/\gamma P_{vM}$, is considerably longer than the time which represents the carrier loss, i.e., t_0 instead of τ in our case.

It follows that usually the trapping has no effect on the relaxation of the photocurrent of a well-constructed ($\tau \gg t_0$) photodiode. In one case only (α trapping, $t_0 \gg \theta$) does the relaxation of J_f extend over a long time, but even then the time is much shorter than for the photoconductivity relaxation under similar conditions.

§74. FAST-RESPONSE PHOTODIODES

In general, the response of a photodiode may be governed by the following factors: the lifetime τ of the minority nonequilibrium carriers, the diffusion time t_0 of the minority carriers toward the p-n junction, the transit time t_j through the p-n junction, the Maxwell relaxation time $\varepsilon/4\pi\sigma$, and, finally, the time constant of the circuit RC, where C is the capacitance of the p-n junction and the rest of the assembly, and R is the resistance, consisting of the load resistance, the resistance of the semiconductor material, the contact resistances, and the internal resistance of the p-n junction (§72). In currently available photodiodes, some of these factors are unimportant.

Let us consider the photodiode regime, in which the response is governed by the kinetics of J_f (§72). In a well-constructed photodiode, the recombination and, therefore, τ do not affect the kinetics of J_f. The quantity $\varepsilon/4\pi\sigma$ for a material with $\sigma \approx 0.1 \ \Omega^{-1} \cdot cm^{-1}$ amounts to $\sim 10^{-11}$ sec and can therefore be neglected. The time to travel across the space-charge region of the junction can be estimated as follows:

$$t_j = \frac{d}{\mu E},\qquad(74.1)$$

where d is the space-charge layer thickness, and μE is the rate of drift of carriers in a field E which exists in the junction. This field is

$$E = \frac{\varphi_c + V}{d},\qquad(74.2)$$

where V is the external voltage applied to the p-n junction in the reverse direction, φ_c is the contact potential difference which governs the field in the junction in the absence of an external voltage.

Using the well-known expression for the thickness of the space-charge layer

$$d = \sqrt{\frac{\varepsilon(\varphi_c + V)}{2\pi en}},\qquad(74.3)$$

we obtain, after substituting Eqs. (74.3) and (74.2) into Eq. (74.1),

$$t_j = \frac{\varepsilon}{2\pi en\mu} = \frac{\varepsilon}{4\pi\sigma}\qquad(74.4)$$

Thus, the order of magnitude of t_j is the same as that of the Maxwell time constant. It is independent of the applied voltage and is governed only by the properties of the material of which the p-n junction is made. Since, in the usual photodiodes, the time constant is considerably greater than the quantity $\varepsilon/4\pi\sigma$, it is obvious that the time to travel across the junction does not restrict the time constant.*

Consequently, the time constants of the usual photodiodes are governed by two factors only: t_0 and RC. This conclusion was referred to earlier in §§ 71 and 72.

The minimum value of RC, which is independent of the external circuit and is governed by the "intrinsic" properties of the photodiode, is $R_b C_j$, where R_b is the resistance of the base and contacts, and C_j is the junction capacitance. The quantity $R_b C_j$ may amount to 10^{-9} sec under real conditions.

Thus, down to $\sim 10^{-9}$ sec, the intrinsic time constant of a photodiode is governed by the diffusion time $t_0 \approx w^2/2D$. By reducing the value of w, we can considerably reduce t_0 and the time constant of the photodiode. For example, by reducing the base thickness to $\sim 10~\mu$, it has been possible to prepare photodiodes with time constants of $\sim 10^{-8}$ sec without reducing their sensitivity. The oscillogram in Fig. 234 shows a square' pulse reproduced by a standard photodiode (the upper curve) and a fast-response photodiode (the lower curve).

---

*On the other hand, the quantity $\varepsilon/4\pi\sigma$ governs the fastest possible response of a photodiode.

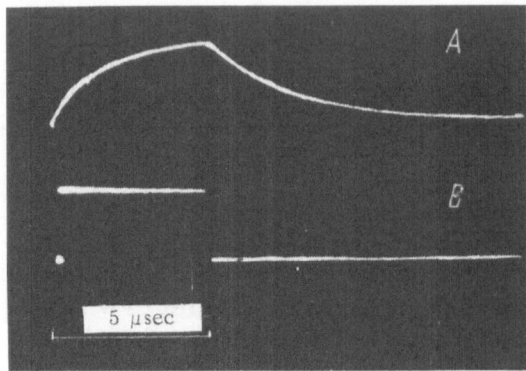

Fig. 234. Reproduction of a square light pulse by: A)
A standard photodiode; B) fast-response diode [35].

The response time was for a long time the only parameter in respect of which photodiodes were worse than vacuum photocells. Germanium photodiodes now surpass vacuum photocells by a factor of several hundred in respect of sensitivity (and the sensitivity extends over a
wide range of the spectrum), require lower voltage sources (in the barrier regime no voltage source is needed at all) and have small dimensions. The development of fast-response photodiodes disposed of the in-
feriority of photodiodes, compared with vacuum photocells, in respect
of their response time. Further reduction of the photodiode time constant would be achieved by a structure in which carriers are generated
directly in the space-charge region of the junction (for example, by using
surface-barrier junctions). One could then achieve time constants of the
order of 10^{-9} sec and even 10^{-10} sec.

§75. OPTIMUM CONDITIONS FOR USING PHOTODIODES
AS DETECTORS OF WEAK SIGNALS

The use of the photodiode regime for the conversion of light signals
has two important advantages compared with the barrier regime. First,
there is the higher voltage sensitivity. Under the barrier regime, the
full increment of current due to illumination J_f may be obtained only
under short-circuit conditions. However, when working in the photodiode regime, the same value of J_f may be obtained with a large load
resistance in the circuit. Consequently, the signal taken from this resistance will be large. The second advantage is the faster response of a
photodiode under the photodiode regime compared with the barrier regime. This follows from the fact that the response under the photodiode
regime is governed by the transit time t_0:

$$t_0 \approx \frac{w^2}{2D}, \tag{75.1}$$

(as before, w denotes the thickness of the photodiode base, and D is the diffusion coefficient of the minority carriers in the photodiode material), while under the barrier regime, with the circuit open, the response depends on the lifetime τ of the nonequilibrium carriers, which is usually considerably longer than t_0. One must point out, however, that the barrier regime has two important advantages over the photodiode regime. One is the absence of voltage sources and the other is the extremely low noise level. The latter (considerably lower than that of the photodiode regime) is mainly due to the fact that the dark current in the barrier regime is zero, while in the photodiode regime this current – whose fluctuations (due to instability of the contacts, etc.) usually determine the noise level – may be considerable.

Obviously, the optimum conditions would be obtained by combining the advantages of the photodiode (high voltage sensitivity, fast response) and the barrier (low noise, absence of voltage sources) regimes. This can be done by operating a photodiode under barrier-regime short-circuit conditions at sufficiently low temperatures. Then, two of the advantages (fast response, * low noise level) are retained. However, the voltage sensitivity is normally low because of the very small load resistance R. This difficulty can be avoided as follows. The short-circuit condition is:†

$$R_0 \gg R. \tag{75.2}$$

If we increase in some way the value of R_0, we can then increase R and, consequently, the voltage sensitivity without violating the conditions (75.2). Usually, the value of R has a limit dictated by other requirements (for example, the requirement that RC must not exceed a certain value in order to retain the fast response). Thus the problem is to find conditions when R_0 is greater than this limiting value of R.

To increase R_0, we can prepare p-n junctions from materials with the widest possible forbidden band. ‡

For example, silicon has a sufficiently wide forbidden band and R_0 of silicon photocells amounts to several megohms at room temperature. Thus, silicon photodiodes can be used for signal conversion under the barrier regime at room temperature.

---

* Under short-circuit conditions the response, as in the case of the photodiode regime, is governed by the transit time t_0.

† The small-signal case is considered. In the general case, this condition should be written in the form $R_{int} \gg R$, where $R_{int} = R_0 \exp(e\varphi/kT)$ (§ 72).

‡ In this connection, we come across the problem of the doping of dielectrics with impurities and the preparation of p-n junctions of sufficient quality.

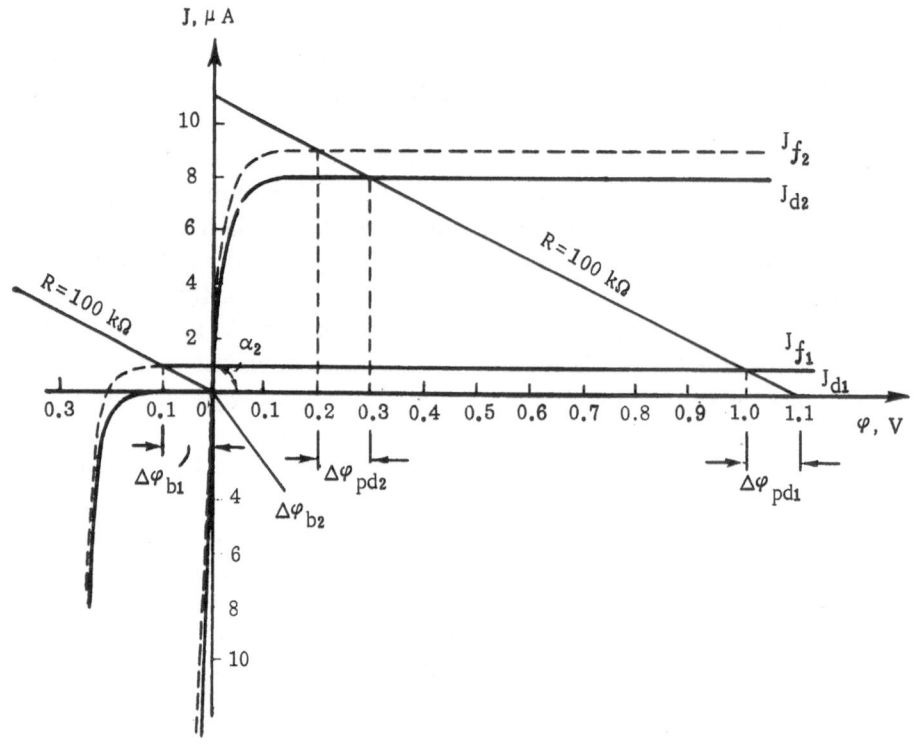

Fig. 235. Current-voltage characteristics calculated for a germanium photodiode at temperatures of +20°C and − 78°C.

Another way of increasing R_0 is to lower the temperature if germanium photodiodes are used. On cooling, we can reach a sufficiently high value of R_0, because of the exponential fall of J_S, and thus increase the load resistance R (without violating the short-circuit condition $R_0 \gg R$) to values which are usually employed in the photodiode regime. *

It is known that J_f is practically independent of temperature (§70) and, therefore, under the conditions considered the voltage sensitivity under the barrier regime should not be different from the sensitivity in the photodiode regime at room temperature. The device should then be also practically noise free.

These conclusions are readily illustrated by considering the current-voltage characteristic of a p-n junction. This characteristic is given by the fundamental photodiode equation (69.14). In Fig. 235, the curves J_{d1} and J_{d2} represent the "dark" characteristics at temperatures of +20°C and −78°C, respectively, while curves J_{f1} and J_{f2} represent the character-

---

* The load resistance in the photodiode regime, selected so that the RC of the circuit does not delay the signals too much, usually amounts to 10^4-10^5 Ω.

istics during illumination at the same temperatures. All these curves were calculated for a photodiode whose dark current at room temperature is 8 μA.

The quantity R_0 is equal to cot α, where α is the angle of the slope of the current-voltage characteristic in darkness when $\varphi = 0$. This angle is close to 90° at room temperature (α_2 in Fig. 235, curve J_{d1} merges with the abscissa).

Under the photodiode regime, the load characteristic intersects the current-voltage characteristic J_d and J_f in the region of their saturation and the magnitude of the signal remains constant at room temperature and at low temperatures ($\Delta\varphi_{pd1} = \Delta\varphi_{pd2}$).

Under the barrier regime, the load characteristic intersects the current-voltage characteristics in the saturation region only at low temperatures (Fig. 235). Then the signal $\Delta\varphi_{b1}$ is found to be equal to the signal $\Delta\varphi_{pd1}$. At room temperature, the intersection occurs in the region where the current-voltage characteristic rises sharply ($\alpha_2 \simeq 90°$) and the signal $\Delta\varphi_{b2}$ is found to be considerably smaller than $\Delta\varphi_{pd2}$.

Thus, at room temperature, the change from the photodiode to the barrier regime reduces the signal due to the smallness of R_0. On increase of R_0 at low temperatures, because the value of J_f does not depend strongly on temperature, the difference between the signals under the photodiode and barrier regimes decreases and, if $R_0 \gg R$, the barrier-regime signals may be as large as those obtainable under the photodiode regime.

When the signals are large, the internal resistance of the photodiode

$$R_{int} = R_0 e^{\frac{e\varphi}{kT}}$$

is smaller than R_0. Then the short-circuit condition $R_{int} \gg R$ is more difficult to satisfy than in the case of small signals ($R_0 \gg R$). However, the noise is not very important in the case of large signals, so that we can use the photodiode regime, reserving the barrier regime for small-signal conditions close to the threshold of the photodiode sensitivity.

Calculations show that

$$\frac{\Delta\varphi_b}{\Delta\varphi_{pd}} = \frac{1}{1 + \frac{R}{R_0}} = \frac{\frac{kT}{e}}{\frac{kT}{e} + J_s R}. \qquad (75.3)$$

This relationship is in good agreement with experiment [27].

Figure 236 shows the rise of $\Delta\varphi_b$ on cooling. It is seen that beginning at about −40°C the barrier-regime signal is equal to the photodiode-regime signal, while the noise level in the former case is at least two orders of magnitude lower than in the latter case.

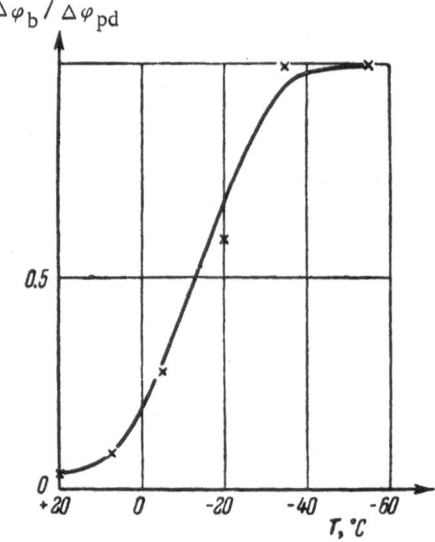

Fig. 236. Dependence of $\Delta\varphi_b$ on temperature for a germanium photodiode: $J_s = 8\ \mu A$, $R = 10^5\ \Omega$. The continuous curve represents calculated values, the crosses experimental values.

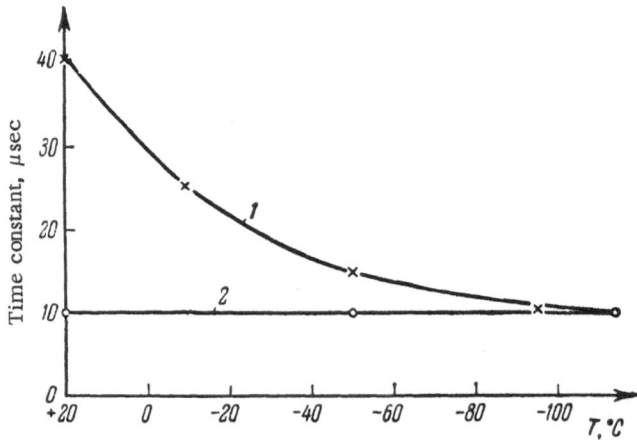

Fig. 237. Comparison of the response of photodiodes under the barrier (1) and photodiode (2) regimes at various temperatures.

Figure 237 demonstrates that for the barrier regime the speed of response decreases on cooling, approaching the response of the photodiode regime.

It follows that the barrier regime under short-circuit conditions is the optimum one for the recording of small signals.

§ 76. PHOTOELECTROMOTIVE FORCES
IN THE IMPURITY EXCITATION REGION

We have shown in § 69 that the photo-emf appears at p-n junctions due to the increase, during illumination, of the minority-carrier density near the junction. *

In the fundamental excitation region, electron-hole pairs are generated and, therefore, minority carriers are always produced.

In the case of excitation with light from the impurity region, one would expect that mainly majority carriers would be liberated and that, consequently, there would be no photo-emf. In fact (Fig. 238a), in an n-type semiconductor (with the Fermi level in the upper half of the forbidden band) the number of electron-occupied levels (and the region of the spectrum suitable for the ionization of these levels and for the consequent appearance of majority carriers – electrons – in the conduction band) is greater than the number of vacant levels (and the corresponding region of the spectrum suitable for the generation of minority carriers).

Thus, the appearance of minority carriers in the case of impurity excitation is, in general, less likely. However, when we consider actual models we can distinguish at least three cases of quite high generation rate of minority carriers by the impurity-region illumination. They are enumerated below.

a) The minority carriers may be generated by optical transitions from the valence band to an impurity level at equilibrium which is not completely occupied with electrons (such a level is close to or higher than the Fermi level, cf. Fig. 238b).

b) The minority carriers may be generated by thermo-optical transitions (Fig. 238c): light transfers electrons from the filled levels to the conduction band. As a result, the equilibrium between the levels and the valence band is upset. The equilibrium is reestablished by thermal electron transitions from the valence band to these levels, a phenomenon which is connected with the appearance of nonequilibrium holes (minority carriers).

c) The minority carriers may be produced by double optical transitions. Here, the sequence is as in (b) but thermal transitions from the lower band are replaced with optical ones due to the absorption of the same impurity-region illumination. Clearly the double optical transitions

---

* The most general considerations show that it is impossible to observe the steady-state photo-emf due to the generation of majority carriers only (with the exception of the rare case of the existence of hot majority carriers with long lifetimes).

This is because in the case of, for example, a semiconductor which is homogeneous before illumination, but in which majority carriers are generated nonuniformly, the diffusion photo-emf in the interior of the semiconductor is compensated exactly by the photo-emf of opposite sign which appears at the electrode-semiconductor contact.

Similar considerations apply to systems which are inhomogeneous before illumination, for example, p-n junctions.

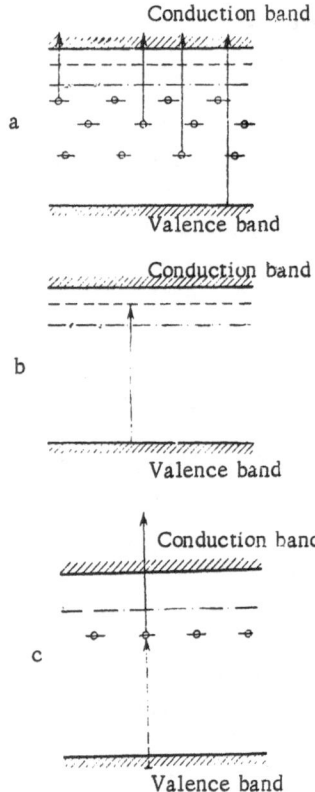

Fig. 238. a) Electron band structure of a semiconductor. The formation of minority carriers (holes) is not very likely in the case of impurity excitation. b) Direct optical generation of minority carriers. c) Generation of minority carriers by two-step (thermo-optical or double optical) transitions.

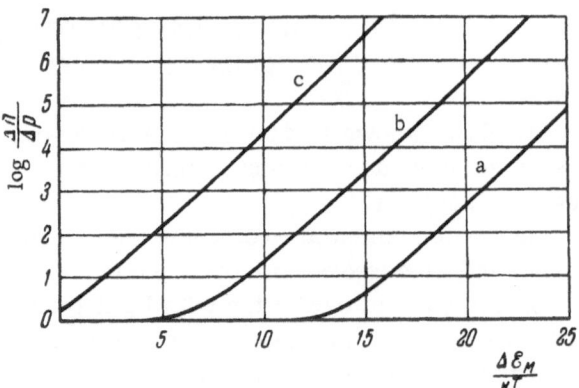

Fig. 239. Dependence, in the case of thermo-optical transitions, of the degree of ambipolar conduction on the separation between the impurity level and the band with which carriers are exchanged by thermal excitation. a) $M = 10^{12}$ cm^{-3}; b) $M = 10^{15}$ cm^{-3}; c) $M = 10^{18}$ cm^{-3}.

are, in principle, possible only if the energy of the quanta is not less than half the forbidden band width.

In §43, we have considered the criterion for the impurity photoconductivity to be unipolar and deduced quantitative relationships [Eqs. (43.5) and (43.9)] between Δn and Δp in the case of thermo-optical and double optical transitions. Numerical estimates made with the aid of these formulas show that in many real cases the role of ambipolar conduction and, consequently, the role of the minority carriers, may be very important.

Figure 239 gives the dependence of $\log \Delta n / \Delta p$ for thermo-optical transitions on the separation between the impurity level and the "minority" band (in our case, the valence band), from which thermal transitions take place. Figure 239 shows that, for example, when $M = m_0 = 10^{15}$ cm^{-3}, $\Delta \mathcal{E} / kT \leq 5$, the minority-carrier density is raised by thermal transitions so that it becomes equal to the density of the majority carriers generated in the conduction band by direct optical ionization of the impurities. Consequently, the impurity-region excitation, like the excitation in the fundamental region, can generate practically equal amounts of nonequilibrium electrons and holes.

The experimental results for cuprous oxide [6] and germanium [41] photocells confirm that the photo-emf can be observed in the impurity

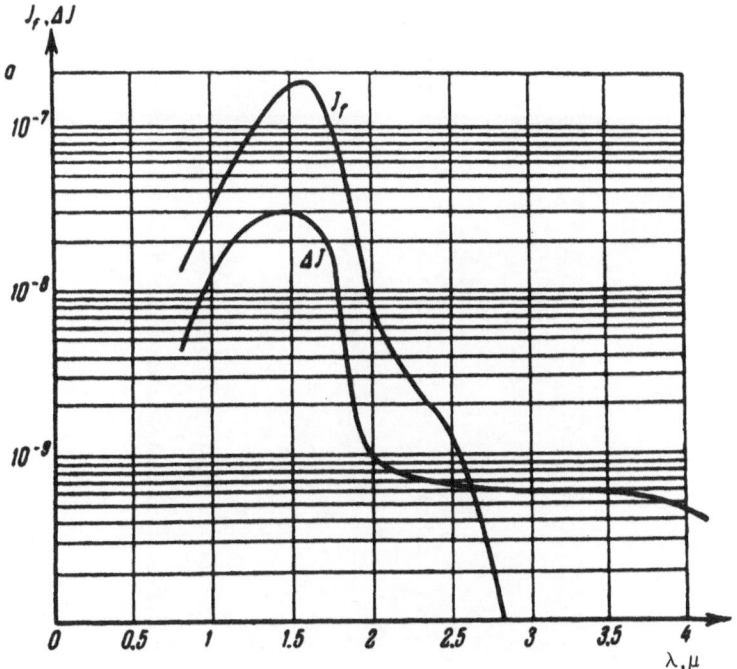

Fig. 240. Spectra of the short-circuit photocurrent and of the photoconductivity of gold-doped germanium samples.

region. Figure 240 shows the short-circuit photocurrent (J_f) spectrum
for a p-n junction prepared by diffusing antimony into p-type germanium
containing $\sim 10^{15}$ cm^{-3} gold atoms (curve 1). This figure also gives the
photoconductivity spectrum of the same material (curve 2). The two
curves show that both the photoconductivity and the current through the
p-n junction extend to the impurity-excitation region (beyond $1.8\ \mu$). The
photoconductivity extends much further into the long-wavelength region
than does the photo-emf. The $J_f(\lambda)$ and $\Delta J(\lambda)$ curves of Fig. 240 are
in good agreement with the foregoing theoretical discussion.

The long-wavelength edge of the photoconductivity is determined by
the transitions from the valence band to the gold levels at $E_V + 0.15$ eV.
The appearance of the minority carriers (electrons) and the photo-emf
is either related to electron transitions to the conduction band from the
partly filled levels at $E_V + 0.15$ eV, or to double optical transitions*
from the valence band to the $E_V + 0.15$ eV levels, and hence to the conduction band. The long-wavelength edge of the photo-emf is then governed by the transitions with the larger jump ($h\nu \approx 0.5$ eV).

---

* Thermo-optical transitions are unlikely at 77°K.

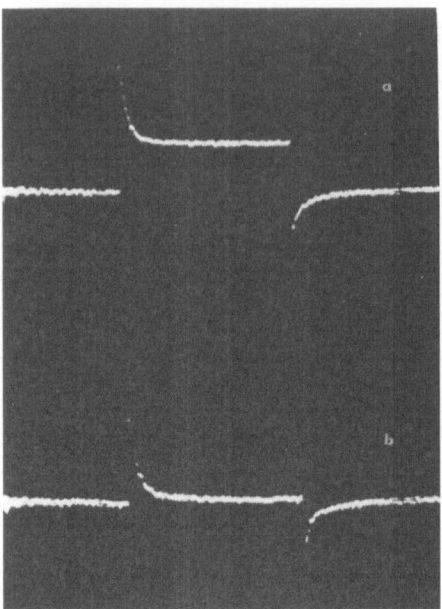

Fig. 241. Relaxation curves of the impurity photo-
emf [42]: a) λ = 2.6 μ , generation of majority and
minority carriers possible; b) λ = 3.2 μ , generation of majority carriers only.

Finally, it should be noted that the impurity photo-emf may be observed also when only the majority carriers are generated under transient conditions. As previously pointed out, the absence of a photo-emf
when the majority carriers are generated is due to the fact that the volume emf is compensated exactly by the contact photo-emf. However, if
the establishment of the photo-emf in the interior and at the contact requires different time (because the conditions for the recombination and
diffusion are different in the interior and near the contact) then, obviously, the compensation under steady-state conditions will not apply during
relaxation.

The oscillograms in Fig. 241 illustrate the relaxation curves of the
photo-emf in the gold-doped samples of germanium recently referred to,
excited with square light pulses. The oscillogram a was obtained when
the generation of both the majority and minority carriers was possible.
It is evident that, in addition to the steady-state emf — related to the generation of minority carriers — there is also a photo-emf "flash" at the
moment of beginning or stopping the illumination. At λ = 3.2 μ (Fig.
241b) when only the generation of the majority carriers is possible, the
steady-state component of the photo-emf disappears and only the flashes
remain. The flashes represent the transient photo-emf due to the generation of the majority carriers.

The spectrum of the transient photo-emf in the long-wavelength region coincides, as expected, with the photoconductivity spectrum.

It can be shown that, in addition to the impurity photo-emf at a p-n junction, under certain conditions the induced impurity photo-emf* may also be observed. Preliminary excitation in the fundamental-band region may produce such overpopulation ("charge exchange") of the impurity levels that the impurity excitation may lead to the generation of minority carriers. Obviously, under these conditions, the impurity photo-emf (induced by the preliminary illumination) may be generated. This was observed at p-n junctions in copper-compensated germanium [44].

Instead of the fundamental-band excitation, one may use injection at a p-n junction to produce charge exchange; like the fundamental-band excitation, the injection generates nonequilibrium electron-hole pairs. Such a current-induced impurity photo-emf was detected in compensated-silicon photodiodes [46].

Charge exchange in systems with p-n junctions is related to certain other effects, such as photodiode "sensitization" by preliminary impurity excitation [45].

§77. "EXTERNAL" PHOTOEMISSION FROM A METAL INTO A SEMICONDUCTOR

The external photoemission from a metal into vacuum is due to the supply by photons of sufficient energy to the metal electrons so that they overcome the barrier represented by the metal-vacuum work function (Fig. 242). Obviously, a similar effect may take place at a metal-semiconductor or metal-dielectric interface. In the former case, the photoeffect may be produced by quanta of much lower energy because the barrier to be overcome by the metal electrons is the difference between the work function for emission from the metal into vacuum and the electron work function of the semiconductor.

Figure 243a shows the energy band structures of a metal and an n-type semiconductor before contact and Fig. 243b shows the same structures after close contact.† The photoemission from a metal into a semiconductor is observed only outside the fundamental excitation region of the semiconductor (i.e., in the impurity region) because the strong effects of the

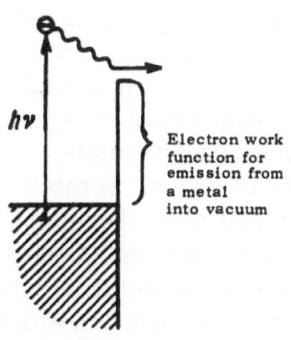

$h\nu$

Electron work function for emission from a metal into vacuum

Fig. 242. External photo-emission from a metal into vacuum.

* Like the induced impurity photoconductivity (see §44).

† The energy band structures are shown for the most interesting case when the work function of the metal is greater than that of the semiconductor. Then the contact of a metal with an n-type semiconductor produces a band curvature representing a barrier layer.

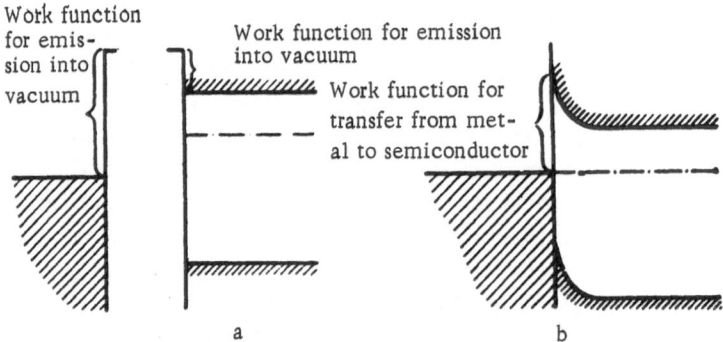

Work function for emission into vacuum

Work function for emission into vacuum

Work function for transfer from metal to semiconductor

a   b

Fig. 243. Energy band structures of a metal and a semiconductor: a) Before contact, and b) after contact.

fundamental excitation mask such photoemission. The external photoemission from a metal into vacuum is usually observed by measuring the current to an additional electrode (anode). The best method for observing the external photoemission from a metal into a semiconductor is to measure the photo-emf in the barrier layer formed in the semiconductor at the contact with the metal. For this reason, external-photoeffect photocells should be made of such metal-semiconductor pairs which form barrier layers, for example, cadmium sulfide in contact with electrolytically deposited copper or another suitable metal.

Figure 244 shows the short-circuit photocurrent spectrum of a photocell with this type of contact [33]. The effect was observed at wavelengths longer than $0.5\ \mu$ (the impurity region) and was governed by the external photoemission from copper into cadmium sulfide. It should be mentioned that photocells using the external photoemission from a metal into a semiconductor have a very fast response.

The kinetics of such photocells [38] may be analyzed in the same way as the kinetics of photocells with p-n junctions (§ 71).

The appearance of a barrier photo-emf in the external photoemission from a metal into an n-type semiconductor may be explained as follows. During illumination, the electrons of the metal which, due to the absorbing of photons, acquire excess energy greater than the potential barrier, can penetrate the semiconductor where a strong contact field drives them into the interior and the semiconductor thus accumulates excess negative charge, leaving the metal positively charged. In this way, the conduction band of the semiconductor is raised and, therefore, the reverse thermal current of the majority carriers increases. Under steady-state conditions, the photoelectron current from the metal is balanced by the reverse current from the semiconductor, a condition which corresponds to the appearance of the photo-emf. If the circuit is closed, the photoelectrons can be returned to the metal, producing a current in the external circuit.

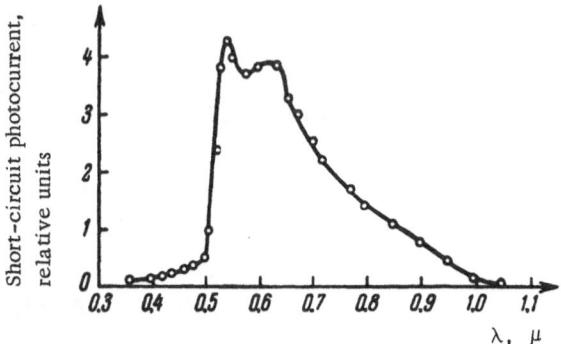

Fig. 244. Short-circuit photocurrent spectrum for copper in contact with cadmium sulfide. The photocurrent
is due to external photoemission from metal into semiconductor.

The analysis shows that there is a very close analogy between the
mechanism of the appearance of the barrier photo-emf in the presence
of a p-n junction and the photo-emf in the case of the external emission
from a metal into a semiconductor. The metal in contact with an n-type
semiconductor seems to take on the role of a p-type semiconductor. The
electrons in the metal which have absorbed photons, and have therefore
acquired excess energy greater than the barrier height, act now as the
nonequilibrium minority carriers. Consequently, the analog of the nonequilibrium minority carrier lifetime in a semiconductor is the time
T during which the excited electron in a metal dissipates the excess energy by collisions until this energy becomes equal to the potential barrier.
Otherwise, the relationships governing the kinetics are analogous to those
given in § 71.

The quantity T is equal to several relaxation times and may amount
to $\sim 10^{-11}$ sec. As shown in § 71, when the nonequilibrium carrier lifetime is $\tau \ll R_{in}C$ (where R_{in} is the internal differential resistance of the
photocell, C is the capacitance of the space-charge layer and of the assembly), the photo-emf kinetics, under the open-circuit conditions, is
governed by the time constant $R_{in}C$; when an external load R_l is in circuit, the kinetics is governed by the quantity R_eC, where $1/R_e = 1/R_l$
$+ 1/R_{in}$, provided $\tau \ll R_eC$. The time t_j that a photoelectron takes to
travel across the space-charge region of a p-n junction is estimated to be
$\sim 10^{-12}$ sec (assuming that the resistivity of the semiconductor is 1 $\Omega \cdot$ cm
and its permittivity is $\varepsilon = 9$). This quantity can be neglected, as in § 71.

If the assumptions about the smallness of T and t_j are justified, the
relaxation of the photocell is then governed only by the quantities $R_{in}C$ or
R_eC, and the dependence of the relaxation time τ_e on the load resistance
should be as follows: when $R \gg R_{in}$, the time τ_e is independent of R_l
and is equal to $R_{in}C$; when R_l is small, τ_e is a linear function of R_l.

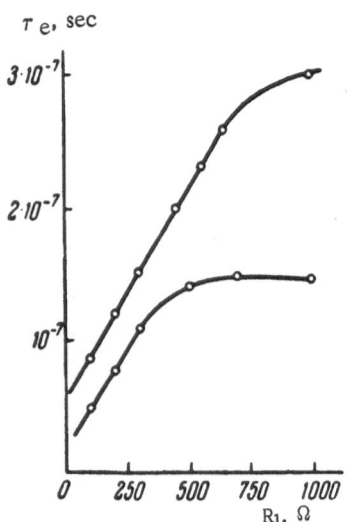

τ_e, sec

Fig. 245. Dependence, on the load resistance, of the time constants of two photocells which use the photo-emission from a metal into a semi-conductor.

Figure 245 shows the experimental dependences of τ_e on R_{in} for two Cu-CdS photocells. The curves agree with the conclusions arrived at above. The time constants of these photocells amount to several tens of nanoseconds when the load resistance is small.*

§ 78. PHOTOTRANSISTORS

Light signals can be converted into electrical ones by phototransistors, which are very sensitive devices due to their current amplification. A phototransistor consists of three regions in contact ($n-p-n$ or $p-n-p$) and is frequently connected up as shown in Fig. 246, i.e., in the same way as photodiodes. The left-hand p-n junction (the emitter) is connected in the forward direction and the right-hand junction (the collector) in the reverse direction.

The current in the external circuit is due to the minority carriers injected at the emitter and passing through the central region (base) to the collector. Because of the presence of the emitter (which distinguishes a phototransistor from a photodiode), the phototransistor current in darkness is higher than the photodiode current. During illumination, the minority carriers generated by light in the base of the transistor (Fig. 246b) diffuse to the emitter and collector junctions and are driven by the contact fields at these junctions into the collector and emitter parts of the transistor, respectively. The nonequilibrium majority carriers (electrons in Fig. 246b) are in a potential well and therefore they stay in the base, altering its charge and, consequently, its potential with respect to the emitter. As a result of the change of the potential difference, the current injected into the base from the emitter increases and, therefore, the collector and external-circuit currents increase as well. Thus, a small increase of the potential difference at the emitter junction during illumination strongly increases the current in the external circuit, thereby ensuring the high sensitivity of phototransistors.

In photodiodes, the maximum quantum yield, defined as the ratio of the photoexcited rise of the saturation current to the number of funda-

---

* There is as yet no certainty that the external photoemission does take place at the Cu-CdS contact. This question requires further study.

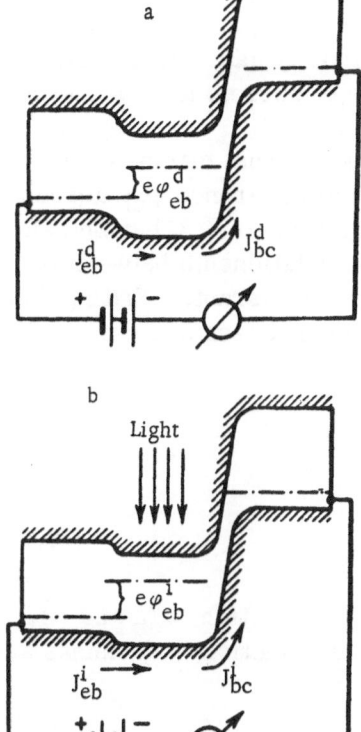

Fig. 246. Energy band structure of a p-n-p phototransistor: a) In darkness; b) during illumination.

mental-region quanta absorbed per unit time, does not exceed unity, but in phototransistors it may reach 100 or more.

Calculations give the following expression for the amplification factor of a phototransistor [11, 26]:

$$k = \frac{\Delta J_{pt}}{\Delta J_{pd}} = \frac{l_D^b}{w} \frac{\alpha}{1 + \frac{\alpha w}{2 l_D^b}},$$

where $\alpha = \sigma_e l_D^e / \sigma_b l_D^b$; σ_e and l_D^e are, respectively, the conductivity and the diffusion length of the minority carriers in the emitter region; σ_b and l_D^b are the same quantities for the base region.

In the special case when $\alpha w / 2 l_D^b \gg 1$, we find that $k = 2 (l_D^b/w)^2 = \tau/t_0$. When $\alpha w / 2 l_D^b \ll 1$, $k = \sigma_e l_D^e / \sigma_b w$. In the former case, which is typical of alloyed phototransistors, the amplification factor may be several hundred for reasonable values of l_D^b/w (~ 10).

We shall consider the moment of stopping the illumination and the subsequent behavior in order to analyze the kinetics of phototransistors [22]. We shall assume that the time constant RC of the external circuit is small and ignore the recombination of the nonequilibrium minority carriers in the base. We shall further assume that the conductivity of p-type germanium is higher than that of n-type germanium and therefore only the hole currents need be considered in the first approximation.

During illumination under steady-state conditions, the current through the emitter junction J_{eb}^i (Fig. 246b) and the equal collector current J_{bc}^i (and, consequently, the measured current in the external circuit) are governed only by the voltage at the emitter junction φ_{eb}^i. When the illumination is stopped, this voltage can change only due to a change of the charge of the nonequilibrium majority carriers (electrons) in the base. The number of such carriers may decrease for two reasons: 1) due to recombination with minority carriers; 2) due to loss to the emitter (which requires the overcoming of a barrier).

The change of φ_{eb} alters immediately (assuming that the emitter capacitance can be neglected) the emitter current J_{eb}. After a transit time $t_0 \approx w^2/2D$, which is required by the minority carriers to diffuse from the emitter to the collector, this change of current affects the collector current, i.e., the external-circuit current.

Thus, the relaxation of the current in the external circuit is governed by t_0 and by the relaxation of the potential difference φ_{eb} across the emitter. We shall return later to the relaxation of φ_{eb} but we can immediately draw some conclusions about the relationship between the rate of relaxation of φ_{eb} and the external-circuit current. Writing down the steady-state current through the emitter in darkness and during illumination in the following form*

$$J_{eb}^d = J_s e^{\frac{e\varphi_{eb}^d}{kT}},\qquad(78.1)$$

$$J_{eb}^i = J_s e^{\frac{e\varphi_{eb}^i}{kT}}\qquad(78.2)$$

(here J_s is the saturation current at the emitter junction, φ_{eb}^d and φ_{eb}^i are the potential differences across the emitter junction in darkness and during illumination, respectively). Introducing

$$\Delta\varphi_{eb} = \varphi_{eb}^i - \varphi_{eb}^d,\qquad(78.3)$$

$$\Delta J_{eb} = J_{eb}^i - J_{eb}^d,\qquad(78.4)$$

we obtain

$$\Delta J_{eb} = J_s e^{\frac{e\varphi_{eb}^d}{kT}}\left(e^{\frac{e\Delta\varphi_{eb}}{kT}} - 1\right) = J_d\left(e^{\frac{e\Delta\varphi_{eb}}{kT}} - 1\right)\qquad(78.5)$$

[here $J_d = J_s \exp(e\varphi_{eb}^d/kT)$ is the saturation current of the phototransistor in darkness].

If we neglect the influence of the emitter capacitance, we find that the emitter current ΔJ_{eb} follows immediately any change of the potential difference across the emitter junction, and that the relationship (78.5) between ΔJ_{eb} and $\Delta\varphi_{eb}$ applies also under transient conditions.

---

*We are neglecting here small "primary" photocurrents to the collector and the emitter due to the minority carriers liberated directly by photons. The formulas (78.1) and (78.2) differ from the usual expression for the current through a p-n junction, $J = J_s [\exp(e\varphi/kT) - 1]$ by the absence of unity in the brackets. This is due to the influence of the collector which, for a thin base, eliminates the unity and alters somewhat the value of J_s.

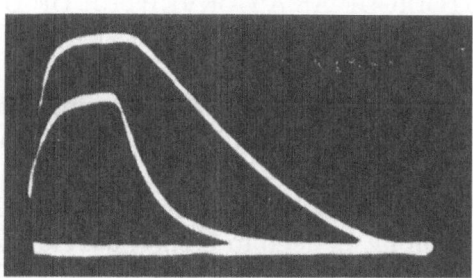

Fig. 247. Relaxation curves for a phototransistor: Upper
curve gives the potential drop between the emitter and
the base; lower curve gives the external-circuit current.

It is then obvious that the relaxation of ΔJ_{eb} should be faster than that of
$\Delta \varphi_{eb}$. In fact, irrespective of the time dependence of $\Delta \varphi_{eb}$, it is evident that

$$\Delta J_{eb}(t) = J_d \left(e^{\frac{e\Delta\varphi_{eb}(t)}{kT}} - 1 \right) \tag{78.6}$$

should decay faster than $\Delta \varphi_{eb}(t)$ because $\Delta \varphi_{eb}(t)$ occurs in the power index of the exponential function in Eq. (78.6).

Only at the final stages of the relaxation, when $\Delta \varphi_{eb} \ll kT/e$ and

$$\Delta J_{eb}(t) = \frac{eJ_d}{kT} \Delta \varphi_{eb}(t), \tag{78.7}$$

do the relaxation rates of ΔJ_{eb} and $\Delta \varphi_{eb}$ become equal.

The experimentally measured change of the collector current ΔJ_{bc}
(if the influence of the emitter and collector capacitances is neglected)
follows the change of the emitter current ΔJ_{eb} with a time delay of t_0. If
this time is short compared with the time constant for a change $\Delta \varphi_{eb}$,
it is evident that the experimentally measured current ΔJ_{bc}, like ΔJ_{eb},
relaxes faster than $\Delta \varphi_{eb}$ (Fig. 247).

Calculations show that at a low excitation level ($\varphi_{eb} \ll kT/e$) the relaxation of φ_{eb} and, consequently, of ΔJ_{eb} [cf. Eq. (78.7)] depends strongly on the ratio of the conductivities of the p- and n-type regions. The
smaller the difference between the conductivities, the faster the relaxation. This is because the majority carriers are lost from the base mainly by fast leakage across the emitter barrier if the p-n junction asymmetry is not too great.

In the case of strongly asymmetric p-n junctions, the leakage is small
and the majority carriers are lost mainly by recombining with the minority carriers in the base. Obviously, under these conditions, the relaxation time constant is simply equal to the pair lifetime.

§ 79. PHOTODIODE AS A CONVERTER OF OPTICAL
INTO ELECTRICAL ENERGY

In the barrier regime, a photodiode transforms optical energy directly into electrical energy. We shall now consider the efficiency of this conversion [12, 14, 16, 21]. *

Figure 248 shows a photodiode connected across a load resistance R_l. The energy of a quantum $h\nu$ is used to form a carrier pair. We shall now consider the subsequent fate of this energy. The electron and the hole may, in general (if $h\nu$ is considerably greater than the forbidden bandwidth), possess energies greater than the average thermal energy of electrons and holes ($\sim kT$). Then the excess energy is rapidly transferred to the lattice (wavy arrows 1 and 1' in Fig. 248). Some of this energy is transformed into heat and therefore lost. Next, the minority carrier (the electron in Fig. 248) crosses the p-n junction and enters the n-type region, again losing some of the energy uselessly (wavy arrow 2). Finally, the electron may pass through the external circuit generating a useful energy in the load, equal to $e\varphi$.† Thus, the external circuit efficiency is $e\varphi/h\nu$ per electron.

However, not all the photons incident on a photocell are absorbed; some of them are reflected or transmitted right through the cell. Moreover, not all the absorbed quanta produce electron-hole pairs because of the possible other competing absorption mechanisms. Therefore, the expression for the efficiency should include the multipliers: r, which is a coefficient representing the reflection and transmission of light, and β, which is the quantum yield. Moreover, due to recombination, not all the minority carriers generated by the absorption of photons reach the p-n junction. We shall allow for this by introducing a collection coefficient κ which represents the fraction of the minority carriers reaching the junction during their lifetime. Finally, we should remember that not all the electrons which cross the junction reach the external circuit. The total number of electrons crossing the junction per unit time is given by the current J_f. Some of them are transferred by the thermal motion across the junction back into the n-type region [the leakage current J_y $= J_S[\exp{(e\varphi/kT)} - 1]$. Thus, the expression for the efficiency should include also the multiplier J/J_f, where J is the useful current through the load.

When all these considerations are taken into account, the expression for the efficiency becomes

---

* We shall consider only the physical aspect of this problem without dealing with the technical aspects connected with the use of silicon "solar cells" [29, 37].

† In the same way, we can describe the passage of a hole from the p-type region through the external circuit into the n-type region where it neutralizes the electron that has crossed the p-n junction.

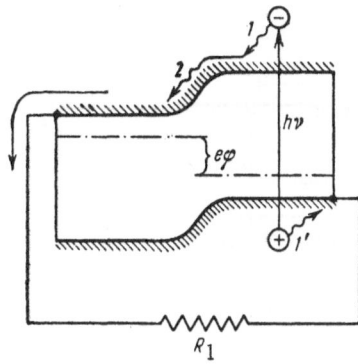

Fig. 248. Circuit of a p-n junction converter of optical into electrical energy.

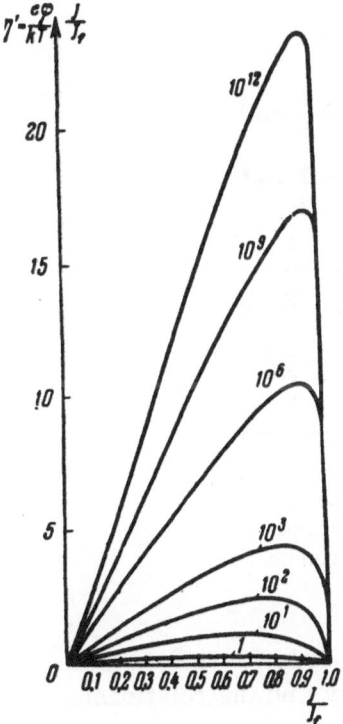

Fig. 249. Dependence of the efficiency on the current in the external circuit. The numbers by the curves denote the values of the ratio J_f/J_s.

$$\eta = r\beta\kappa \frac{J}{J_f} \frac{e\varphi}{h\nu}. \qquad (79.1)$$

Substituting into Eq. (79.1) the expression for φ from Eq. (69.15) and introducing the notation $\alpha = r\beta\kappa$, we obtain

$$\eta = \alpha \frac{kT}{h\nu} \frac{J}{J_f} \ln\left(\frac{J_f - J}{J_s} + 1\right). \qquad (79.2)$$

It follows from Eq. (79.2) that η depends on the current J, i.e., on the load resistance. Other conditions being equal, there is an optimum load (or an optimum value of the current J) which gives maximum efficiency. We shall now determine this optimum condition.

Equating $d\eta/dJ$ to zero, we find that

$$\ln\left(\frac{J_f - J_m}{J_s} + 1\right) = \frac{J_m}{J_f + J_s - J_m}, \qquad (79.3)$$

where J_m is the current corresponding to the maximum efficiency. At a low excitation level

$$\frac{J_f - J_m}{J_s} \ll 1, \qquad (79.4)$$

the left-hand part of Eq. (79.3) transforms into $(J_f - J_m)/J_s$ and using Eq. (79.4), we find

$$J_m = \frac{J_f}{2}, \qquad (79.5)$$

i.e., under these conditions we should, by varying the load resistance R_l, make the current J equal to half the short-circuit current. It is easily seen that the optimum load resistance is then

$$R = \frac{\varphi_m}{J_m} = \frac{kT}{eJ_s} = R_0. \qquad (79.6)$$

In the general case of an arbitrary illumination level, we have to solve

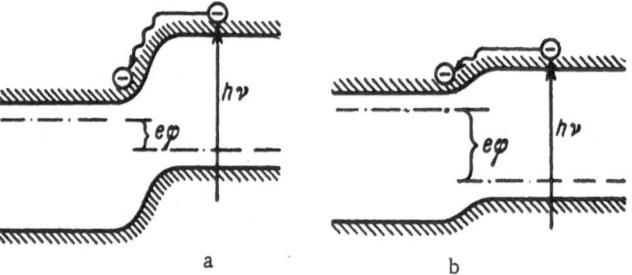

Fig. 250. Energy losses at low (a) and high (b) excitation levels.

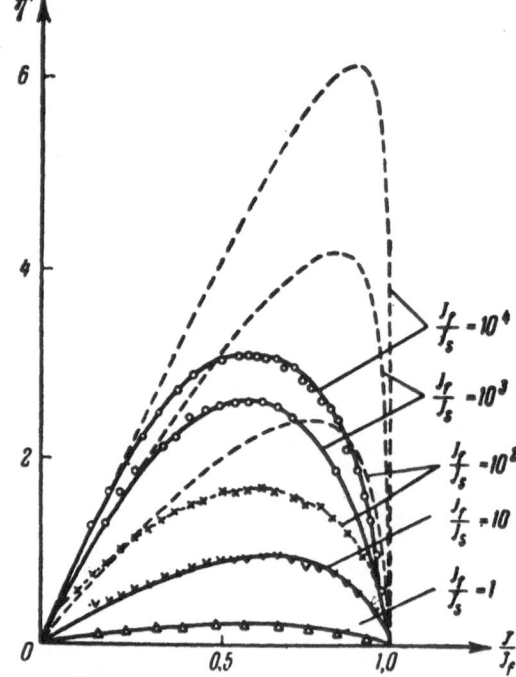

Fig. 251. Dependence of the efficiency on the external-circuit current for germanium photodiodes.
Room temperature. Dashed curves are theoretical,
continuous curves — experimental.

graphically the transcendental equation (79.3) in order to find J_m and
φ_m. Using Eqs. (79.3) and (69.15) we can easily find the relationship
between J_m and φ_m

$$J_m = \frac{J_s\left(1+\frac{J_f}{J_s}\right)\frac{e\varphi_m}{kT}}{1+\frac{e\varphi_m}{kT}} = \frac{1+\frac{J_f}{J_s}}{1+\frac{e\varphi_m}{kT}}\frac{\varphi_m}{R_0}. \tag{79.7}$$

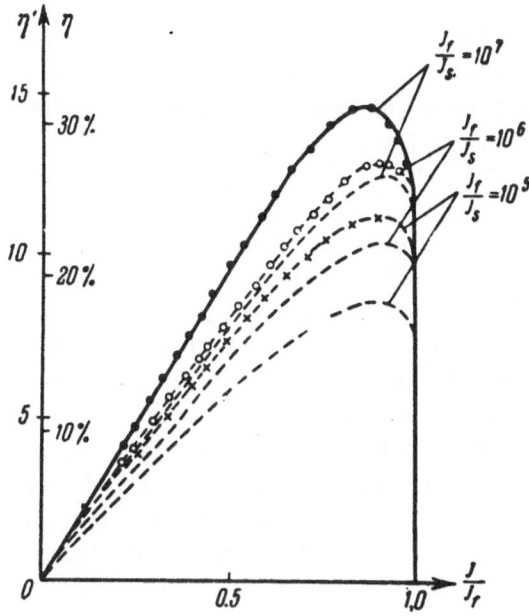

Fig. 252. The same as Fig. 251 but for −77°C.

It follows from Eq. (79.7) that when the illumination intensity is increased and J_f / J_S, as well as the slowly rising $e\varphi_m/kT$, become greater than unity, J_m approaches J_f.

Figure 249 shows the theoretical dependence of the efficiency on the external-circuit current (J/J_f) plotted in accordance with Eq. (79.2). The various curves in Fig. 249 represent various values of the ratio J_f / J_S (these values are given next to the curves). Each of the curves has a maximum corresponding to the optimum condition. With increase of the ratio J_f / J_S, i.e., with increase of the illumination intensity, to which J_f is directly proportional, the value of the efficiency increases and the maximum shifts toward higher currents J. At low illumination intensities (J_f / J_S small) the maximum corresponds to the current $J = J_f /2$ in the external circuit. With increase of J_f / J_S, this maximum shifts toward the value $J = J_f$.

The increase of the efficiency with increase of the illumination intensity can be explained quite easily (Fig. 250). The generation of a minority carrier needs an energy $h\nu$, which is then used in two stages to heat the lattice (wavy arrows) and to carry out work $e\varphi$ in the external circuit. The higher the illumination intensity, and the higher the potential difference φ, the larger the useful fraction of the energy. This is seen clearly by a comparison of Figs. 250a and 250b.

It should be stressed that, with increase of the illumination intensity and the value of φ, the return transfer of carriers (leakage current) increases as well, but calculations and Fig. 249 show that the overall result is such that the efficiency increases with the illumination intensity.

Figure 251 presents the results of an experimental check of dependence of the efficiency on the load (J_f / J_s). This figure indicates that at low intensities ($J_f / J_s = 1$, $J_f / J_s = 10$), the experimental points fit well the theoretical curves but at high intensities there is agreement only at low values of the current J (i.e., at high load resistances). The discrepancy at higher values of J_f / J_s is due to the effect of the series-connected resistance of the interior of germanium and the contacts. The voltage drop across this resistance, which is responsible for the deviation from the theoretical dependence, should have the strongest effect at high absolute values of the external-circuit current, which corresponds to high values of J_f / J_s and J/J_f. To check whether this explanation is correct, we can use high values of J_f / J_s and J/J_f without increasing the absolute value of the current in the external circuit of the photocell. For this purpose, we can lower the temperature keeping the (weak) illumination constant. Thus, keeping J_f fixed, we can lower J_s.

Figure 252 gives curves obtained at dry-ice temperatures. Here, the position of the maximum is approximately in agreement with the theory, even for high values of J_f / J_s (10^5, 10^6).

LITERATURE

CHAPTER I

1. B. I. Davydov, ZhTF $\underline{7}$, 2212 (1937).
2. V. P. Zhuze and S. M. Ryvkin, DAN SSSR $\underline{58}$, 1629 (1947).
3. V. E. Lashkarev, ZhÉTF $\underline{19}$, 876 (1949).
4. V. E. Lashkarev, I. R. Potapenko, and G. A. Fedorus, ZhÉTF $\underline{19}$, 887 (1949).
5. S. M. Ryvkin, ZhÉTF $\underline{20}$, 139 (1950).
6. N. A. Tolstoi and P. P. Feofilov, UFN $\underline{41}$, 44 (1950).
7. A. F. Ioffe, Physics of Semiconductors [in Russian] (AN SSSR, 1957).

CHAPTER II

1. B. Schonwald, Ann. Physik $\underline{15}$, 395 (1932).
2. S. M. Ryvkin, ZhÉTF $\underline{20}$, 139 (1950).

CHAPTER III

1. I. Runge and R. Sewig, Z. Physik $\underline{62}$, 726 (1930).
2. P. G. Tager, Kerr Cell [in Russian] (Giz. "Iskusstvo, " Moscow-Leningrad, 1937).
3. N. A. Tolstoi and P. P. Feofilov, DAN SSSR $\underline{58}$, 389 (1947).
4. I. R. Potapenko, ZhTF $\underline{18}$, 1356 (1948).
5. V. E. Lashkarev and I. R. Potapenko, Izv. AN SSSR ser. fiz. $\underline{13}$, 566 (1949).
6. N. A. Tolstoi and P. P. Feofilov, UFN $\underline{41}$, 44 (1950).
7. S. M. Ryvkin, ZhÉTF $\underline{20}$, 139 (1950).
8. V. E. Lashkarev, Izv. AN SSSR ser. fiz. $\underline{14}$, 199 (1950).
9. S. M. Ryvkin, Izv. AN SSSR ser. fiz. $\underline{15}$, 721 (1951).
10. N. A. Tolstoi, Izv. AN SSSR ser. fiz. $\underline{15}$, 712 (1951).
11. G. C. Monch, Optik $\underline{10}$, 8 (1953).

12. S. M. Ryvkin, ZhTF 25, 1471 (1955).

13. D. K. Barsukov and A. R. Regel', ZhTF 25, 1472 (1955).

14. A. M. Bonch-Bruevich and V. A. Molchanov, ZhTF 25, 1653 (1955).

15. N. A. Tolstoi, DAN SSSR 102, 935 (1955).

16. N. A. Tolstoi, N. N. Tkachuk, M. Ya. Tsenter, Z. S. Mansurova, and A. V. Burlakov, Optika i spektroskopiya 1, 719 (1956).

17. M. Garbuny, T. P. Vogl, and J. R. Hansen, Rev. Sci. Instr. 28, 826 (1957).

18. N. A. Tolstoi, Optika i spektroskopiya 4, 279 (1958).

19. J. A. Hull, IRE Nat. Convention Record 6, 228 (1958).

20. Yu. V. Popov, I. I. Andrianov, and I. A. Tel'tovskii, Optiko-mekhanicheskaya promyshlennost' 1, 30 (1959).

21. N. N. Ogurtsova and A. V. Podmoshenskii, Optika i spektro-skopiya 4, 539 (1958).

22. V. A. Romanov, PTÉ No. 6, 70 (1959).

23. A. Whetston, Rev. Sci. Instr. 30, 6, 447 (1959).

24. M. S. Marshak, PTÉ No. 3, 5 (1962).

25. L. G. Paritskii and S. M. Ryvkin, FTT 2, 547 (1960).

26. Kh. S. Valeev, Yu. P. Vorontsov, and M. G. Morozov, PTÉ 2, 122 (1960).

27. A. A. Meier, PTÉ No. 2, 182 (1961).

28. S. M. Ryvkin and L. N. Malakhov, DAN SSSR 85, 765 (1952).

29. B. R. Holeman and C. Hilsum, J. Phys. Chem. Solids 22, 19 (1961).

30. F. M. Berkovskii, N. B. Strokan, and G. V. Khozov, PTÉ No. 2, 165 (1962).

31. L. G. Paritskii and S. M. Ryvkin, FTT 3, 2245 (1961).

CHAPTER IV

1. F. F. Vol'kenshtein, Electrical Conductivity of Semiconductors [in Russian] (Gostekhizdat, 1947).

2. N. F. Mott and R. W. Gurney, Electronic Processes in Ionic Crystals [Russian translation] (IL, 1950).

3. S. M. Ryvkin, DAN SSSR 22, 482 (1950).

4. F. S. Goucher, Phys. Rev. 78, 816 (1950).

5. S. I. Pekar, Investigations of Electron Theory of Crystals [in Russian] (Gostekhizdat, 1951).

6. H. Brooks, "Theory of electrical properties of germanium and silicon" [Russian translation] in: Problems of Modern Physics (IL, 1957), No. 8, p. 74.

7. R. Bray, "Exclusion of minority carriers from germanium" [Russian translation] in: Problems of Semiconductor Physics (IL, 1957), p. 221.

8. G. G. E. Low, "Deviations from equilibrium carrier densities in semiconductors" [Russian translation] in: Problems of Semiconductor Physics (IL, 1957), p. 216.
9. S. Koc, Czech. J. Phys. $\underline{6}$, 668 (1956).
10. T. S. Moss, "Absorption and photoconductivity of indium antimonide" [Russian translation] in: New Semiconducting Materials (1958), p. 125.
11. E. Antončik, Czech. J. Phys. $\underline{7}$, 674 (1957).
12. N. Sclar and E. Burstein, J. Phys. Chem. Solids $\underline{2}$, 1 (1957).
13. V. S. Vavilov and K. I. Britsyn, ZhÉTF $\underline{34}$, 521 (1958); $\underline{34}$, 1354 (1958).
14. B. Lax, UFN $\underline{70}$, 111 (1960).
15. V. S. Vavilov and K. I. Britsyn, FTT $\underline{1}$, 1629 (1959).
16. B. M. Vul, V. S. Vavilov, and A. P. Shotov, "On the problem of impact ionization in semiconductors" [in Russian] in: Collection Dedicated to G. S. Landsberg (AN SSSR, 1959), p. 95.
17. T. S. Moss, Optical Properties of Semiconductors (London, 1959).
18. B. M. Vul, É. I. Zavaritskaya, and L. V. Keldysh, DAN SSSR $\underline{135}$, 1361 (1960).
19. W. Shockley, Czech. J. Phys., ser. B $\underline{11}$, 81 (1961).

CHAPTER V

1. V. E. Lashkarev, ZhÉTF $\underline{19}$, 876 (1949); V. E. Lashkarev and G. A. Fedorus, Izv. AN SSSR ser. fiz. $\underline{16}$, 81 (1952).
2. V. P. Zhuze and S. M. Ryvkin, DAN SSSR $\underline{68}$, 673 (1949); Izv. AN SSSR ser. fiz. $\underline{16}$, 93 (1952).
3. R. N. Hall, Phys. Rev. $\underline{87}$, 387 (1952); W. Shockley and W. T. Read, "Statistics of recombination of holes and electrons" [Russian translation] in: Semiconducting Electronic Devices (IL, 1953).
4. Hofman, "Carrier lifetime and trap models from the point of view of the law of mass action" in: Halbleiterprobleme (1955), Vol. 2, p. 106.
5. S. G. Kalashnikov, ZhTF $\underline{26}$, 241 (1956).
6. A. V. Rzhanov, ZhTF $\underline{26}$, 1389 (1956).
7. J. Okada, J. Phys. Soc. Japan $\underline{12}$, 1338 (1956).
8. É. I. Adirovich, G. M. Guro, V. F. Kuleshov, and V. A. Chuenko, Tr. Fiz. in-ta im. P. N. Lebedeva AN SSSR $\underline{8}$, 129 (1956).
9. É. I. Adirovich and G. M. Guro, DAN SSSR $\underline{108}$, 417 (1956).
10. M. I. Iglitsyn, Yu. A. Kontsevoi, and A. I. Sidorov, ZhTF $\underline{27}$, 2461 (1957).
11. V. E. Lashkarev, V. G. Litovchenko, I. M. Omel'yanovskaya, R. N. Bondarenko, and V. I. Strikha, ZhTF $\underline{27}$, 2437 (1957).

12. V. G. Alekseeva, S. G. Kalashnikov, L. P. Kalpach, I. V. Karpova, and A. I. Morozov, ZhTF $\underline{27}$, 1931 (1957).

13. D. Sandiford, Phys. Rev. $\underline{105}$, 524 (1957).

14. D. Clarke, J. Electronics and Control $\underline{3}$, 375 (1957).

15. V. L. Bonch-Bruevich, Izv. AN SSSR ser. fiz. $\underline{21}$, 87 (1957).

16. V. L. Bonch-Bruevich, ZhÉTF $\underline{32}$, 1470 (1957).

17. S. M. Ryvkin, "Recombination in semiconductors" [in Russian] in: Semiconductors in Science and Technology (1958), Vol. 2, Chap. 22.

18. D. Sandiford, Proc. Phys. Soc. (London) $\underline{71}$, 1002 (1958).

19. G. Bemski, Proc. Inst. Radio Engrs. $\underline{46}$, 990 (1958).

20. T. V. Mashovets and S. M. Ryvkin, FTT Sbornik II, 70 (1959).

21. J. Okada, J. Phys. Soc. Japan $\underline{14}$, 1550 (1959).

22. V. G. Alekseeva, I. V. Karpova, and S. G. Kalashnikov, FTT $\underline{1}$, 529 (1959).

23. S. M. Ryvkin and I. B. Strokan, DAN SSSR $\underline{124}$, 1034 (1959).

24. M. Lax, J. Phys. Chem. Solids $\underline{8}$, 66 (1959).

25. G. I. Galkin, I. S. Rytova, and V. S. Vavilov, FTT $\underline{2}$, 2025 (1960).

26. G. M. Guro, UFN $\underline{72}$, 711 (1960).

27. Ya. E. Pokrovskii and K. I. Svistunova, FTT $\underline{3}$, 757 (1961).

CHAPTER VI

1. V. E. Lashkarev, ZhÉTF $\underline{19}$, 876 (1949).

2. A. Rose, "Recombination processes in insulators and semiconductors" [Russian translation] in: Problems of Semiconductor Physics (IL, 1957).

3. R. H. Bube, Proc. Inst. Radio Engrs. $\underline{43}$, 1836 (1955).

4. A. Rose, Proc. Inst. Radio Engrs. $\underline{43}$, 1850 (1955).

5. E. A. Niekisch, Ann. Physik $\underline{15}$, 279 (1955); Z. Physik $\underline{161}$, 38 (1961).

6. K. W. Böer and H. Vogel, Z. physik. Chem. $\underline{206}$ (1956); Ann. Physik $\underline{17}$, 10 (1955).

7. J. Hornbeck and J. R. Haynes, "Trapping of minority carriers in silicon" [Russian translation] in: Problems of Semiconductor Physics (IL, 1957).

8. R. G. Schulman, Phys. Rev. $\underline{102}$, 1451 (1956).

9. V. L. Bonch-Bruevich, ZhTF $\underline{28}$, 67 (1958).

10. S. M. Ryvkin, "Recombination in semiconductors" [in Russian] in: Semiconductors in Science and Technology (1958), Vol. 2.

11. M. Zerbst and W. Heywang, Z. Naturforsch. $\underline{14a}$, 645 (1959).

12. A. G. Mironov, FTT $\underline{1}$, 522 (1959).

13.   G. M. Guro, UFN 72, 711 (1960).
14.   L. G. Paritskii and S. M. Ryvkin, FTT 2, 547 (1960).
15.   S. M. Ryvkin and I. D. Yaroshetskii, FTT 2, 1966 (1960).
16.   A. A. Grinberg, L. G. Paritskii, and S. M. Ryvkin, FTT 2, 1545 (1960).
17.   L. G. Paritskii and S. M. Ryvkin, FTT 3, 2245 (1961).
18.   M. I. Boiko, FTT 2, 1835 (1960).
19.   V. E. Lashkarev, E. A. Sal'kov, and M. K. Sheinkman, FTT 3, 1973 (1961).

CHAPTER VII

1.   W. C. Dunlap, Phys. Rev. 97, 614 (1955).
2.   W. Shockley and J. T. Last, Phys. Rev. 107, 392 (1957).
3.   P. T. Landsberg, Proc. Phys. Soc. (London) B69, 1056 (1957).
4.   P. T. Landsberg, Proc. Phys. Soc. (London) B70, 282 (1957).
5.   C. T. Sah and W. Shockley, Phys. Rev. 109, 1103 (1958).
6.   G. Bemski, Proc. Inst. Radio Engrs. 46, 990 (1958).
7.   T. V. Mashovets, ZhTF 28, 1140 (1958).
8.   V. E. Khartsiev, ZhTF 28, 1651 (1958).
9.   N. G. Zhdanova, S. G. Kalashnikov, and A. I. Morozov, FTT 1, 535 (1959).
10.   Yu. A. Kontsevoi, FTT 1, 1289 (1959).
11.   K. D. Glinchuk, E. G. Miselyuk, and N. N. Fortunatova, FTT 1, 1345 (1959).
12.   S. G. Kalashnikov and K. P. Tissen, FTT 1, 545 (1959).
13.   S. G. Kalashnikov and K. P. Tissen, FTT 1, 1754 (1959).
14.   G. K. Wertheim, Phys. Rev. 115, 37 (1959).
15.   S. G. Kalashnikov and A. I. Morozov, FTT 1, 1294 (1959).
16.   R. Newman and W. W. Tyler, UFN 72, 587 (1960).
17.   S. G. Kalashnikov and K. Konstantsinesku, FTT 1, 1763 (1959).
18.   A. A. Grinberg, FTT Sbornik II, 192 (1959).
19.   L. Johnson and H. Levinstein, Phys. Rev. 117, 1191 (1960).
20.   S. G. Kalashnikov and K. P. Tissen, FTT 2, 2743 (1960).
21.   M. I. Iglitsyn and Yu. A. Kontsevoi, FTT 2, 1148 (1960).
22.   E. G. Miselyuk and K. D. Glinchuk, "Investigation of carrier recombination in germanium containing gold and silver impurities" in: Proc. Intern. Conf. Semicond. Prague, 1960.
23.   S. G. Kalashnikov and A. I. Morozov, FTT 2, 2813 (1960).
24.   V. G. Alekseeva, I. V. Karpova, and S. G. Kalashnikov, FTT 3, 964 (1961).
25.   E. G. Landsberg and S. G. Kalashnikov, FTT 3, 1566 (1961).

CHAPTER VIII

1. J. R. Haynes, "New radiation due to recombination of holes and electrons in germanium" [Russian translation] in: Problems of Semiconductor Physics (IL, 1958).
2. W. van Roosbroeck and W. Shockley, "Radiative recombination of electrons and holes in germanium" [Russian translation] in: Problems of Semiconductor Physics (IL, 1958).
3. R. Braunstein, Phys. Rev. 99, 1892 (1955).
4. T. S. Moss and T. H. Hawkins, Phys. Rev. 101, 1609 (1956).
5. E. Burstein and P. H. Egli, "Physics of semiconductors (a review)" [Russian translation] in: Physics of Semiconductors (IL, 1957).
6. L. N. Sosnovskii, Izv. AN SSSR ser. fiz. 21, 70 (1957).
7. F. Kane, J. Phys. Chem. Solids 1, 249 (1957).
8. P. T. Landsberg and A. R. Beattie, J. Phys. Chem. Solids 8, 73 (1959).
9. P. H. Brill and R. F. Schwarz, J. Phys. Chem. Solids 8, 75 (1959).
10. V. S. Vavilov, UFN 68, No. 2 (June, 1959).
11. J. R. Haynes, M. Lax, and W. F. Flood, J. Phys. Chem. Solids 8, 392 (1959).
12. R. N. Zitter, A. J. Strauss, and A. E. Attard, Phys. Rev. 115, 226 (1959).
13. J. I. Panhove, Phys. Rev. Letters 4, 20 (1960).
14. P. T. Landsberg, Abhandl. Deut. Akad. Wiss. 7, 57 (1960).
15. D. Curie, Luminescence in Crystals [Russian translation] (IL, 1961).
16. F. M. Gashimzade and V. E. Khartsiev, FTT 3, 1453 (1961).
17. D. N. Nasledov, A. A. Rogachev, S. M. Ryvkin, and B. V. Tsarenkov, FTT 4, 1062 (1962).
18. A. A. Rogachev and S. M. Ryvkin, FTT 4, 1676 (1962).
19. D. N. Nasledov, A. A. Rogachev, S. M. Ryvkin, V. E. Khartsiev, and B. V. Tsarenkov, FTT 4, 3346 (1962).
20. N. G. Basov, B. M. Vul, and Yu. M. Popov, ZhÉTF, 37, 587 (1959).
21. N. G. Basov, O. N. Krokhin, and Yu. M. Popov, UFN 72, 161 (1960).
22. R. N. Hall, G. E. Fenner, J. D. Kingsley, T. J. Soltys, and R. O. Carlson, Phys. Rev. Letters 9, 366 (1962).
23. T. M. Quist, R. H. Rediker, R. J. Keyer, W. E. Krag, B. Lax, A. L. McWhorter, and H. J. Zeigler, Appl. Phys. Letters 1, 91 (1962).

24. M. I. Nathan, W. P. Dumke, G. Burns, F. H. Dill, Jr, and G. Zasher, Appl. Phys. Letters 1, 62 (1962).

25. N. Holonyak, Jr and S. F. Bevacqua, Appl. Phys. Letters 1, 82 (1962).

26. V. S. Bagaev, I. G. Basov, B. M. Vul, B. D. Kopylovskii, O. I. Krokhin, E. P. Markin, Yu. M. Popov, A. I. Khvoshchev, and A. P. Shotov, DAN SSSR 150, 275 (1963).

CHAPTER IX

1. E. Burstein, G. Picus, and N. Sclar, Photoconductivity Conference, Atlantic City (1954), p. 353.

2. E. Burstein, J. W. Davisson, E. E. Bell, W. Turner, and H. G. Lipson, "Infrared photoconductivity of germanium due to neutral impurities" [Russian translation] in: Electrical Properties of Germanium and Silicon (IL, 1956), p. 158.

3. R. Newman, Phys. Rev. 94, 278 (1954).

4. R. Newman and W. W. Tyler, "Properties of germanium with iron impurity" [Russian translation] in: Problems of Semiconductor Physics (1957), p. 34.

5. R. Newman, H. H. Woodbury, and W. W. Tyler, Phys. Rev. 102, No. 3, 613 (1956).

6. H. Y. Fan, Reports on Progress in Physics 29, 107 (1956).

7. J. S. Blakemore, Can. J. Phys. 34, 938 (1956).

8. B. T. Kolomiets, Izv. AN SSSR ser. fiz. 21, 643 (1957).

9. C. B. Collins, R. O. Carlson, and C. J. Gallagher, Phys. Rev. 105, 1168 (1957).

10. H. Y. Fan and K. Lark-Horowitz, Effects of Radiation on Materials (1958), p. 159.

11. F. Stockmann, E. E. Klontz, J. MacKay, H. Y. Fan, and K. Lark-Horowitz, Z. Physik 153, 331 (1958).

12. R. Newman and W. W. Tyler, UFN 72, 587 (1960).

13. E. N. Arkad'eva, L. G. Paritskii, and S. M. Ryvkin, FTT 2, 1161 (1960).

14. É. I. Adirovich, FTT 2, 2384 (1960).

15. S. M. Ryvkin, L. G. Paritskii, R. Yu. Khansevarov, and I. D. Yaroshetskii, FTT 3, 252 (1961).

16. A. V. Rzhanov and A. F. Plotnikov, FTT 3, 1557 (1961).

17. E. N. Arkad'eva, R. S. Kasymova, and S. M. Ryvkin, FTT 3, 2417 (1961).

18. V. S. Vavilov and A. F. Plotnikov, FTT 3, 2455 (1961).

19. S. M. Ryvkin, R. Yu. Khansevarov, and I. D. Yaroshetskii, FTT 3, 3211 (1961).

20. E. N. Arkad'eva, FTT 4, 3048 (1962).

CHAPTER X

1.   E. A. Taft and M. H. Hebb, J. Opt. Soc. Am. 42, 249 (1952).

2.   J. Lambe and C. C. Klick, Phys. Rev. 98, 909 (1955).

3.   R. H. Bube, Phys. Rev. 101, 1668 (1956).

4.   K. W. Böer, S. Oberlander, and J. Voigt, Ann. Physik 2, 130 (1958).

5.   A. A. Trofimenko and G. A. Fedorus, Ukr. fiz. zhurn. 3, No. 4 (1958).

6.   E. N. Arkad'eva and S. M. Ryvkin, FTT 1, 1460 (1959).

7.   I. I. Boiko, É. I. Rashba, and A. P. Trofimenko, FTT 2, 109 (1960).

8.   E. N. Arkad'eva, L. G. Paritskii, and S. M. Ryvkin, FTT 2, 1160 (1960).

9.   E. N. Arkad'eva and S. M. Ryvkin, FTT 2, 1889 (1960).

10. L. G. Paritskii and S. M. Ryvkin, FTT 3, 2245 (1961).

11. E. N. Arkad'eva, R. S. Kasymova, and S. M. Ryvkin, FTT 3, 2417 (1961).

12. E. N. Arkad'eva, L. G. Paritskii, and S. M. Ryvkin, FTT 4, No. 6 (1962).

13. Yu. L. Ivanov and S. M. Ryvkin, FTT 4, 1482 (1962).

14. N. Sclar and E. Burstein, J. Phys. Chem. Solids 2, 1 (1957).

15. S. H. Koenig and G. R. Gunther-Mohr, J. Phys. Chem. Solids 2, 268 (1957).

16. B. M. Vul, E. I. Zavaritskaya, and V. A. Chuenkov, Proc. Intern. Conf. Semicond. Phys., Prague, 1960.

17. S. M. Ryvkin, V. P. Dobrego, B. M. Konovalenko, and I. D. Yaroshetskii, FTT 4, 1911 (1962).

18. S. M. Ryvkin, V. P. Dobrego, B. M. Konovalenko, and I. D. Yaroshetskii, Report Intern. Conf. Phys. Semicond., Exeter, 1962.

19. A. L. McWhorter and R. H. Rediker, Proc. Intern. Conf. Semicond. Phys., Prague, 1960.

20. N. F. Mott and W. D. Twose, Adv. Phys. 10, 107 (1961).

21. A. L. McWhorter, Bull. Am. Phys. Soc. II 4, 186 (1959).

22. V. P. Dobrego and S. M. Ryvkin, FTT 4, 553 (1962).

23. V. P. Dobrego and S. M. Ryvkin, "Hopping photoconductivity" [in Russian] in: Abstracts of Papers Presented at the Third All-Union Conference on Photoelectric Phenomena in Semiconductors, Kiev (KGU, July 1963); FTT 6, 1203 (1964).

24. Sh. M. Kogan, T. M. Lifshits, and V. I. Sidorov, ZhÉTF 46, 395 (1964).

CHAPTER XI

1. É. I. Adirovich and G. M. Guro, DAN 108, 417 (1956); G. M. Guro, UFN 72, 711 (1960).
2. É. I. Adirovich, G. M. Guro, V. F. Kuleshov, and V. A. Chuenkov, Tr. Fiz. in-ta im. P. I. Lebedeva 8, 129 (1956).
3. A. Rose, "Recombination processes in insulators and semiconductors" [Russian translation] in: Problems of Semiconductor Physics (IL, 1957).
4. S. M. Ryvkin, FTT 2, 2410 (1960).

CHAPTER XII

1. B. Gudden and R. W. Pohl, Z. Physik 6, 248 (1921).
2. B. I. Davydov, ZhÉTF 9, 451 (1939).
3. N. F. Mott and R. W. Gurney, Electronic Processes in Ionic Crystals [Russian translation] (IL, 1950).
4. A. N. Gubanov, ZhTF 24, 933 (1954).
5. A. I. Gubanov, Theory of the Rectifying Effect in Semiconductors [in Russian] (Gostekhizdat, 1956).
6. S. M. Ryvkin, ZhTF 26, 2439 (1956).

CHAPTER XIII

1. A. V. Ioffe and A. F. Ioffe, ZhÉTF 5, 111 (1935).
2. J. Haynes and W. Shockley, Phys. Rev. 81, 835 (1951).
3. F. Goucher, Phys. Rev. 81, 475 (1951).
4. V. E. Lashkarev, Izv. AN SSSR ser. fiz. 16, 186 (1952).
5. L. B. Valdes, Proc. Inst. Radio Engrs. 40, 1420 (1952).
6. R. Lawrance and A. Gibson, Proc. Phys. Soc. (London) 65, 994 (1952).
7. A. Giordano et al., Phys. Rev. 88, 1368 (1952).
8. M. Prince, Phys. Rev. 92, 681 (1953).
9. W. van Roosbroeck, Phys. Rev. 91, 182 (1953).
10. W. Shockley, Electrons and Holes in Semiconductors [Russian translation] (IL, 1953).
11. S. M. Ryvkin, ZhTF 24, 2136 (1954).
12. G. Adam, Physica 20, 1037 (1954).
13. S. M. Ryvkin and R. V. Khar'yuzov, ZhTF 25, 563 (1955).
14. É. I. Adirovich, G. M. Guro, V. F. Kuleshov, and V. A. Chuenkov, Tr. Fiz. in-ta im. P. I. Lebedeva 8, 129 (1956).
15. É. I. Rashba and K. B. Tolpygo, Ukr. fiz. zhurn. 1, 29 (1956).

16. O. V. Sorokin, ZhTF 26, 2473 (1956).
17. S. M. Ryvkin and Yu. A. Makhalov, ZhTF 27, 441 (1957).
18. G. Pikus "Contact phenomena in semiconductors" [in Russian] in: Semiconductors in Science and Technology (1958), Vol. 1.
19. G. M. Guro, UFN 72, 711 (1960).

CHAPTER XIV

1. I. K. Kikoin and M. M. Noskov, Physik. Z. Sowjetunion 5, 586 (1934); 6, 478 (1934). I. K. Kikoin, DAN SSSR 3, 418 (1934).
2. Ya. I. Frenkel', Physik. Z. Sowjetunion 5, 597 (1934); 8, 185 (1935).
3. L. D. Landau and E. M. Lifshits, Physik. Z. Sowjetunion 9, 477 (1936); E. M. Lifshits, Physik. Z. Sowjetunion 9, 641 (1936).
4. V. E. Lashkarev, ZhÉTF 18, 915 (1948).
5. K. B. Tolpygo, Tr. In-ta fiziki AN USSR 3, 52 (1952).
6. A. I. Ansel'm, ZhTF 22, 1146 (1952).
7. P. Aigrain and H. Bulliard, Compt. rend. 236, 595 (1953); 236, 672 (1953).
8. T. S. Moss, Proc. Phys. Soc. (London) B66, 993 (1953).
9. T. S. Moss, L. Pincherle, and A. M. Woodward, Proc. Phys. Soc. (London) B66, 743 (1953).
10. W. Shockley, Electrons and Holes in Semiconductors [Russian translation] (IL, 1953).
11. H. Bulliard, Phys. Rev. 94, 1564 (1954).
12. A. I. Ansel'm, ZhTF 24, 2064 (1954).
13. A. P. Komar, N. M. Reinov, and S. S. Shalyt, DAN SSSR 96, 47 (1954).
14. H. Bulliard, Ann. Phys. 9, 52 (1954).
15. S. W. Kurnick, A. J. Strauss, and R. N. Zitter, Phys. Rev. 94, 1791 (1954).
16. L. H. Hall, Phys. Rev. 97, 1741 (1955).
17. W. van Roosbroeck, Phys. Rev. 101, 1713 (1956).
18. S. Kurnick and R. Zitter, J. Appl. Phys. 27, 278 (1956).
19. L. Pincherle, Photoconductivity Conference, 1956, p. 307.
20. I. K. Kikoin and Yu. A. Bykhovskii, DAN SSSR 109, 735 (1956); DAN SSSR 116, 381 (1957).
21. O. Caretta and J. Crosvalet, "Photomagnetoelectric effect in semiconductors, " in: Progress in Semiconductors (1956), Vol. 1a, p. 165.
22. B. Ya. Moizhes, ZhTF 27, 495 (1957).
23. B. Ya. Moizhes and Yu. N. Obraztsov, ZhTF 27, 1446 (1957).
24. T. M. Buck and F. Z. McKim, Phys. Rev. 106, 904 (1957).

25. R. N. Zitter, Phys. Rev. 112, 852 (1958).
26. S. G. Kalashnikov and E. G. Landsberg, ZhTF 28, 1387 (1958).
27. G. M. Guro, ZhTF 28, 1036 (1958).
28. B. Ya. Moizhes, FTT 1, 1239 (1959).
29. A. Amith, Phys. Rev. 116, 793 (1959).
30. A. G. Mironov, FTT 1, 525 (1959).
31. S. M. Ryvkin, Yu. L. Ivanov, A. A. Grinberg, S. R. Novikov, and N. D. Potekhina, FTT 1, 1372 (1959).
32. W. van Roosbroeck, Phys. Rev. 119, 636 (1960).
33. E. G. Landsberg, FTT 2, 848 (1960).
34. F. A. Brand, A. N. Baker, and H. Mette, Phys. Rev. 119, 922 (1960).
35. A. A. Grinberg, FTT 2, 836 (1960).
36. Yu. I. Ravich, FTT 2, 2366 (1960).
37. S. M. Ryvkin, A. A. Grinberg, Yu. L. Ivanov, S. R. Novikov, and N. D. Potekhina, FTT 2, 575 (1960).
38. A. A. Grinberg, FTT 3, 94 (1961).
39. A. A. Grinberg, S. R. Novikov, and S. M. Ryvkin, DAN SSSR 136, 329 (1961).
40. Yu. I. Ravich, FTT 3, 1601 (1961).
41. A. A. Grinberg and S. M. Ryvkin, FTT 3, 2470 (1961).

CHAPTER XV

1.   A. V. Ioffe, DAN SSSR 27, 547 (1940); ZhÉTF 15, 721 (1945).
2.   L. Landau and E. Lifshits, Physik. Z. Sowjetunion 9, 477 (1936).
3.   B. I. Davydov, ZhTF 7, 2212 (1937).
4.   Yu. P. Maslakovets, ZhÉTF 10, 393 (1940).
5.   V. E. Lashkarev, ZhÉTF 18, No. 10 (1948).
6.   V. E. Lashkarev and K. M. Kosonogova, ZhÉTF 18, No. 10 (1948).
7.   S. M. Ryvkin, ZhÉTF 18, 1521 (1948); 19, 286 (1949).
8.   V. E. Lashkarev, DAN SSSR 70, 813 (1950).
9.   K. B. Tolpygo, ZhÉTF 23, 340 (1952).
10. J. Shive, J. Opt. Soc. Am. 43, 239 (1953).
11. W. Shockley, Electrons and Holes in Semiconductors [Russian translation] (IL, 1953).
12. R. L. Cummerow, "Photoeffect in a p-n junction" [Russian translation] in: Semiconducting Converters of Radiant Energy (IL, 1959).
13. Zh. I. Alferov, B. M. Konovalenko, S. M. Ryvkin, V. M. Tuchkevich, and A. I. Uvarov, ZhTF 25, 11 (1955).
14. S. M. Ryvkin, ZhTF 25, 21 (1955).
15. É. I. Adirovich and V. G. Kolotilova, ZhÉTF 29, 770 (1955).
16. V. S. Vavilov, Atomnaya énergiya 1, 107 (1956).

17. V. E. Lashkarev and V. A. Romanov, Radiotekhnika i elektronika
     2, 1144 (1956).
18. S. M. Ryvkin, ZhTF 27, 1676 (1957).
19. J. Tauc, Revs. Modern Phys. 29, 308 (1957).
20. Yu. P. Maslakovets, Yu. A. Vodakov, G. P. Naumov, and G. A.
     Lomakina, ZhTF 27, 1594 (1957).
21. V. K. Subashiev and M. S. Sominskii, "Semiconducting photocells"
     [in Russian] in: Semiconductors in Science and Technology (AN
     SSSR, 1958), Vol. 2.
22. S. M. Ryvkin and N. B. Strokan, ZhTF 28, 1169 (1958).
23. S. M. Ryvkin, N. B. Strokan, and L. L. Makovskii, ZhTF 28,
     1871 (1958).
24. S. M. Ryvkin, N. B. Strokan, V. M. Tuchkevich, and V. E.
     Chelnokov, ZhTF 28, 1165 (1958).
25. D. E. Sawyer and R. H. Rediker, Proc. Inst. Radio Engrs. 46,
     1122 (1958).
26. N. D. Potekhina, FTT 1, 1509 (1959).
27. N. A. Vitovskii, P. I. Maleev, and S. M. Ryvkin, Radiotekhnika
     i elektronika 4, 330 (1959).
28. L. Ya. Pervova, Radiotekhnika i elektronika 4, 330 (1959).
29. V. S. Vavilov, G. N. Galkin, and V. M. Malovetskaya, "Investiga-
     tion of silicon photocells as solar energy converters" [in Russian]
     in: Proceedings Conf. Photoelec. and Optical Effects in Semicond.,
     AN UkrSSR, Kiev, 1959.
30. V. M. Tuchkevich and V. E. Chelnokov, "Silicon photocells" [in
     Russian] in: Proceedings Conf. Photoelec. and Optical Effects in
     Semicond. AN UkrSSR, Kiev, 1959.
31. D. N. Nasledov and B. V. Tsarenkov, FTT Sbornik (1959).
32. K. B. Tolpygo, "Kinetics of photo-emf's in homogeneous semi-
     conductors" [in Russian] in: Proceedings Conf. Photoelec. and Opti-
     cal Effects in Semicond. AN UKrSSR,Kiev, 1959.
33. R. Williams and R. Bube, J. Appl. Phys. 36, No. 6 (1960).
34. A. A. Grinberg and N. B. Strokan, FTT 2, 1536 (1960).
35. S. M. Ryvkin, R. F. Konopleva, L. V. Maslova, O. A. Matveev,
     N. B. Strokan, D. V. Tarkhin, and G. V. Khozov, FTT 2, 2199
     (1960).
36. G. M. Avakyants, Phenomenological Theory of Semiconductors [in
     Russian] AN Uzbek SSR, Tashkent, 1960.
37. V. K. Subashiev and É. M. Pedyash, FTT 2, 213 (1960).
38. L. G. Paritskii, A. A. Rogachev, and S. M. Ryvkin, FTT 3, 1613
     (1961).
39. F. M. Berkovskii, S. M. Ryvkin, and N. B. Strokan, FTT 3, 230
     (1961).
40. F. M. Berkovskii, S. M. Ryvkin, and N. B. Strokan, FTT 3, 3535
     (1961).

41.   F. M. Berkovskii and S. M. Ryvkin, FTT $\underline{4}$, 366 (1962).

42.   F. M. Berkovskii and S. M. Ryvkin, FTT $\underline{4}$, 376 (1962).

43.   K. Thiessen and G. Jungk, Phys. Status Solidi $\underline{2}$, 4, 473 (1962).

44.   F. M. Berkovskii and S. M. Ryvkin, FTT $\underline{5}$, 381 (1963).

45.   F. M. Berkovskii, R. S. Kasymova, and S. M. Ryvkin, FTT $\underline{5}$, 525 (1963).

46.   F. M. Berkovskii and S. M. Ryvkin, FTT $\underline{5}$, 2023 (1963).

INDEX

399